Lecture Notes
in Computational Science
and Engineering

34

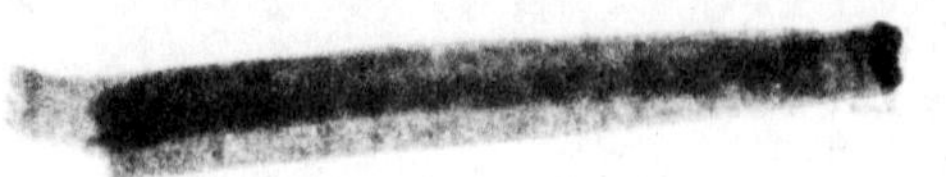

Springer
Berlin
Heidelberg
New York
Hong Kong
London
Milan
Paris
Tokyo

Volker John

Large Eddy Simulation of Turbulent Incompressible Flows

Analytical and Numerical Results for a Class of LES Models

Springer

Volker John
Institute of Analysis and Numerical Mathematics
Department of Mathematics
Otto-von-Guericke-University Magdeburg
39106 Magdeburg, Germany
e-mail: john@mathematik.uni-magdeburg.de

Cataloging-in-Publication Data applied for

A catalog record for this book is available from the Library of Congress.

Bibliographic information published by Die Deutsche Bibliothek
Die Deutsche Bibliothek lists this publication in the Deutsche Nationalbibliografie; detailed bibliographic data is available in the Internet at <http://dnb.ddb.de>.

Mathematics Subject Classification (2000): 76-02, 76F65, 65M60, 65M55

ISSN 1439-7358
ISBN 3-540-40643-3 Springer-Verlag Berlin Heidelberg New York

Springer-Verlag Berlin Heidelberg New York
a member of BertelsmannSpringer Science + Business Media GmbH
http://www.springer.de

Printed in Germany

Cover Design: Friedhelm Steinen-Broo, Estudio Calamar, Spain
Cover production: *design & production*
Typeset by the author using a Springer TEX macro package

Printed on acid-free paper 46/3142/LK - 5 4 3 2 1 0

For Anja and Josephine.

Preface

The numerical simulation of turbulent flows is undoubtedly very important in applications. The richness of scales inherent in turbulent flows makes it impossible to solve the governing equations, the Navier-Stokes equations, on present-day computers and even on computes in the foreseeable future. Turbulence models are the tool to modify the Navier-Stokes equations such that equations arise which can be numerically approximated using present-day hardware and software. One kind of modelling turbulence is large eddy simulation (LES) which aims to compute the large flow structures accurately and models the interactions of the small flow structures to the large ones.

This monograph considers a class of LES models whose derivation is mainly based on mathematical (and not on physical) arguments and in addition the Smagorinsky model. One main goal of this monograph is to present all mathematical analysis which is known for these models. The second goal is to give a detailed description of the implementation of these LES models into a finite element code. Since the probably best model within the considered class of LES models is rather new, it still requires comprehensive numerical tests. Therefore, the last main topic of this monograph is the presentation of first numerical studies with this new LES model which investigate, e.g., how good a space averaged flow field is approximated.

The writing of this monograph would have been impossible without the support of friends and colleagues. A particular thank goes to William (Bill) J. Layton (University of Pittsburgh). The participation at the scientific cooperation of Lutz Tobiska with Bill and his group enabled me to enlarge my fields of research considerably. It was Bill who brought me into contact with LES and together with whom a number of results presented in this monograph were obtained. The Deutsche Akademische Austauschdienst (D.A.A.D.) made it possible for me to pay three longer research visits at the University of Pittsburgh within the past two years which were essential for the work at this monograph. I like to thank Annette and Bill Layton also for their hospitality during these visits. The computational results were obtained with the code MooNMD which was developed in our group. My special thanks go to Gunar

Matthies who often answered questions which arose in the implementation of the algorithms. I like to thank Lutz Tobiska who gave useful suggestions for improving the monograph. For helpful discussions on subjects of this monograph, I would like to thank, besides the already mentioned colleagues, Adrian Dunca, Traian Iliescu and Friedhelm Schieweck. Useful suggestions for the preparation of the final version of this monograph came from Michael Griebel, Max D. Gunzburger and Tobias Knopp. I like to thank also Walfred Grambow for his efforts to provide the computer resources which were necessary to do the numerical simulations.

Last but not least, I like to thank my beloved wife Anja for her constant encouragement and support. Her efforts to solve the daily problems of our life were the basis of finding sufficient time to work at this monograph in the past two and a half years.

Colbitz,
July 2003

Volker John

Contents

1

Introduction

1.1 Short Remarks on the Nature and Importance of Turbulent Flows

Turbulent flows occur in many processes in nature as well as in many industrial applications. A storm, for instance, is accompanied with a wind which has a high velocity and can change its direction and also its speed abruptly. The take-off of an aircraft leads to turbulent air above the runway which has to be calmed down before the start of the next aircraft. This necessary interval between the take-offs of the aircrafts limits the number of starts in a given time period. There are many more examples of turbulent flows, e.g., in oceanography and in wind channels for the design of cars. Often, turbulent flows go along with chemical reactions like in gas engines. A wide variety of other examples for turbulent flows can be found, e.g., in the book by Lesieur [Les97, Chapter I].

There are many important and interesting physical phenomena which are connected with turbulent flows. In particular, turbulent flows possess a richness of scales, that means there are large flow structures and also very small ones. E.g., a hurricane has a number of very large eddies but also millions of small eddies. Since this monograph will mainly deal with models for turbulent flows, their analysis and their implementation, we refer for more details on the physics of turbulent flows, e.g., to the books by Frisch [Fri95], Lesieur [Les97] or Pope [Pop00].

1.2 Remarks on the Direct Numerical Simulation (DNS) and the $k - \varepsilon$ Model

The understanding and predicting of many processes in nature and in industry requires the simulation of turbulent flows. We consider in this monograph turbulent incompressible flows which are governed by the incompressible Navier-Stokes equations. These equations describe the conservation of momentum and the conservation of mass of a fluid in a domain $\Omega \in \mathbb{R}^d$, $d = 2, 3$, see (3.5). It is, in general, not possible to find an analytical representation of a solution of the Navier-Stokes equations. This necessitates the computation of a numerical approximation.

The ideal approach is to compute this numerical approximation directly by using a discretisation of the Navier-Stokes equations, like the Galerkin finite element discretisation described in Chapter 7. This approach, called direct numerical simulation (DNS) tries to solve the Navier-Stokes equations such that all persisting eddies are resolved. If the Reynolds number Re of a flow becomes very large, small flow structures will develop. But, due to viscous forces in the flow, very small eddies die rather quickly. The size of the persisting eddies is given by the Kolmogorov law [Kol42]. It is $\mathcal{O}(Re^{-3/4})$ in 3d. The persisting eddies in 2d are of size $\mathcal{O}(Re^{-1/2})$, Kraichnan [Kra67]. To capture all persisting eddies in 3d, one would need a mesh size of $h \approx Re^{-3/4}$. The total number of mesh cells in a uniform mesh in a hexahedron is then $N = h^{-3} \approx Re^{9/4}$. That means, the performance of a DNS is possible on present day computers only to very limited values of the Reynolds number. Good computers can handle today about $N = 10^7$ mesh cells which gives the restriction $Re \approx 10^{28/9} \approx 1292$. However, a reasonable number for industrial applications is $Re = 10^6$. The situation in 2d is less dramatic. For a DNS, one needs a mesh size of $h \approx Re^{-1/2}$ which leads to $N = h^{-2} \approx Re$ mesh cells in a square.

Another difficulty for a DNS is that the boundary and initial conditions must have the precision which is required by the smallest scales of the flow. Whereas this might cause no problems for academic test examples, for real applications like geophysical flows this seems to be impossible. Besides the lack of data on every point at the boundary, also such aspects as wall roughness and wall vibration have to be considered. A large Reynolds number flow is inherently unstable and the uniqueness of a (weak) solution is not yet proved generally in three dimensions for given initial and boundary conditions. Even small perturbations in these conditions may excite small scales. Thus, the impossibility of prescribing precise initial and boundary conditions causes that the resulting flow will have random character.

In situations where a DNS is not possible, one needs a modification of the underlying equations by a turbulence model. This model should lead to a computable flow, using the given discretisation, which still contains important properties of the original turbulent flow. Many turbulence models are based on physical insight and observational evidence. One of the most famous of them is the $k-\varepsilon$ model, which is described, e.g., in the book by Mohammadi and Pironneau [MP94]. The $k-\varepsilon$ model consists of two transport equations, one for the kinetic energy of the turbulence k and one for the rate of dissipation of turbulent energy ε. These equations are coupled to the flow equations and they are coupled mutually. The exact boundary conditions of the two additional equations are not clear. Often, wall laws are used. This model produces satisfactory results for boundary layer type of flows. The $k-\varepsilon$ model includes four empirical constants whose values are determined such that simple flow situations are recovered. It is not clear if these constants are universal. The derivation of the $k - \varepsilon$ model is based on the idealised assumption of homogeneous isotropic turbulence. A list of further hypotheses is given in [MP94, p. 53]. The practical experience with the $k - \varepsilon$ model is that it performs sometimes surprisingly well, even for flows which do not satisfy the assumptions which are made in the derivation of the model, e.g., for recirculating flows which is a very natural situation of a turbulent flow. There are also situations where unsatisfactory results are produced, for an overview see [MP94, p. 73]. Even after three decades of intensive use and study of this model, only partial results from the mathematical point of view, e.g., existence of solutions, are available. Nevertheless, the $k - \varepsilon$ model is one of the most common models used in the engineering community and a lot of experiences on its behaviour are available.

We will consider in this monograph another way of modelling turbulent flows – the large eddy simulation (LES).

1.3 Large Eddy Simulation (LES)

LES is currently a very popular approach for turbulent flow simulation. The basic idea starts by decomposing the quantities which describe the flow (velocity, pressure, body forces) into two parts: one part containing the large flow structures (large scales) and the remainder containing the small scales.

This approach makes sense in applications. Considering again the example of a hurricane, one is primarily interested in predicting the behaviour of the large eddies. These eddies are the most dangerous ones and it is important to predict their way and their velocity in order to take steps to protect men and properties. The actual behaviour of the millions of small eddies is not of interest. But these small eddies of course influence the behaviour of the large eddies. There is a direct interaction between the small and the large eddies. In addition, the interaction of the small eddies among each other influences the large eddies. That means, a prediction of the behaviour of the large eddies is not possible without taking into account the interactions coming from the small eddies.

The distinction of what are the large scales and the small scales might be given by the application. But very often, the available hardware and software determine if a certain scale can be computed and thus belongs to the large scales. The computation uses in general a discretisation scheme in space which is based on a computational mesh on the domain Ω. The given mesh width (distance of the degrees of freedom) is a restriction on the size of the eddies which can be computed since flow structures which are smaller than the mesh width cannot be represented by the discretisation. The fineness of the mesh which can be used depends on the memory available and also on the expected computing time for the flow simulation.

In LES, the large flow structures are always defined by averaging the velocity, pressure, etc. in space. A usual way is to define this spatial average by convolving these quantities with an appropriate filter function with filter width $\delta > 0$. This filter function should filter out the small scale structures, which are the structures of size smaller than $\mathcal{O}(\delta)$, or, equivalently, it should damp the high wave numbers. The goal of LES is to compute only the large flow structures accurately !

After having decided how to define the large scale flow structures, one needs equations for them. The approach to obtain such equations starts by simply averaging the Navier-Stokes equations, i.e. the Navier-Stokes equations are also convolved with the filter function. Then, one assumes that convolution and differentiation commute and these operators are interchanged. This leads to the space averaged Navier-Stokes equations, see (3.11). However, the commutation assumption is in general not valid, e.g., if Ω is a bounded domain. One commits a so-called commutation error which is simply neglected in practical computations. A first analysis of this commutation error is presented in Chapter 3.

The space averaged Navier-Stokes equations are not yet equations for the large flow structures. The non-linear convective term of the Navier-Stokes equations leads to a tensor, the so-called Reynolds stress tensor, whose entries are a priori not related to the large scale velocity and pressure. Thus, the space averaged Navier-Stokes equations are not closed, there are more unknowns than equations. A main issue in LES is to model the Reynolds stress tensor in terms of the large scale velocity and pressure.

This modelling process is addressed in Chapter 4. There are a variety of models in the literature. We will concentrate in this monograph on models which are based mainly on mathematical (and not on physical) considerations. In this approach, the Reynolds stress tensor is transformed to the Fourier space or wave number space. There, the Fourier transform of the filter function defining the spatial average, in our case the Gaussian filter, is approximated by a simpler function. There are two proposals for such an approximation in the literature:

- a second order Taylor polynomial approximation, Leonard [Leo74] and Clark et al. [CFR79]. This model is called in this monograph *Taylor LES model*. In the literature one finds also the name gradient model.
- a second order rational Padé approximation, Galdi and Layton [GL00]. This model is called *rational LES model.*

After having replaced the Fourier transform of the filter function by its approximation, terms which are in some sense of higher order are neglected. Applying the inverse Fourier transform leads to the first part of the model of the Reynolds stress tensor.

It turns out that neither the second order Taylor polynomial nor the second order rational Padé approximation give a model for the subgrid scale term, which describes how the interaction of the small eddies among each other influences the large eddies, compare Section 4.3. There is a clear numerical evidence that a model for this term is necessary, see Chapters 10 and 11. A fourth order rational approximation of the Fourier transform of the Gaussian filter gives a model for the subgrid scale term, but a number of difficulties arise in the implementation of this model into a code based on finite element methods. One proposal, which was already used by Clark et al. [CFR79] for the Taylor LES model, is to model the subgrid scale term by the Smagorinsky model [Sma63]. This is a simple LES model based on physical considerations. From the mathematical point of view, a non-linear artificial viscosity is added. Besides the Smagorinsky model, there are also other proposals for modelling the subgrid scale tensor by Iliescu and Layton [IL98].

Replacing the subgrid scale stress tensor by a model, one obtains equations for the large scale velocity and pressure. These equations have to be equipped with an initial condition for the large scale velocity and with boundary conditions. The initial condition is simply obtained by convolving the initial velocity of the Navier-Stokes equations with the filter function. The question of accurate boundary conditions for the large flow structures is largely unresolved. There is currently an active research in this topic. One proposal, to use no penetration and slip with friction boundary conditions, is considered in more detail in this monograph.

An important feature of LES, from our point of view, is the possibility of supporting LES models with mathematical analysis. A sound mathematical support is necessary for better understanding the advantages and drawbacks of turbulence models and it is also helpful for assessing the accuracy of computational results. Based on the better mathematical insight into a turbulence model, it can be perhaps improved. The rational LES model by Galdi and Layton [GL00] can be considered as an example for such an improvement. Based on the observation that the Taylor polynomial approximates the Fourier transform of the Gaussian filter completely wrong for high wave numbers, Figure 4.1, Galdi and Layton proposed a more consistent approximation by a rational function, Figure 4.2.

1.4 Contents of this Monograph

The flow models which occur in this monograph are summarised in Figure 1.1. This monograph provides a comprehensive study of three LES models:

- the Smagorinsky model,
- the Taylor LES model,
- the rational LES model.

The Smagorinsky model is included since it is the LES model which is currently studied best by mathematical and numerical analysis. It will be considered also as one proposal for modelling the subgrid scale term in the Taylor LES model and the rational LES model.

The scope of this monograph reaches from the very beginning of LES, the derivation of the space averaged Navier-Stokes equations, to numerical investigations. We will give here a short overview about the contents of this monograph and the most important results.

The governing equations

$$\begin{aligned} \mathbf{w}_t - \nabla\cdot((2\nu+\nu_T)\mathbb{D}(\mathbf{w})) + (\mathbf{w}\cdot\nabla)\,\mathbf{w} & \\ +\nabla r + \nabla\cdot\frac{\delta^2}{2\gamma}\left(A\left(\nabla\mathbf{w}\nabla\mathbf{w}^T\right)\right) &= \mathbf{f} \text{ in } (0,T]\times\Omega, \\ \nabla\cdot\mathbf{w} &= 0 \text{ in } [0,T]\times\Omega \end{aligned}$$

The models

- *Navier-Stokes equations*

$$\nu_T = 0, \quad A = 0,$$

- *Smagorinsky model*

$$\nu_T = c_S\delta^2\left\|\mathbb{D}(\mathbf{w})\right\|_F, \quad A = 0,$$

- *Taylor LES model*

$$\nu_T = c_S\delta^2\left\|\mathbb{D}(\mathbf{w})\right\|_F, \quad A = I \text{ (identity)},$$

- *rational LES model with auxiliary problem and Smagorinsky subgrid scale model*

$$\nu_T = c_S\delta^2\left\|\mathbb{D}(\mathbf{w})\right\|_F, \quad A = \left(I - \frac{\delta^2}{4\gamma}\Delta\right)^{-1} \quad \text{(auxiliary problem)}$$

  auxiliary problem equipped with homogeneous Neumann boundary conditions
- *rational LES model with auxiliary problem and Iliescu-Layton subgrid scale model*

$$\nu_T = c_S\delta\left\|\mathbf{w} - \left(I - \frac{\delta^2}{4\gamma}\Delta\right)^{-1}\mathbf{w}\right\|_2, \quad A = \left(I - \frac{\delta^2}{4\gamma}\Delta\right)^{-1}$$

  auxiliary problems equipped with homogeneous Neumann boundary conditions
- *rational LES model with convolution and Smagorinsky subgrid scale model*

$$\nu_T = c_S\delta^2\left\|\mathbb{D}(\mathbf{w})\right\|_F, \quad A = g_\delta\, * \text{ (convolution with Gaussian filter)}$$

- *rational LES model with convolution and Iliescu-Layton subgrid scale model*

$$\nu_T = c_S\delta\left\|\mathbf{w} - \left(I - \frac{\delta^2}{4\gamma}\Delta\right)^{-1}\mathbf{w}\right\|_2, \quad A = g_\delta *$$

  second order partial differential equation equipped with homogeneous Neumann boundary conditions

Fig. 1.1. The flow models occurring in this monograph

LES Models

- The derivation of the above mentioned LES models, which is mainly based on mathematical arguments, is presented in Chapter 4. It is easy to realise that the rational approach by Galdi and Layton [GL00] improves the traditional Taylor polynomial approach by Leonard [Leo74] and Clark et al. [CFR79].
- Several models for the subgrid scale term $\overline{\mathbf{u}'\mathbf{u}'^T}$ are presented in Section 4.3. The Smagorinsky model [Sma63] and a model by Iliescu and Layton [IL98] are considered in more detail in the analysis and the numerical studies. A very important parameter in these models is a scaling factor c_S.
- The slip with friction and no penetration boundary condition is proposed as boundary condition for the large eddies in Section 5.2. Heuristic arguments suggest that this boundary condition is more appropriate for the large eddies than, e.g., the no slip boundary condition.

Analysis in LES

A main topic of this monograph is the presentation of, to our knowledge, all available analytical results for the LES models mentioned above.

- The first rigorous analysis of a commutation error, which is joint work with A. Dunca and W.L. Layton [DJL03b], is presented in Chapter 3. For an arbitrary but fixed time, the convergence of the commutation error with respect to the filter width $\delta \to 0$ is studied. It is shown that the commutation error does in general not converge to zero in $L^p\left(\mathbb{R}^d\right)$, Theorem 3.11. This result implies that the error of simply dropping the commutation error term is $\mathcal{O}(1)$ if the strong form of the space averaged Navier-Stokes equations is discretised, e.g., as in finite difference methods. The convergence of the commutation error to zero for $\delta \to 0$ could be proved for the $H^{-1}(\Omega)$ norm, Theorem 3.15, and for a weak form of the commutation error, Theorem 3.17. This implies that the commutation error vanishes asymptotically as $\delta \to 0$ for discretisations which are based on a variational formulation of the space averaged Navier-Stokes equations, e.g., like finite element methods. Estimates of the order of convergence are given.
- The unique existence of a weak solution of the Smagorinsky model, Theorems 6.12 and 6.14, and the Taylor LES model with Smagorinsky subgrid scale term with c_S sufficiently large, Theorem 6.21, can be proved. The proofs, which are mainly based on a paper by Ladyzhenskaya [Lad67], are presented in Chapter 6.
- The first finite element error analysis for LES models, based on a joint work with W.J. Layton [JL02], is presented in Chapter 8. In this chapter, the continuous-in-time finite element method is considered. For a variant of the Smagorinsky model, an error estimate with constants independent

of the Reynolds number *Re* and with minimal assumptions on the regularity of the solution of the continuous problem can be proved, Theorem 8.17. With stronger assumptions on the regularity of the solution of the continuous problem, a finite element error estimate with constants independent of *Re* can be proved also for the original Smagorinsky model, Theorem 8.18. The main ideas of the finite element error analysis can be carried over to the Taylor LES model with Smagorinsky subgrid scale term and c_S sufficiently large. The final error estimate, obtained in a joint work with T. Iliescu and W.J. Layton [IJL02], is given in Theorem 8.22.

Numerical Algorithms and their Implementation

- A discretisation of the LES models based on
  - second order implicit discretisations in time,
  - inf-sup stable and higher order finite element discretisations in space

  is descriped in Chapter 7. In particular, the treatment of the non-linear terms coming from the space averaged Navier-Stokes equations and the LES models is discussed.
- It is shown that the slip with friction and no penetration boundary conditions can be incorporated easily into a finite element code.
- Solvers for the arising non-symmetric linear saddle point problems are described in detail, together with aspects of their implementation, in Chapter 9. The core of the solvers is a coupled multigrid method with local smoothers of Vanka type.
- Besides the standard multigrid method, which uses the same discretisation on each multigrid level, the so-called multiple discretisation multilevel method for higher order finite element discretisations is presented in Section 9.3.3. The multiple discretisation multilevel method uses stabilised lowest order non-conforming finite element discretisations on all coarse levels. The goal of this method is to combine the accuracy of higher order finite element discretisations with the efficiency of multilevel methods for lowest order non-conforming discretisations.
- It is proposed to use the multigrid methods rather as preconditioners in a flexible GMRES method (FGMRES) than to use them as solvers. The use as preconditioner in FGMRES improves in general the efficiency and the robustness of the solution process.

Numerical Studies of the LES Models

- It is studied in Chapter 10 if the considered LES models fulfil a necessary condition for the acceptability of turbulence models, namely if the computed solutions possess bounded total kinetic energy.
  - The solutions computed with the Taylor LES model with Smagorinsky subgrid scale model and a standard choice of the scaling factor c_S blow

up in finite time. Thus, the Taylor LES model proves to be not suited for the simulation of turbulent flows.
  - The rational LES model with Smagorinsky subgrid scale model or Iliescu-Layton subgrid scale model computes solutions with bounded total kinetic energy with standard choices of the scaling factor in the subgrid scale models.
  - The total kinetic energy of the solutions computed with the rational LES model and the different types of the subgrid scale model differs considerably.
- Chapter 11 studies how good space averaged flow fields are approximated by numerical solutions obtained with the considered LES models. The test examples are mixing layer problems in two and three dimensions.
  - In 2d, it is possible to compute a reference solution with a DNS. Among the considered LES models, the best results are obtained with the rational LES model with Iliescu-Layton subgrid scale term. These results are considerably better than the results computed with the Smagorinsky model and the rational LES model with Smagorinsky subgrid scale term.
  - It was not possible to compute a reference solution in 3d with our computer resources. Thus, the evaluation of the results obtained with the LES models becomes much harder. Based on the behaviour of a characteristic parameter of the flow, which is similar to the 2d results, it can be concluded that the solutions obtained with the rational LES model and both types of the subgrid scale term are better than the solution of the Smagorinsky model in the initial phase of the flow.
  - The scaling factor c_S of the subgrid scale term has often a great influence on the computed results.
- Based on all computational results, the rational LES model with Iliescu-Layton subgrid scale term (4.31) performs best among all considered LES models.

This monograph is devoted mainly to the LES models mentioned above. Thus, many interesting topics in turbulent flow modelling and simulation, even in LES, are not addressed. E.g., LES models whose mathematical analysis is beyond the current state of art or other ways of modelling turbulent flows, like the $k-\varepsilon$ model, scale similarity models, multiscale models, ..., are mentioned only shortly or not at all. We refer for these models to the books on turbulent flow simulation appeared in recent years, e.g., Mohammadi and Pironneau [MP94] or Sagaut [Sag01], and to papers in relevant journals. Many of the turbulence models are still in its development, like the rational LES model considered here, such that new results are published every year. Also for the LES models considered in this monograph, there are a number of open problems which demand further investigations, see Chapter 12.

2

Mathematical Tools and Basic Notations

This chapter provides the mathematical tools which are used in this monograph and introduces basic notations. For a concise presentation of this preliminary material, theorems and inequality are given without proofs.

2.1 Function Spaces

Lebesgue Spaces

Let $\Omega \subset \mathbb{R}^d$, $d \in \{1,2,3\}$, be a domain (open set) and let $p \in [1,\infty]$. The Lebesgue space $L^p(\Omega)$ is the space of all measurable functions v on Ω for which

$$\begin{aligned} \|v\|_{L^p(\Omega)} &:= \left(\int_\Omega |v(\mathbf{x})|^p \, d\mathbf{x}\right)^{1/p} < \infty \quad \text{if } p \in [1,\infty), \\ \|v\|_{L^\infty(\Omega)} &:= \operatorname*{ess\,sup}_{\mathbf{x}\in\Omega} |v(\mathbf{x})| < \infty \quad \text{if } p = \infty. \end{aligned} \tag{2.1}$$

Property (2.1) means that there is a positive constant C such that $|v(\mathbf{x})| \le C$ for almost all $\mathbf{x} \in \Omega$. The spaces $L^p(\Omega)$ are Banach spaces for $p \in [1,\infty]$. $L^2(\Omega)$ is even a Hilbert space with respect to the inner product

$$(v,w) = \int_\Omega v(\mathbf{x})w(\mathbf{x})d\mathbf{x}.$$

Let $p \in (1,\infty)$ and let q be the conjugate exponent of p given by

$$\frac{1}{p} + \frac{1}{q} = 1. \tag{2.2}$$

The conjugate exponent of 1 is ∞ and vice versa. The dual space of $L^p(\Omega)$, $p \in (1,\infty)$, is $L^q(\Omega)$ with the dual pairing

$$\langle v, w \rangle = \int_\Omega v(\mathbf{x})w(\mathbf{x})d\mathbf{x}, \quad v \in L^p(\Omega), \; w \in L^q(\Omega).$$

The spaces $L^p(\Omega)$ are reflexive if and only if $p \in (1, \infty)$.

The space $L_0^2(\Omega)$ is the subspace of all functions from $L^2(\Omega)$ which have integral mean zero:

$$\int_\Omega v(\mathbf{x})d\mathbf{x} = 0.$$

For more details and properties of Lebesgue spaces, we refer to Adams [Ada75, Chapter II].

Sobolev Spaces

Let m and α_i, $1 \le i \le d$, be non-negative integers and $\alpha = (\alpha_1, \ldots, \alpha_d)$ a multi-index, such that $|\alpha| = \sum_{i=1}^d \alpha_i = m$. The partial derivative $D^\alpha v$ of the function v (defined in $\Omega \subset \mathbb{R}^3$) is given by

$$D^\alpha v(\mathbf{x}) := \frac{\partial^m v}{\partial x^{\alpha_1} \partial y^{\alpha_2} \partial z^{\alpha_3}}(\mathbf{x}), \quad \mathbf{x} = (x, y, z) \in \mathbb{R}^3.$$

The Sobolev space $W^{m,p}(\Omega)$ is the space of all functions for which

$$\|v\|_{W^{m,p}(\Omega)} := \left(\sum_{0 \le |\alpha| \le m} \|D^\alpha v\|_{L^p(\Omega)}^p \right)^{1/p} < \infty \quad \text{if } p \in [1, \infty),$$
$$\|v\|_{W^{m,\infty}(\Omega)} := \max_{0 \le |\alpha| \le m} \|D^\alpha v\|_{L^\infty(\Omega)} < \infty \quad \text{if } p = \infty.$$

The spaces $W^{m,p}(\Omega)$ are Banach spaces and the spaces $W^{m,2}(\Omega)$ are Hilbert spaces. Like often in the literature, the notation $H^m(\Omega)$ instead of $W^{m,2}(\Omega)$ is used in this monograph. It is $W^{0,p}(\Omega) = L^p(\Omega)$.

Sobolev spaces with $m \in \mathbb{R}$ can be defined in different ways, e.g., by interpolation between Sobolev spaces with integer m. Usually, these spaces will not appear explicitly in this monograph. Only the $W^{m,p}(\Omega)$ norm of functions, $m \in \mathbb{R}$, has to be estimated in some proofs. This is done with the interpolation estimate (2.15). On the right hand side of this estimate there are norms of the function in Sobolev spaces with integer m.

Let Ω be a bounded domain. The space $W_0^{1,p}(\Omega)$ contains all functions of $W^{1,p}(\Omega)$ which vanish on the boundary $\partial\Omega$ of Ω (in the sense of traces):

$$W_0^{1,p}(\Omega) = \left\{ v \; : \; v \in W^{1,p}(\Omega), v|_{\partial\Omega} = 0 \right\}.$$

In the case $p = 2$, we use $H_0^1(\Omega)$ instead of $W_0^{1,2}(\Omega)$. The dual space of $W_0^{1,p}(\Omega)$, $p \in (1, \infty)$, is $W^{-1,q}(\Omega)$, where q is the conjugate exponent of p given by (2.2). The space $W^{-1,q}(\Omega)$ is equipped with the norm

$$\|v\|_{W^{-1,q}(\Omega)} := \sup_{w \in W_0^{1,p}(\Omega), \|w\|_{W^{1,p}(\Omega)} \neq 0} \frac{\int_\Omega v(\mathbf{x})w(\mathbf{x})d\mathbf{x}}{\|w\|_{W^{1,p}(\Omega)}}. \tag{2.3}$$

For further properties of Sobolev spaces, we refer to Adams [Ada75, Chapter III].

Spaces of Continuous Functions and Hölder Continuous Functions

The space of continuous functions on Ω which have continuous partial derivatives $D^\alpha v$ up to order $|\alpha| \leq m \in \mathbb{N} \cup \{0\}$ is denoted by $C^m(\Omega)$. The space $C^\infty(\Omega)$ consists of all functions which are infinitly often differentiable in Ω.

The subspace of all functions of $C^m(\Omega)$ which have compact support in Ω is denoted by $C_0^m(\Omega)$. The support of a function $v : \Omega \to \mathbb{R}$ is defined by

$$\operatorname{supp}(v) = \overline{\{\mathbf{x} \in \Omega \ : \ v(\mathbf{x}) \neq 0\}},$$

where the overline denotes the closure of a set. If a function v has compact support, then the distance of $\operatorname{supp}(v)$ and $\partial\Omega$ is positive. In particular, v and all existing derivatives of v vanish on $\partial\Omega$. The space $C_0^\infty(\Omega)$ is dense in $L^p(\Omega)$, $p \in [1, \infty)$.

Let $\overline{\Omega}$ be the closure of Ω. The space $C^m(\overline{\Omega})$ is the subspace of $C^m(\Omega)$ of all functions whose partial derivatives up to order m can be extended continuously to $\partial\Omega$. The space $C^m(\overline{\Omega})$ is equipped with the norm

$$\|v\|_{C^m(\overline{\Omega})} := \max_{0 \leq |\alpha| \leq m} \sup_{\mathbf{x} \in \Omega} |D^\alpha v(\mathbf{x})|.$$

The space of Hölder continuous functions $C^{0,\beta}(\overline{\Omega})$, $\beta \in (0, 1]$, is the subspace of $C^0(\overline{\Omega})$ which consists of those functions v for which a Hölder condition with the exponent β is satisfied, i.e. there is a constant C such that

$$|v(\mathbf{x}) - v(\mathbf{y})| \leq C \|\mathbf{x} - \mathbf{y}\|_2^\beta, \quad \forall\, \mathbf{x}, \mathbf{y} \in \Omega, \tag{2.4}$$

where $\|\cdot\|_2$ denotes the Euclidean norm in $\mathbb{R}^d$. For $\beta = 1$, (2.4) is also called Lipschitz condition. The spaces $C^{0,\beta}(\overline{\Omega})$ are Banach spaces.

Spaces in a Time-Space Domain

Let $\Omega \subset \mathbb{R}^d$ be a domain, (a, b) a (real) time interval and v be a function defined on $(a, b) \times \Omega$. Let V be a Banach space of functions defined on Ω with the associated norm $\|\cdot\|_V$. We denote by $L^p(a, b; V)$, $p \in [1, \infty]$, the space of functions $v : (a, b) \to V$ such that

$$\|v\|_{L^p(a,b;V)} := \left(\int_a^b \|v(t)\|_V^p \right)^{1/p} < \infty \quad \text{if } p \in [1, \infty),$$

$$\|v\|_{L^\infty(a,b;V)} := \operatorname*{ess\,sup}_{t \in (a,b)} \|v(t)\|_V < \infty \quad \text{if } p = \infty.$$

Let $I \subset \mathbb{R}$ be an interval which is not necessarily open. For m being a non-negative integer, we denote by $C^m(I;V)$ the space of continuous functions from I to V which are differentiable in I up to the order m.

Spaces of Vector Valued Functions

To simplify the notations, function spaces for vector valued functions are denoted in this monograph in the same way as function spaces for scalar functions. E.g., if $\mathbf{v}$ is a function with d components, then we write $\mathbf{v} \in L^2(\Omega)$ instead of $\mathbf{v} \in \left(L^2(\Omega)\right)^d$ if all components of $\mathbf{v}$ belong to $L^2(\Omega)$. Likewise, the inner product in $\left(L^2(\Omega)\right)^d$ is denoted also by $(\cdot,\cdot)$.

2.2 Some Tools from Analysis and Functional Analysis

Inequalities

This section contains a number of inequalities which are used in this monograph.

- an inequality for sums:

$$\sum_{i=1}^{n} a_i \le \left(\sum_{i=1}^{n} a_i^{1/p}\right)^p, \quad a_i \ge 0, p \ge 1. \tag{2.5}$$

- Young's inequality:

$$ab \le \frac{t}{p}a^p + \frac{t^{-q/p}}{q}b^q, \quad \frac{1}{p}+\frac{1}{q} = 1, \quad 1 < p,q < \infty, \quad t > 0. \tag{2.6}$$

- generalised Hölder's inequality: Let $u \in L^p(\Omega)$, $v \in L^q(\Omega)$, $w \in L^r(\Omega)$ with $1 \le p,q,r \le \infty$ and $p^{-1}+q^{-1}+r^{-1} = 1$. Then $uvw \in L^1(\Omega)$ and

$$\int_\Omega |uvw|\, d\mathbf{x} \le \|u\|_{L^p(\Omega)} \|v\|_{L^q(\Omega)} \|w\|_{L^r(\Omega)}. \tag{2.7}$$

- Poincaré's inequality (or Poincaré-Friedrichs' inequality), see, e.g., Girault and Raviart [GR86, p. 3]: If Ω is connected and bounded, then there exists a constant C such that

$$\|v\|_{H_1(\Omega)} \le C \|\nabla v\|_{L^2(\Omega)} \quad \forall\, v \in H_0^1(\Omega) \tag{2.8}$$

or equivalently

$$\|v\|_{L^2(\Omega)} \le C \|\nabla v\|_{L^2(\Omega)} \quad \forall\, v \in H_0^1(\Omega).$$

Poincaré's inequality holds also for functions $v \in H^1(\Omega)$ with $v = 0$ on $\Gamma_0 \subset \Gamma$ with $\operatorname{meas}(\Gamma_0) > 0$. Poincaré's inequality stays valid for vector-valued functions $\mathbf{v}$ if Ω is bounded with a locally Lipschitz boundary, $\mathbf{v} \in W^{1,q}(\Omega)$, $1 \le q < \infty$ and $\mathbf{v}\cdot\mathbf{n} = 0$ on $\partial\Omega$, see Galdi [Gal94, Section II.4].

- Let $\int_\Omega d\mathbf{x} < \infty$ and $1 \le p \le q \le \infty$. If $u \in L^q(\Omega)$, then $u \in L^p(\Omega)$ and

$$\|u\|_{L^p(\Omega)} \le \left(\int_\Omega d\mathbf{x}\right)^{1/p-1/q} \|u\|_{L^q(\Omega)}, \tag{2.9}$$

  see Adams [Ada75, Theorem 2.8].
- Korn's inequalities relate $L^p(\Omega)$ norms of the deformation tensor $\mathbb{D}(\mathbf{v})$ to the same norms of the gradient $\nabla\mathbf{v}$ for $1 < p < \infty$. There is a $C > 0$ such that for $1 < p < \infty$

$$\|\mathbf{v}\|^p_{W^{1,p}(\Omega)} \le C\left(\|\mathbf{v}\|^p_{L^p(\Omega)} + \|\mathbb{D}(\mathbf{v})\|^p_{L^p(\Omega)}\right) \quad \forall \mathbf{v} \in W^{1,p}(\Omega).$$

  Further, if in addition $|\cdot|$ is a semi norm on $L^p(\Omega)$ which is a norm for the constants, then

$$\|\nabla\mathbf{v}\|_{L^p(\Omega)} \le C\left(|\mathbf{v}| + \|\mathbb{D}(\mathbf{v})\|_{L^p(\Omega)}\right) \quad \forall \mathbf{v} \in W^{1,p}(\Omega). \tag{2.10}$$

- The Gagliardo-Nirenberg inequality states that provided $\partial\Omega$ satisfies a weak regularity condition, which holds in particular for polygonal and polyhedral domains, and $\mathbf{v} = \mathbf{0}$ on $\Gamma_0 \subset \partial\Omega$ with $\text{meas}(\Gamma_0) > 0$, then

$$\|\mathbf{v}\|_{L^q(\Omega)} \le C\,\|\nabla\mathbf{v}\|^a_{L^p(\Omega)}\,\|\mathbf{v}\|^{1-a}_{L^r(\Omega)} \quad \forall \mathbf{v} \in W^{1,p}(\Omega) \tag{2.11}$$

  with $1 \le r \le q \le \infty$, $p \ge 3$ and

$$a = \left(\frac{1}{r} - \frac{1}{q}\right)\left(\frac{1}{3} - \frac{1}{p} + \frac{1}{r}\right)^{-1}.$$

  It follows $0 \le a \le 1$. In particular, taking $q = 6, p = 3, r = 2$ gives

$$\|\mathbf{v}\|_{L^6(\Omega)} \le C\,\|\nabla\mathbf{v}\|^{2/3}_{L^3(\Omega)}\,\|\mathbf{v}\|^{1/3}_{L^2(\Omega)}. \tag{2.12}$$

Theorems in Banach and Hilbert Spaces

This section summarises a number of theorems which will be applied in the analytical parts of this monograph.

- Embedding theorems for Sobolev spaces are used frequently in the analysis presented in this monograph. They can be found in the book by Adams [Ada75]. For convenience, we will give here some embeddings. Let Ω be a bounded domain with locally Lipschitz boundary and let $d \in \{2,3\}$. Then there exist the following embeddings

$$\begin{aligned} &H^1(\Omega) \to L^6(\Omega), \;\; W^{1,3}(\Omega) \to L^6(\Omega), \; H^{1/3}(\Omega) \to L^{18/7}(\Omega), \\ &H^2(\Omega) \to L^\infty(\Omega), \; H^2(\Omega) \to C^{0,\beta}(\overline{\Omega}) \end{aligned} \tag{2.13}$$

with $\beta \in (0,1)$ if $d = 2$ and $\beta = 1/2$ if $d = 3$.
The embedding theorems are used to estimate the norm of functions in a given space by the norm in another space in the following way. Let V be a Banach space such that an embedding $W^{m,p}(\Omega) \to V$ holds. Then, there is a constant C depending on Ω such that

$$\|v\|_V \le C \|v\|_{W^{m,p}(\Omega)}$$

for all functions $v \in W^{m,p}(\Omega)$.

- If an embedding $V \to W$ is in addition compact, then every uniformly bounded sequence $\{v_i\} \in V$ has a subsequence which converges in W. A compact embedding theorem, the so-called Rellich-Kondrachov theorem, can be found in Adams [Ada75]. For convenience, we will give here the compact embeddings which are used. If Ω is a bounded domain in $\mathbb{R}^d, d = 2,3$, with locally Lipschitz boundary, then the embeddings

$$W_0^{1,3}(\Omega) \to L^2(\Omega), \quad H^2(\Omega) \to C^0\left(\overline{\Omega}\right) \tag{2.14}$$

are compact.
- An interpolation theorem for norms of Sobolev spaces is given in Adams [Ada75, Theorem 4.17]. If Ω is bounded, the boundary is locally Lipschitz continuous and if $1 \le p < \infty$, then there exists a constant C such that for $0 \le j \le m$ and for all $v \in W^{m,p}(\Omega)$

$$\|v\|_{W^{j,p}(\Omega)} \le C \|v\|_{W^{m,p}(\Omega)}^{j/m} \|v\|_{L^p(\Omega)}^{(m-j)/m}. \tag{2.15}$$

- Every bounded sequence $\{x_n\}$ in a reflexive Banach space has a weakly convergent subsequence.
- Every bounded set in a Hilbert space is relatively compact.

Gronwall's lemma

Gronwall's lemma is a major tool for the analysis of time dependent problems. We will give two versions of this lemma and refer to Emmrich [Emm99] for proofs and a discussion of the differences of the versions.

Lemma 2.1. Gronwall's lemma in integral form: *Let $T \in \mathbb{R}^+ \cup \infty$, $f, g, \in L^\infty(0,T)$ and $\lambda \in L^1(0,T)$, $\lambda(t) \ge 0$ for almost all $t \in [0,T]$. Then*

$$f(t) \le g(t) + \int_0^t \lambda(s) f(s)\, ds \quad \text{a.e. in } [0,T]$$

implies for almost all $t \in [0,T]$

$$f(t) \le g(t) + \int_0^t \exp\left(\int_s^t \lambda(\tau)\, d\tau\right) \lambda(s) g(s)\, ds.$$

If $g \in W^{1,1}(0,T)$, it follows

$$f(t) \le \exp\left(\int_0^t \lambda(\tau)\, d\tau\right)\left(g(0) + \int_0^t \exp\left(-\int_0^s \lambda(\tau)\, d\tau\right)\right) g'(s)\, ds.$$

Moreover, if g is a monotonically increasing continuous function, it holds

$$f(t) \le \exp\left(\int_0^t \lambda(\tau)\, d\tau\right) g(t).$$

Gronwall's lemma in differential form: *Let $T \in \mathbb{R}^+ \cup \infty$, $f \in W^{1,1}(0,T)$ and $g, \lambda \in L^1(0,T)$. Then*

$$f'(t) \le g(t) + \lambda(t) f(t) \quad \text{a.e. in } [0,T]$$

implies for almost all $t \in [0,T]$

$$f(t) \le \exp\left(\int_0^t \lambda(\tau)\, d\tau\right) f(0) + \int_0^t \exp\left(\int_s^t \lambda(\tau)\, d\tau\right) g(s)\, ds.$$

2.3 Convolution and Fourier Transform

The convolution is the main tool for the definition of the large flow structures in LES, see Chapter 3. The Fourier transform will be used for the derivation of LES models which are studied in this monograph in Chapter 4.

The convolution of two scalar functions f and g is defined by

$$(f * g)(y) = \int_{\mathbb{R}} f(y-x)\, g(x)\, dx = \int_{\mathbb{R}} f(x)\, g(y-x)\, dx = (g * f)(y),$$

provided that the integrals exist for almost all $y \in \mathbb{R}$. The Fourier transform of a scalar function f is defined by

$$\mathcal{F}(f)(y) = \int_{\mathbb{R}} f(x)\, e^{-ixy} dx \tag{2.16}$$

and the inverse Fourier transform of $F(y)$ by

$$\mathcal{F}^{-1}(F)(x) = \frac{1}{2\pi} \int_{\mathbb{R}} F(y)\, e^{ixy} dy. \tag{2.17}$$

It holds

$$\mathcal{F}(f * g) = \mathcal{F}(f)\, \mathcal{F}(g), \quad \mathcal{F}(fg) = \mathcal{F}(f) * \mathcal{F}(g).$$

If f is differentiable and $\lim_{|x|\to\infty} f(x) = 0$, integration by parts yields

$$y\mathcal{F}(f)(y) = -i\mathcal{F}(f')(y).$$

This formulae implies other relations which are used in the derivation of the LES models in Chapter 4

$$\|\mathbf{y}\|_2^2 \mathcal{F}(\mathbf{f}) = -\mathcal{F}(\Delta \mathbf{f}),$$
$$\frac{1}{\|\mathbf{y}\|_2^2} \mathcal{F}(\mathbf{f}) = -\mathcal{F}\left(\Delta^{-1}(\mathbf{f})\right),$$
$$\frac{1}{1 + c\|\mathbf{y}\|_2^2} \mathcal{F}(\mathbf{f}) = \mathcal{F}\left(\left(I - c\Delta\right)^{-1}(\mathbf{f})\right).$$

The $L^r(\Omega)$ norm of $f * g$, $1 \le r \le \infty$, can be estimated by Young's inequality for convolutions (sometimes also called Hölder's inequality for convolutions), e.g., see Hörmander [Hör90, Section IV.4.5]. Let $1 \le p, q \le \infty$, $\frac{1}{p} + \frac{1}{q} \ge 1$ and $\frac{1}{r} = \frac{1}{p} + \frac{1}{q} - 1$. For $f \in L^p(\mathbb{R}^d)$ and $g \in L^q(\mathbb{R}^d)$ is $f * g \in L^r(\mathbb{R}^d)$ and Young's inequality for convolution

$$\|f * g\|_{L^r(\Omega)} \le \|f\|_{L^p(\Omega)} \|g\|_{L^q(\Omega)} \tag{2.18}$$

holds. If $\frac{1}{p} + \frac{1}{q} - 1 = 0$ then $f * g \in L^r(\mathbb{R}^d)$ is continuous and bounded.

The derivation of the space averaged Navier-Stokes equations in a bounded domain requires the convolution of a distribution and a function from $C^\infty(\mathbb{R}^d)$. The technical details of this operation are explained in Section 3.3.

2.4 Notations for Matrix-Vector Operations

In general, scalar quantities are denoted by small letters, vectors by bold small letters and matrices or tensors by capital letters. For operations with matrices or vectors, we use the standard symbols. Let $\mathbf{x} = (x_i)_{i=1}^d$ and $\mathbf{y} = (y_i)_{i=1}^d$ be two (column) vectors and $A = (a_{ij})_{i,j=1}^d$ and $B = (b_{ij})_{i,j=1}^d$ two $d \times d$ matrices. Then the following notations are used:

- The dot product (scalar product, inner product) of two vectors is

$$\mathbf{x} \cdot \mathbf{y} = \sum_{i=1}^{d} x_i y_i.$$

- The dyadic product of two vectors of length d is a $d \times d$ matrix:

$$\mathbf{x}\mathbf{y}^T = (x_i y_j)_{i,j=1}^d.$$

  This operation is in the literature sometimes denoted by $\otimes$.
- The 'dot product' of two matrices is a scalar

$$A : B = \sum_{i,j=1}^{d} a_{ij} b_{ij}.$$

The Frobenius norm of the matrix A is given by

$$\|A\|_F = \left(\sum_{i,j=1}^{d} a_{ij}^2 \right)^{1/2} = (A : A)^{1/2} = \left(\mathrm{tr}(AA^T) \right)^{1/2},$$

where $\mathrm{tr}(\cdot)$ is the trace of a matrix (the sum of its diagonal entries). Often, the Frobenius norm of A is denoted by $|A|$ in the literature.

There is one exception from the strict use of standard matrix-vector notations in the monograph. The divergence operator of a matrix is applied row-wise even if the notations suggested a column-wise application, e.g., for $d = 2$:

$$\nabla \cdot A = \begin{pmatrix} \dfrac{\partial}{\partial x_1} \\ \dfrac{\partial}{\partial x_2} \end{pmatrix} \cdot \begin{pmatrix} a_{11} & a_{12} \\ a_{21} & a_{22} \end{pmatrix} = \begin{pmatrix} \dfrac{\partial a_{11}}{\partial x_1} + \dfrac{\partial a_{12}}{\partial x_2} \\ \dfrac{\partial a_{21}}{\partial x_1} + \dfrac{\partial a_{22}}{\partial x_2} \end{pmatrix}.$$

3

The Space Averaged Navier-Stokes Equations and the Commutation Error

To compute the space averaged velocity $\overline{\mathbf{u}}$ and pressure $\overline{p}$, equations for these quantities are needed. These equations have to be derived from the governing equations for $\mathbf{u}$ and p, i.e. from the Navier-Stokes equations. The simple approach consists in applying the filter which defines $(\overline{\mathbf{u}}, \overline{p})$ also to the Navier-Stokes equations. Then, under the assumption that differentiation and filtering commute, the basic equations of LES, the space averaged Navier-Stokes equations, are obtained. However, it turns out that an additional modelling step is necessary to derive equations for $(\overline{\mathbf{u}}, \overline{p})$ from the space averaged Navier-Stokes equations. This modelling step is discussed in Chapter 4.

This chapter, which is based on a joint work with A. Dunca and W.J. Layton, [DJL03b], deals with the mathematical analysis of the derivation of the space averaged Navier-Stokes equations. There are important situations in which the assumed commutation of filtering and differentiation is not true, e.g., if Ω is a bounded domain. In commuting these operators nevertheless, a so-called commutation error is committed. After deriving the commutation error, Section 3.3, we present in the following sections an analysis of the commutation error in various norms for an arbitrary but fixed time. In practical computations, the commutation error term is always neglected, expecting that it is small and vanishes if the filter width δ tends to zero. In Section 3.5 it is shown that this is not always true ! The commutation error is asymptotically negligible in $L^p(\mathbb{R}^d)$ (i.e., it vanishes as the averaging radius $\delta \to 0$) if and only if the normal stress vanishes almost everywhere on the boundary. With other words, it is asymptotically negligible in $L^p(\mathbb{R}^d)$ if and only if the fluid and the boundary exert exactly zero force on each other. The expected convergence of the commutation error as $\delta \to 0$ can be shown in the $H^{-1}(\Omega)$ norm, Section 3.6, and for a weak form of the commutation error, Section 3.7.

3.1 The Incompressible Navier-Stokes Equations

The conservation of mass for flow problems without sources of mass and for incompressible fluids, that means fluids with constant density ρ, is given by

$$\nabla \cdot \mathbf{v} = 0 \quad \text{in } \Omega. \tag{3.1}$$

The conservation of linear momentum has the form

$$(\mathbf{v}_{t'} + (\mathbf{v} \cdot \nabla)\mathbf{v}) - \nabla \cdot \mathbb{S}(\mathbf{v}, P) = \mathbf{F} \quad \text{in } (0, T] \times \Omega. \tag{3.2}$$

Here, we denote by $\mathbf{v}$ the velocity, by $\mathbf{v}_{t'}$ the derivative of the velocity with respect to the time and by $\mathbb{S}(\mathbf{v}, P)$ the stress tensor

$$\mathbb{S}(\mathbf{v}, P) = 2\nu\mathbb{D}(\mathbf{v}) - \frac{P}{\rho}\mathbb{I}, \tag{3.3}$$

where ν is the kinematic viscosity of the fluid, $\mathbb{D}(\mathbf{v})$ is the deformation tensor (or strain rate tensor) of $\mathbf{v}$

$$\mathbb{D}(\mathbf{v}) = \frac{\nabla \mathbf{v} + \nabla \mathbf{v}^T}{2}, \tag{3.4}$$

P is the pressure, $\mathbf{F}$ an acceleration acting on the fluid and $\mathbb{I}$ the unit tensor. The deformation tensor of $\mathbf{v}$ is the symmetric part of the gradient of the velocity. The skew symmetric part of the gradient of the velocity

$$\frac{\nabla \mathbf{v} - \nabla \mathbf{v}^T}{2}$$

is called spin or vorticity tensor. The gradient of the velocity is a tensor with the components

$$(\nabla \mathbf{v})_{ij} = \frac{\partial v_i}{\partial x_j}, \quad i, j = 1, \ldots, d.$$

The domain $\Omega \subset \mathbb{R}^d, d = 2, 3$ is assumed to be independent of the time. The system of equations (3.1), (3.2) is called incompressible Navier-Stokes equations. It has to be equipped with appropriate initial and boundary conditions.

The functions in system (3.1), (3.2) are not dimensionless. To change this, one introduces the following quantities

- L – a characteristic length scale for the problem,
- U – a characteristic velocity scale for the problem,
- T – a characteristic time scale for the problem.

By the transformation of variables

$$\mathbf{x} = \frac{\mathbf{x}'}{L}, \quad \mathbf{u} = \frac{\mathbf{v}}{U}, \quad t = \frac{t'}{T}$$

one obtains from (3.1), (3.2)

$$\begin{aligned} \frac{L}{UT}\mathbf{u}_t - \frac{2\nu}{UL}\nabla\cdot\mathbb{D}(\mathbf{u}) + (\mathbf{u}\cdot\nabla)\mathbf{u} + \nabla\frac{P}{\rho U^2} &= \frac{L}{U^2}\mathbf{F} \ \text{ in } (0,T]\times\Omega, \\ \nabla\cdot\mathbf{u} &= 0 \qquad \text{in } [0,T]\times\Omega, \end{aligned}$$

where all derivatives are with respect to the new variables. One defines

$$p = \frac{P}{\rho U^2}, \quad Re = \frac{UL}{\nu}, \quad St = \frac{L}{UT}, \quad \mathbf{f} = \frac{L}{U^2}\mathbf{F}$$

and obtains the Navier-Stokes equations in dimensionless form

$$\begin{aligned} St\mathbf{u}_t - \frac{2}{Re}\nabla\cdot\mathbb{D}(\mathbf{u}) + (\mathbf{u}\cdot\nabla)\mathbf{u} + \nabla p &= \mathbf{f} \ \text{ in } (0,T]\times\Omega, \\ \nabla\cdot\mathbf{u} &= 0 \ \text{ in } [0,T]\times\Omega. \end{aligned}$$

The constant Re is called Reynolds number and the constant St Strouhal number. These numbers allow the classification and comparison of different flows. Two stationary flows are similar if their Reynolds numbers coincide and two time dependent flows are similar if they have the same Reynolds number and the same Strouhal number.

To simplify the notations, we use the characteristic quantities $L = 1\ m$, $U = 1\ m/s$, and $T = 1\ s$ such that the dimensionless Navier-Stokes equations have the form

$$\begin{aligned} \mathbf{u}_t - 2\nu\nabla\cdot\mathbb{D}(\mathbf{u}) + (\mathbf{u}\cdot\nabla)\mathbf{u} + \nabla p &= \mathbf{f} \text{ in } (0,T]\times\Omega, \\ \nabla\cdot\mathbf{u} &= 0 \text{ in } [0,T]\times\Omega, \end{aligned} \tag{3.5}$$

with the dimensionless viscosity $\nu = Re^{-1}$. This transform is used also in the numerical solution of the equations. In addition, an initial condition for the velocity

$$\mathbf{u}(0,\cdot) = \mathbf{u}_0 \text{ in } \Omega \tag{3.6}$$

with a given function $\mathbf{u}_0$ and boundary conditions on $\partial\Omega = \Gamma$ have to be prescribed.

The diffusive term and the non-linear convection term can be transformed with the help of the divergence constraint:

$$-2\nu\nabla\cdot\mathbb{D}(\mathbf{u}) = -\nu\Delta\mathbf{u}, \quad (\mathbf{u}\cdot\nabla)\mathbf{u} = \nabla\cdot(\mathbf{u}\mathbf{u}^T), \tag{3.7}$$

where the transformation of the convective term follows immediately from

$$\nabla\cdot(\mathbf{u}\mathbf{v}^T) = (\nabla\cdot\mathbf{v})\mathbf{u} + (\mathbf{v}\cdot\nabla)\mathbf{u}.$$

3.2 The Space Averaged Navier-Stokes Equations in the Case $\Omega = \mathbb{R}^d$

The aim of LES is to compute a space averaged flow accurately. The classical approach is to choose first an appropriate averaging operator which defines a space averaged velocity $\overline{\mathbf{u}}$ and a space average pressure $\overline{p}$. Then, equations for $(\overline{\mathbf{u}}, \overline{p})$ have to be derived which are based on the Navier-Stokes equations (3.5). In the first step of the derivation, the so-called space averaged Navier-Stokes equations are obtained.

The space averaged velocity $\overline{\mathbf{u}}$ and pressure $\overline{p}$ are defined by applying an averaging operator with respect to the spatial variable, a so-called filter, to $(\mathbf{u}, p)$. We will indicate the filter of a function by a bar. This gives the decomposition

$$\mathbf{u} = \overline{\mathbf{u}} + \mathbf{u}', \quad p = \overline{p} + p', \tag{3.8}$$

where $(\overline{\mathbf{u}}, \overline{p})$ are called mean or large scale component and $(\mathbf{u}', p')$ turbulent or subgrid scale component.

We assume that the filter has the following two properties:

- The filter is a linear operator

$$\overline{\mathbf{u} + \lambda \mathbf{v}} = \overline{\mathbf{u}} + \lambda \overline{\mathbf{v}}. \tag{3.9}$$

- Derivatives and averages commute, e.g.,

$$\overline{\left(\frac{\partial \mathbf{u}}{\partial x_i}\right)} = \left(\frac{\partial \overline{\mathbf{u}}}{\partial x_i}\right), \; i = 1, \ldots, d, \quad \overline{\left(\frac{\partial \mathbf{u}}{\partial t}\right)} = \left(\frac{\partial \overline{\mathbf{u}}}{\partial t}\right). \tag{3.10}$$

In LES, the filter operator is often given by a convolution with an appropriate filter function, see Section 3.4 for a concrete example. Hence, the filter operator is linear. If $\Omega = \mathbb{R}^d$ and if the functions $\mathbf{u}, p, \mathbf{f}$ in (3.5) are sufficiently smooth in space and time, then filtering and differentiation also commute. The crucial assumption is that $\Omega = \mathbb{R}^d$. In fact, the regularity of the functions plays a minor role since convolution and differentiation commute even for distributions.

Let $\Omega = \mathbb{R}^d$. Applying a filter which has the properties (3.9) and (3.10) to the Navier-Stokes equations (3.5) and the initial condition (3.6) gives the space averaged Navier-Stokes equations (or Reynolds equations)

$$\begin{aligned} \overline{\mathbf{u}}_t - 2\nu \nabla \cdot \mathbb{D}(\overline{\mathbf{u}}) + \nabla \cdot \left(\overline{\mathbf{u}\mathbf{u}^T}\right) + \nabla \overline{p} &= \overline{\mathbf{f}} \quad \text{in } (0, T] \times \mathbb{R}^d, \\ \nabla \cdot \overline{\mathbf{u}} &= 0 \quad \text{in } [0, T] \times \mathbb{R}^d, \\ \overline{\mathbf{u}}(0, \cdot) &= \overline{\mathbf{u}}_0 \text{ in } \mathbb{R}^d, \end{aligned} \tag{3.11}$$

where the linearity of the filter implies

$$\overline{\mathbf{u}\mathbf{u}^T} = \overline{\overline{\mathbf{u}}\,\overline{\mathbf{u}}^T} + \overline{\overline{\mathbf{u}}\mathbf{u}'^T} + \overline{\mathbf{u}'\overline{\mathbf{u}}^T} + \overline{\mathbf{u}'\mathbf{u}'^T}. \tag{3.12}$$

The first term in (3.12) is called large scale advective term. It describes the convection of the large eddies driven by themselves. The second and third term are the cross terms describing the interaction of the large scale and subgrid scale components. The last tensor is the subgrid scale term which describes how the small eddies extract energy from the flow.

Remark 3.1. The closure problem in the space averaged Navier-Stokes equations. Since the dyadic product of a d dimensional vector with itself is a symmetric matrix, $\overline{\mathbf{u}\mathbf{u}^T}$ is symmetric, too. The unknowns in $\overline{\mathbf{u}\mathbf{u}^T}$ are a priori not related to $(\overline{\mathbf{u}}, \overline{p})$. Thus, on the one hand there are in (3.11) $d+1$ equations and on the other hand there are $d+1$ unknown space averaged values and $d(d+1)/2$ unknown quantities in $\overline{\mathbf{u}\mathbf{u}^T}$. Thus, a closure problem arises and the term $\overline{\mathbf{u}\mathbf{u}^T}$ or the individual terms in (3.12) need to be modelled. Modelling $\overline{\mathbf{u}\mathbf{u}^T}$ in terms of $\overline{\mathbf{u}}$ is the main issue in LES. This topic will be addressed in Chapter 4. □

Remark 3.2. Averaging in time. The concept of LES models does not include averaging in time. Temporal averaging is used in the derivation of Reynolds Averaged Navier-Stokes (RANS) models, e.g., see Ferziger and Perić [FP99, Section 9.4] for more details. □

3.3 The Space Averaged Navier-Stokes Equations in a Bounded Domain

Remark 3.3. The space averaged Navier-Stokes equations in bounded domains in practical computations. Usually, (3.11) is also used in practical computations in bounded domains Ω, i.e. $\mathbb{R}^d$ is simply replaced by Ω. However, if Ω is a bounded domain, the commutation of filtering and differentiation requires special attention. In general, an extra term occurs and omitting this term leads to the so-called commutation error. This commutation error will be studied in the following sections. □

We will consider in the remainder of this chapter the Navier-Stokes equations in a bounded domain where the velocity is equipped with homogeneous Dirichlet boundary conditions. In order to apply a convolution operator, one has to extend first all functions outside the domain. These functions will fulfil the Navier-Stokes equations in a distributional sense. Then, the convolution operator can be applied, filtering and differentiation commute and space averaged Navier-Stokes equations are obtained.

Let Ω be a bounded domain in $\mathbb{R}^d$, $d = 2, 3$, with Lipschitz boundary $\partial\Omega$, outward pointing unit normal $\mathbf{n}_{\partial\Omega}$ and $(d-1)$ dimensional measure $|\partial\Omega| < \infty$. We consider the incompressible Navier-Stokes equations with homogeneous Dirichlet boundary conditions

$$
\begin{aligned}
\mathbf{u}_t - 2\nu\nabla\cdot\mathbb{D}(\mathbf{u}) + \nabla\cdot\left(\mathbf{u}\mathbf{u}^T\right) + \nabla p &= \mathbf{f} \quad \text{in } (0,T]\times\Omega, \\
\nabla\cdot\mathbf{u} &= 0 \quad \text{in } [0,T]\times\Omega, \\
\mathbf{u} &= \mathbf{0} \quad \text{on } [0,T]\times\partial\Omega, \\
\mathbf{u}(0,\cdot) &= \mathbf{u}_0 \text{ in } \Omega, \\
\int_\Omega p d\mathbf{x} &= 0 \quad \text{in } (0,T].
\end{aligned}
\tag{3.13}
$$

The stress tensor is given by

$$\mathbb{S}(\mathbf{u}, p) = 2\nu\mathbb{D}(\mathbf{u}) - p\mathbb{I}$$

and the normal stress or Cauchy stress vector or traction vector on $\partial\Omega$ is defined by $\mathbb{S}(\mathbf{u}, p)\,\mathbf{n}_{\partial\Omega}$.

Our analysis will require that (3.13) possesses a unique solutions $(\mathbf{u}, p)$ which is sufficiently regular, such that the normal stress has a well defined trace on the $\partial\Omega$ which belongs to some Lebesgue space defined on $\partial\Omega$. We assume that

$$\begin{aligned} \mathbf{u} \in H^2(\Omega) \cap H_0^1(\Omega),\ p \in H^1(\Omega) \cap L_0^2(\Omega) &\ \text{ for a.e. } t \in [0, T], \\ \mathbf{u} \in H^1((0, T)) &\ \text{ for a.e. } \mathbf{x} \in \overline{\Omega}. \end{aligned} \tag{3.14}$$

Lemma 3.4. *If (3.14) holds then $\mathbb{S}(\mathbf{u}, p)\,\mathbf{n}_{\partial\Omega}$ belongs to $H^{1/2}(\partial\Omega)$. In particular, for almost every $t \in [0, T]$, $\mathbb{S}(\mathbf{u}, p)\,\mathbf{n}_{\partial\Omega} \in L^q(\partial\Omega)$ with $1 \le q < \infty$ if $d = 2$ and $1 \le q \le 4$ if $d = 3$ and*

$$\|\mathbb{S}(\mathbf{u}, p)\,\mathbf{n}_{\partial\Omega}\|_{L^q(\partial\Omega)} \le C\left(\nu\,\|\mathbf{u}\|_{H^2(\Omega)} + \|p\|_{H^1(\Omega)}\right). \tag{3.15}$$

Proof. This follows from the usual trace theorem and embedding theorems, e.g., see Galdi [Gal94, Chapter II, Theorem 3.1]. □

In writing down an equation to which the convolution with a filter function can be applied, $\mathbf{f}$ must be extended off Ω and then $(\mathbf{u}, p)$ must be extended compatible with the extension of $\mathbf{f}$. For $\overline{\mathbf{f}}$ to be easily computable, $\mathbf{f}$ is extended by $\mathbf{0}$ off Ω. Thus, $(\mathbf{u}, p)$ must be extended by zero off Ω, too. This extension is reasonable since $\mathbf{u} = \mathbf{0}$ on $\partial\Omega$. An extension of $\mathbf{u}$ off Ω as an $H^2(\mathbb{R}^d)$ function exists but is unknown, in particular since $\mathbf{u}$ is not known. Using this extension, instead of $\mathbf{u} \equiv \mathbf{0}$ on $\mathbb{R}^d \setminus \Omega$, would make the extension of $\mathbf{f}$ unknowable and hence $\overline{\mathbf{f}}$ uncomputable in the space averaged momentum equation. Thus, define

$$\mathbf{u} = \mathbf{0}, \quad \mathbf{u}_0 = \mathbf{0}, \quad p = 0, \quad \mathbf{f} = \mathbf{0} \quad \text{if } \mathbf{x} \notin \overline{\Omega}.$$

The extended functions posses the following regularities

$$\begin{aligned} \mathbf{u} \in H_0^1(\mathbb{R}^d),\ p \in L_0^2(\mathbb{R}^d) &\ \text{ for a.e. } t \in [0, T], \\ \mathbf{u} \in H^1((0, T)) &\ \text{ for a.e. } \mathbf{x} \in \mathbb{R}^d. \end{aligned} \tag{3.16}$$

From (3.14) and (3.16) follow that the first order weak derivatives of the extended velocity $\mathbf{u}_t$, $\nabla\mathbf{u}$,$\nabla \cdot \mathbf{u}$ and $\nabla \cdot (\mathbf{u}\mathbf{u}^T)$ are well defined on $\mathbb{R}^d$, taking their indicated values in Ω and being identically zero off Ω.

Since $\mathbf{u} \notin H^2(\mathbb{R}^d)$, $p \notin H^1(\mathbb{R}^d)$, the terms $\nabla \cdot \mathbb{D}(\mathbf{u})$ and ∇p must be defined in the sense of distributions. To this end, let $\varphi \in C_0^\infty(\mathbb{R}^d)$. Since $p \equiv 0$ on $\mathbb{R}^d \setminus \Omega$ for all times, we get

$$\begin{aligned}(\nabla p)(\varphi)(t) &:= -\int_{\mathbb{R}^d} p(t,\mathbf{x})\,\nabla\varphi(\mathbf{x})\,d\mathbf{x} \\ &= \int_{\Omega}\varphi(\mathbf{x})\,\nabla p(t,\mathbf{x})\,d\mathbf{x} - \int_{\partial\Omega}\varphi(\mathbf{s})\,p(t,\mathbf{s})\,\mathbf{n}_{\partial\Omega}(\mathbf{s})\,ds.\end{aligned} \tag{3.17}$$

In the same way, one obtains

$$\begin{aligned}&\nabla\cdot\mathbb{D}(\mathbf{u})(\varphi)(t) \\ &:= -\int_{\mathbb{R}^d}\mathbb{D}(\mathbf{u})(t,\mathbf{x})\,\nabla\varphi(\mathbf{x})\,d\mathbf{x} \\ &= \int_{\Omega}\varphi(\mathbf{x})\,\nabla\cdot\mathbb{D}(\mathbf{u})(t,\mathbf{x})\,d\mathbf{x} - \int_{\partial\Omega}\varphi(\mathbf{s})\,\mathbb{D}(\mathbf{u})(t,\mathbf{s})\,\mathbf{n}_{\partial\Omega}(\mathbf{s})\,ds.\end{aligned} \tag{3.18}$$

Both distributions have compact support. From (3.17) and (3.18), it follows that the extended functions $(\mathbf{u},p)$ fulfil the following distributional form of the momentum equation

$$\begin{aligned}&\left(\mathbf{u}_t - 2\nu\nabla\cdot\mathbb{D}(\mathbf{u}) + \nabla\cdot\left(\mathbf{u}\mathbf{u}^T\right) + \nabla p\right)(\varphi)(t) \\ &= \mathbf{f}(\varphi)(t) + \int_{\partial\Omega}\mathbb{S}(\mathbf{u},p)(t,\mathbf{s})\,\mathbf{n}_{\partial\Omega}(\mathbf{s})\,\varphi(\mathbf{s})\,ds.\end{aligned} \tag{3.19}$$

The correct space averaged Navier-Stokes equations are now derived by convolving (3.19) with a filter function $g(\mathbf{x}) \in C^\infty\left(\mathbb{R}^d\right)$. Applying the convolution with g to (3.19) gives a function in $C^\infty(\Omega)$ and moreover convolution and differentiation commute on $\mathbb{R}^d$, Hörmander [Hör90, Theorem 4.1.1]. This yields immediately

$$\begin{aligned}&g * \left[\left(\mathbf{u}_t - 2\nu\nabla\cdot\mathbb{D}(\mathbf{u}) + \nabla\cdot\left(\mathbf{u}\mathbf{u}^T\right) + \nabla p\right)(\varphi)\right] \\ &= \overline{\mathbf{u}}_t - 2\nu\nabla\cdot\mathbb{D}(\overline{\mathbf{u}}) + \nabla\cdot\left(\overline{\mathbf{u}\mathbf{u}^T}\right) + \nabla\overline{p}\end{aligned} \tag{3.20}$$

in $(0,T]\times\mathbb{R}^d$. Let $H(\varphi)$ be a distribution with compact support which has the form

$$H(\varphi) = -\int_{\mathbb{R}^d} f(\mathbf{x})\,D^\alpha\varphi(\mathbf{x})\,d\mathbf{x},$$

where D^α is the derivative of φ with the multi-index α. Then, $H*g \in C^\infty\left(\mathbb{R}^d\right)$, see Rudin [Rud91, Theorem 6.35], where

$$\overline{H}(\mathbf{x}) = (H*g)(\mathbf{x}) := H(g(\mathbf{x}-\cdot)) = -\int_{\mathbb{R}^d} f(\mathbf{y})\,D^\alpha g(\mathbf{x}-\mathbf{y})\,d\mathbf{y}. \tag{3.21}$$

Applying (3.21) to (3.17) gives

$$\begin{aligned}\nabla\overline{p}(t,\mathbf{x}) &= g * ((\nabla p)(\varphi))(t,\mathbf{x}) \\ &= -\int_{\mathbb{R}^d} p(t,\mathbf{y})\,\nabla g(\mathbf{x}-\mathbf{y})\,d\mathbf{y} \\ &= \int_{\Omega}\nabla p(t,\mathbf{y})\,g(\mathbf{x}-\mathbf{y})\,d\mathbf{y} - \int_{\partial\Omega} g(\mathbf{x}-\mathbf{s})\,p(t,\mathbf{s})\,\mathbf{n}_{\partial\Omega}(\mathbf{s})\,ds.\end{aligned} \tag{3.22}$$

Convolving (3.18) in the same way yields

$$\begin{aligned}
&\nabla \cdot \mathbb{D}(\,\overline{\mathbf{u}}\,)\,(t,\mathbf{x}) \qquad\qquad (3.23)\\
&= \int_{\Omega} \nabla \cdot \mathbb{D}(\mathbf{u})\,(t,\mathbf{y})\,g\,(\mathbf{x}-\mathbf{y})\,d\mathbf{y} - \int_{\partial\Omega} g\,(\mathbf{x}-\mathbf{s})\,\mathbb{D}(\mathbf{u})\,(t,\mathbf{s})\,\mathbf{n}_{\partial\Omega}\,(\mathbf{s})\,ds.
\end{aligned}$$

Since $\mathbf{f}$ vanishes outside Ω for $t \in [0,T]$, we have

$$\overline{\mathbf{f}}\,(t,\mathbf{x}) = \int_{\Omega} \left(\mathbf{u}_t - 2\nu\nabla\cdot\mathbb{D}(\mathbf{u}) + \nabla\cdot\left(\mathbf{u}\mathbf{u}^T\right) + \nabla p\right)(t,\mathbf{y})g\,(\mathbf{x}-\mathbf{y})\,d\mathbf{y}. \quad (3.24)$$

Combining (3.20), (3.22), (3.23) and (3.24) yields the space averaged momentum equation

$$\begin{aligned}
&\overline{\mathbf{u}}_t - 2\nu\nabla\cdot\mathbb{D}(\,\overline{\mathbf{u}}\,) + \nabla\cdot\left(\overline{\mathbf{u}\mathbf{u}^T}\right) + \nabla\,\overline{p}\\
&= \overline{\mathbf{f}} + \int_{\partial\Omega} g\,(\mathbf{x}-\mathbf{s})\,\mathbb{S}\,(\mathbf{u},p)\,(t,\mathbf{s})\,\mathbf{n}_{\partial\Omega}\,(\mathbf{s})\,ds \quad \text{in } (0,T]\times\mathbb{R}^d. \quad (3.25)
\end{aligned}$$

Definition 3.5. *Let $g \in C^\infty\left(\mathbb{R}^d\right)$ be a filter function with filter width δ. The commutation error $A_\delta\left(\mathbb{S}\,(\mathbf{u},p)\right)$ in the space averaged Navier-Stokes equations is defined to be*

$$A_\delta\left(\mathbb{S}\,(\mathbf{u},p)\right)(t,\mathbf{x}) := \int_{\partial\Omega} g\,(\mathbf{x}-\mathbf{s})\,\mathbb{S}\,(\mathbf{u},p)\,(t,\mathbf{s})\,\mathbf{n}_{\partial\Omega}\,(\mathbf{s})\,ds.$$

Remark 3.6. Analysis of the commutation error for arbitrary but fixed time. In the analysis of the commutation error, we will consider an arbitrary but fixed time $t \in (0,T]$ such that the dependence of $A_\delta\left(\mathbb{S}\,(\mathbf{u},p)\right)$ on the time can be neglected. □

Remark 3.7. Laplacian form of the viscous term. If the viscous term in the Navier-Stokes equations is written as $\nu\Delta\mathbf{u}$ instead of $2\nu\nabla\cdot\mathbb{D}(\mathbf{u})$, the resulting space averaged momentum equation is given by replacing $2\nu\mathbb{D}(\,\overline{\mathbf{u}}\,)$ in (3.25) by $\nu\nabla\,\overline{\mathbf{u}}$ and $2\nu\mathbb{D}(\mathbf{u})\,(t,\mathbf{s})$ by $\nu\nabla\mathbf{u}\,(t,\mathbf{s})$ in the stress tensor. □

The correct space averaged Navier-Stokes equations arising from the Navier-Stokes equations on a bounded domain thus posses an extra boundary integral, $A_\delta\left(\mathbb{S}\,(\mathbf{u},p)\right)$. Omitting this integral results in a commutation error. Including this integral in the space averaged momentum equation introduces a new modelling question since it depends on the unknown normal stress on $\partial\Omega$ of $(\mathbf{u},p)$ and not of $(\,\overline{\mathbf{u}}\,,\,\overline{p}\,)$.

Remark 3.8. Analytical studies of the commutation error in the literature. Analytical studies of the commutation error can be found, e.g., in Fureby and Tabor [FT97], Ghosal and Moin [GM95], or Vasilyev et al. [VLM98]. The analysis in these papers is based on Taylor series expansions and it is done in one space dimension. The Taylor series expansions require a smoothness of functions, which is in general not given in practice. A special analysis of the end points a, b of a one dimensional domain (a, b), which is necessary in our opinion, is missing. The final claim in [FT97, GM95, VLM98] is that the commutation error is of second order with respect to δ in (a, b). This result is used to justify that the commutation error is neglected in practice.

Despite the statement that the commutation error can be neglected, almost all computations with LES models on bounded domains, which can be found in the literature, use a special treatment of the region near the boundary. Very popular are wall models, e.g., see Sagaut [Sag01, Section 9.2.2] for an overview. Another approach is to apply a filter with a spatial variation of the filter length which tends to zero at the boundary, e.g., Ghosal and Moin [GM95]. This approach requires extra resolution and another commutation error due to the non-constant filter width occurs. There has been proposed also a new approach in LES which does not use filtering at all - the variational multiscale method by Hughes et al. [HMJ00]. We think that one reason of the necessity of the special treatment of the near boundary region is the neglect of the commutation error. □

3.4 The Gaussian Filter

Filters are an important tool in the development of LES models. An overview of commonly used filters in LES is given by Aldama [Ald90, p. 19ff]. We will use throughout this monograph the so-called Gaussian filter and present it here in detail.

A filter of a function $f(t, \mathbf{x})$ with respect to the space variable can be defined by convolution

$$\overline{f}(t, \mathbf{x}) = (F * f)(t, \mathbf{x}) = \int_{\mathbb{R}^d} F(\mathbf{x} - \mathbf{z}) f(t, \mathbf{z})\, d\mathbf{z}, \tag{3.26}$$

where F is a suitable filter function. Applying the Fourier transform to (3.26) gives

$$\mathcal{F}\left(\overline{f}\right)(t, \mathbf{y}) = \left(\mathcal{F}(F)\,\mathcal{F}(f)\right)(t, \mathbf{y}),$$

where $\mathbf{y}$ is the dual variable (wave number). If $\mathcal{F}(F)(t, \mathbf{y}) = 0$ for $\|\mathbf{y}\|_2 > y_c$, where y_c is a cut-off wave number, then all high wave number components of $f(t, \mathbf{x})$ are filtered out by convolving f with F. This is an ideal situation. For practical applications it is sufficient that the high wave numbers are damped out fast enough. Thus, the most essential requirement on a filter function is that its Fourier transform decreases rapidly for high wave numbers. Then, a cut-off wave number can be chosen in applications such that the loss of information can be neglected.

We assume that the filter function $F(\mathbf{x})$ can be represented as tensor product of one dimensional filter functions

$$F(\mathbf{x}) = \prod_{i=1}^{d} F_i(x_i).$$

Then, the Fourier transform of the filter function is

$$\mathcal{F}(F)(\mathbf{y}) = \prod_{i=1}^{d} \mathcal{F}(F_i)(y_i).$$

Let the positive constant δ be the characteristic filter width or averaging radius or scale of the filter, i.e. all eddies with size lower than $\mathcal{O}(\delta)$ should be filtered out. It is clear, the smaller δ is the larger becomes the cut-off wave number y_c and the less eddies are filtered out.

The filter function for the Gaussian filter is given by

$$F_i(x_i) = \sqrt{\frac{\gamma}{\pi\delta^2}} \exp\left(-\frac{\gamma}{\delta^2} x_i^2\right)$$

and its Fourier transform is

$$\mathcal{F}(F_i)(y_i) = \exp\left(-\frac{\delta^2}{4\gamma} y_i^2\right), \tag{3.27}$$

where γ is a constant. This constant is chosen to be $\gamma = 6$ which equilibrates the filtering effect of the Gaussian filter and the filtering effect of a certain discrete approximation in a model problem, see Aldama [Ald90, Section 3.8] for a detailed discussion. The Gaussian filter function and its Fourier transform are of the same form. Aldama [Ald90] proposes to use $y_c = 2\pi/\delta$ as a cut-off wave number in applications since $\mathcal{F}(F_i)(y_i)$ decays rapidly and it is small outside $[-2\pi/\delta, 2\pi/\delta]$.

The Gaussian filter g_δ in $\mathbb{R}^d$ is given by

$$g_\delta(\mathbf{x}) = \left(\frac{6}{\delta^2\pi}\right)^{d/2} \exp\left(-\frac{6}{\delta^2} \|\mathbf{x}\|_2^2\right), \tag{3.28}$$

see Figure 3.1. Its Fourier transform is

$$\mathcal{F}(g_\delta)(\mathbf{y}) = \exp\left(-\frac{\delta^2}{24} \|\mathbf{y}\|_2^2\right). \tag{3.29}$$

The Gaussian filter fits into the framework of Section 3.3. It has the following properties which are easy to verify:

- regularity: $g_\delta \in C^\infty(\mathbb{R}^d)$, $\mathcal{F}(g_\delta) \in C^\infty(\mathbb{R}^d)$,
- positivity: $0 < g_\delta(\mathbf{x}) \le \left(\frac{6}{\delta^2\pi}\right)^{\frac{d}{2}}$, $0 < \mathcal{F}(g_\delta)(\mathbf{y}) \le 1$
- integrability: $\|g_\delta\|_{L^p(\mathbb{R}^d)} < \infty$, $1 \le p \le \infty$, in particular $\|g_\delta\|_{L^1(\mathbb{R}^d)} = 1$,
- symmetry: $g_\delta(\mathbf{x}) = g_\delta(-\mathbf{x})$, $\mathcal{F}(g_\delta)(\mathbf{y}) = \mathcal{F}(g_\delta)(-\mathbf{y})$,

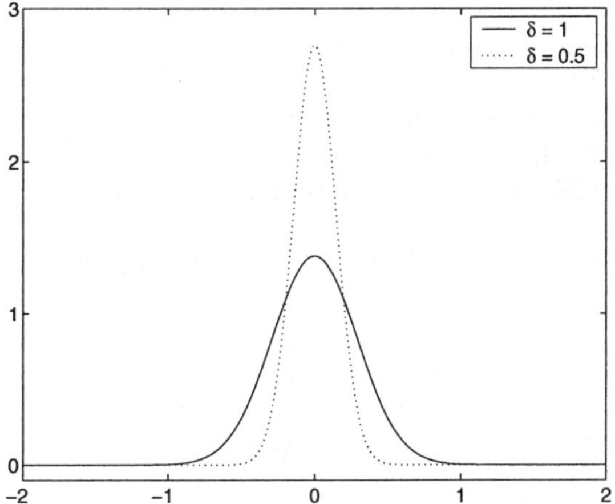

Fig. 3.1. The Gaussian filter in one dimension for different δ

- monotonicity: $g_\delta(\mathbf{x}) \geq g_\delta(\mathbf{y})$ if $\|\mathbf{x}\|_2 \leq \|\mathbf{y}\|_2$.

Lemma 3.9.

i) Let $\varphi \in L^p\left(\mathbb{R}^d\right)$, *then for* $1 \leq p < \infty$

$$\lim_{\delta \to 0} \|g_\delta * \varphi - \varphi\|_{L^p(\mathbb{R}^d)} = 0.$$

ii) Let $\varphi \in L^\infty\left(\mathbb{R}^d\right)$. *If* φ *is uniformly continuous on a set* ω, *then* $g_\delta * \varphi \to \varphi$ *uniformly on* ω *as* $\delta \to 0$.

iii) If $\varphi \in C_0^\infty\left(\mathbb{R}^d\right)$, *then for* $1 \leq p < \infty$, $0 \leq r < \infty$

$$\lim_{\delta \to 0} \|g_\delta * \varphi - \varphi\|_{W^{r,p}(\mathbb{R}^d)} = 0.$$

Proof. The proof of the first two statements can be found, e.g., in Folland [Fol95, Theorem 0.13]. The third statement is an immediate consequence of the first one. □

For convenience of notations, the Gaussian filter with a scalar argument x is understood in this chapter to be

$$g_\delta(x) := \left(\frac{6}{\delta^2 \pi}\right)^{\frac{d}{2}} \exp\left(-\frac{6x^2}{\delta^2}\right).$$

3.5 Error Estimate of the Commutation Error Term in the $L^p\left(\mathbb{R}^d\right)$ Norm

In this section, it is shown that the commutation error $A_\delta(\mathbb{S}(\mathbf{u}, p))$ is a function in $L^p\left(\mathbb{R}^d\right)$, $1 \leq p \leq \infty$, and that $A_\delta(\mathbb{S}(\mathbf{u}, p))$ vanishes in $L^p\left(\mathbb{R}^d\right)$ as

$\delta \to 0$ if and only if the normal stress is identically zero almost everywhere on $\partial\Omega$. This condition means that the wall has zero influence on the wall-bounded turbulent flow. Thus, it is not expected to be satisfied in any interesting flow problem ! If the commutation error term is simply dropped and then the strong form of the space averaged Navier-Stokes equations is discretised, as by, e.g., a finite difference method, this result shows that the error committed is $\mathcal{O}(1)$!

In view of Definition 3.5, Lemma 3.4 and Remark 3.6, it is necessary to study terms of the form

$$\int_{\partial\Omega} g_\delta(\mathbf{x}-\mathbf{s})\,\psi(\mathbf{s})\,ds \tag{3.30}$$

with $\psi \in L^q(\partial\Omega)$, $1 \le q \le \infty$. We will first show, that (3.30) is a function in $L^p(\mathbb{R}^d)$ with $1 \le p \le \infty$.

Theorem 3.10. *Let $\psi \in L^q(\partial\Omega)$, $1 \le q \le \infty$, then (3.30) belongs to $L^p(\mathbb{R}^d)$, $1 \le p \le \infty$.*

Proof. By the Cauchy-Schwarz inequality, one obtains with $r^{-1}+q^{-1}=1$, $q>1$,

$$\begin{aligned}
&\left|\int_{\partial\Omega} g_\delta(\mathbf{x}-\mathbf{s})\,\psi(\mathbf{s})\,ds\right| \\
&\le \left(\int_{\partial\Omega} g_\delta^r(\mathbf{x}-\mathbf{s})\,ds\right)^{1/r} \|\psi\|_{L^q(\partial\Omega)} \\
&= \left(\int_{\partial\Omega} \left(\frac{6}{\delta^2\pi}\right)^{rd/2} \exp\left(-\frac{6r}{\delta^2}\|\mathbf{x}-\mathbf{s}\|_2^2\right) ds\right)^{1/r} \|\psi\|_{L^q(\partial\Omega)}\,.
\end{aligned}$$

By the triangle inequality and Young's inequality, it follows

$$2\|\mathbf{x}-\mathbf{s}\|_2^2 \ge \|\mathbf{x}\|_2^2 - 2\|\mathbf{s}\|_2^2,$$

which implies

$$\exp\left(-\frac{6r\|\mathbf{x}-\mathbf{s}\|_2^2}{\delta^2}\right) \le \exp\left(3r\frac{-\|\mathbf{x}\|_2^2+2\|\mathbf{s}\|_2^2}{\delta^2}\right),$$

and

$$\begin{aligned}
&\left|\int_{\partial\Omega} g_\delta(\mathbf{x}-\mathbf{s})\,\psi(\mathbf{s})\,ds\right| \\
&\le \left(\frac{6}{\delta^2\pi}\right)^{d/2} \|\psi\|_{L^q(\partial\Omega)} \left(\int_{\partial\Omega} \exp\left(\frac{6r\|\mathbf{s}\|_2^2}{\delta^2}\right) ds\right)^{1/r} \exp\left(-\frac{3\|\mathbf{x}\|_2^2}{\delta^2}\right) \\
&< \infty,
\end{aligned} \tag{3.31}$$

since $\partial\Omega$ has finite measure in $\mathbb{R}^{d-1}$ and the exponential is a bounded function. This proves the statement for $L^\infty(\mathbb{R}^d)$. The proof for $p \in [1,\infty)$ is obtained by raising both sides of (3.31) to the power p, integrating on $\mathbb{R}^d$ and using

$$\int_{\mathbb{R}^d} \exp\left(-\frac{3p\,\|\mathbf{x}\|_2^2}{\delta^2}\right) d\mathbf{x} < \infty.$$

If $q = 1$, we have for $1 \le p < \infty$

$$\begin{aligned}\int_{\mathbb{R}^d}\left|\int_{\partial\Omega} g_\delta(\mathbf{x}-\mathbf{s})\,\psi(\mathbf{s})\,d\mathbf{s}\right|^p d\mathbf{x} &\le \int_{\mathbb{R}^d} \sup_{\mathbf{s}\in\partial\Omega} g_\delta^p(\mathbf{x}-\mathbf{s})\,d\mathbf{x}\ \|\psi\|^p_{L^1(\partial\Omega)} \\ &= \int_{\mathbb{R}^d} g_\delta^p(d(\mathbf{x},\partial\Omega))\,d\mathbf{x}\ \|\psi\|^p_{L^1(\partial\Omega)}.\end{aligned}$$

We choose a ball $B(\mathbf{0},R)$ with radius R such that $d(\mathbf{x},\partial\Omega) > \|\mathbf{x}\|_2/2$ for all $\mathbf{x} \notin B(\mathbf{0},R)$. Then, the integral on $\mathbb{R}^d$ is split into a sum of two integrals. The first integral is computed on $B(\mathbf{0},R)$. This is finite since the integrand is a continuous function on $\overline{B}(\mathbf{0},R)$. The second integral on $\mathbb{R}^d \setminus B(\mathbf{0},R)$ is also finite because

$$\int_{\mathbb{R}^d\setminus B(\mathbf{0},R)} g_\delta^p(d(\mathbf{x},\partial\Omega))\,d\mathbf{x} \le \int_{\mathbb{R}^d} g_\delta^p\left(\frac{\|\mathbf{x}\|_2}{2}\right) d\mathbf{x}$$

and the integrability of the Gaussian filter. This concludes the proof for $p < \infty$. For $p = \infty$, we have

$$\begin{aligned}\operatorname*{ess\,sup}_{\mathbf{x}\in\mathbb{R}^d}\left|\int_{\partial\Omega} g_\delta(\mathbf{x}-\mathbf{s})\,\psi(\mathbf{s})\,d\mathbf{s}\right| &\le \operatorname*{ess\,sup}_{\mathbf{x}\in\mathbb{R}^d}\operatorname*{ess\,sup}_{\mathbf{s}\in\partial\Omega} g_\delta(\mathbf{x}-\mathbf{s})\,\|\psi\|_{L^1(\partial\Omega)} \\ &\le g_\delta(\mathbf{0})\,\|\psi\|_{L^1(\partial\Omega)} < \infty.\end{aligned}$$

□

In the next theorem, we study the behaviour of (3.30) in the $L^p(\mathbb{R}^d)$ norm for $\delta \to 0$.

Theorem 3.11. *Let $\psi \in L^p(\partial\Omega)$, $1 \le p \le \infty$. A necessary and sufficient condition for*

$$\lim_{\delta\to 0}\left\|\int_{\partial\Omega} g_\delta(\mathbf{x}-\mathbf{s})\,\psi(\mathbf{s})\,d\mathbf{s}\right\|_{L^p(\mathbb{R}^d)} = 0, \tag{3.32}$$

$1 \le p \le \infty$, is that ψ vanishes almost everywhere on $\partial\Omega$.

Proof. It is obvious that the condition is sufficient.

Let (3.32) hold. From Hölder's inequality, we obtain for an arbitrary function $\varphi \in C_0^\infty(\mathbb{R}^d)$

$$\lim_{\delta\to 0}\left|\int_{\mathbb{R}^d}\varphi(\mathbf{x})\left(\int_{\partial\Omega} g_\delta(\mathbf{x}-\mathbf{s})\,\psi(\mathbf{s})\,d\mathbf{s}\right)d\mathbf{x}\right|$$

$$\leq \lim_{\delta\to 0}\|\varphi\|_{L^q(\mathbb{R}^d)}\left\|\int_{\partial\Omega} g_\delta(\mathbf{x}-\mathbf{s})\,\psi(\mathbf{s})\,d\mathbf{s}\right\|_{L^p(\mathbb{R}^d)} = 0 \qquad (3.33)$$

where $p^{-1}+q^{-1}=1$. By Fubini's theorem and the symmetry of the Gaussian filter, we have

$$\lim_{\delta\to 0}\int_{\mathbb{R}^d}\varphi(\mathbf{x})\left(\int_{\partial\Omega} g_\delta(\mathbf{x}-\mathbf{s})\,\psi(\mathbf{s})\,d\mathbf{s}\right)d\mathbf{x}$$

$$=\lim_{\delta\to 0}\int_{\partial\Omega}\psi(\mathbf{s})\left(\int_{\mathbb{R}^d} g_\delta(\mathbf{x}-\mathbf{s})\,\varphi(\mathbf{x})\,d\mathbf{x}\right)d\mathbf{s}=\int_{\partial\Omega}\psi(\mathbf{s})\,\varphi(\mathbf{s})\,d\mathbf{s}.$$

The last step is a consequence of Lemma 3.9 since $\varphi\in L^\infty\left(\mathbb{R}^d\right)$ and φ is uniformly continuous on the compact set $\partial\Omega$. Thus, from (3.33) follows

$$0=\left|\int_{\partial\Omega}\psi(\mathbf{s})\,\varphi(\mathbf{s})\,d\mathbf{s}\right|$$

for every $\varphi\in C_0^\infty\left(\mathbb{R}^d\right)$. This is true if and only if ψ vanishes almost everywhere on $\partial\Omega$. □

We will now bound the $L^p\left(\mathbb{R}^d\right)$ norm of (3.30) in terms of δ. The next lemma proves a geometric property which is needed later.

Lemma 3.12. *Let $\Omega\subset\mathbb{R}^d$, $d=2,3$, be a bounded domain with Lipschitz boundary $\partial\Omega$. Then there exists a constant $C>0$ such that*

$$\left|\{\mathbf{x}\in\mathbb{R}^d\ :\ d(\mathbf{x},\partial\Omega)\leq y\}\right|\leq C\left(y+y^d\right) \qquad (3.34)$$

for every $y\geq 0$, where $|\cdot|$ denotes the measure in $\mathbb{R}^d$.

Proof. For simplicity, we present the proof for Ω being a simply connected domain. The analysis can be extended to the case that $\partial\Omega$ consists of a finite number of non-connected parts.

We will start with the case $d=2$. We fix a point $\mathbf{x}_0$ on $\partial\Omega$ and an orientation of the boundary. Next, we construct $\mathbf{x}_1$ such that the length of the curve between $\mathbf{x}_0$ and $\mathbf{x}_1$ is y. Continuing this construction, we obtain a sequence $(\mathbf{x}_i)_{0\leq i\leq N}$ such that for every $0\leq i<N$ the length of curve between $\mathbf{x}_i$ and $\mathbf{x}_{i+1}$ is y. The length of the curve between $\mathbf{x}_N$ and $\mathbf{x}_0$ is less or equal than y, see Figure 3.2. The number of intervals is $N+1$ with $N<|\partial\Omega|/y\leq N+1$. Obviously, we have

$$\{\mathbf{x}\in\mathbb{R}^d\ :\ d(\mathbf{x},\partial\Omega)\leq y\}=\bigcup_{\mathbf{x}\in\partial\Omega}\overline{B}(\mathbf{x},y).$$

But for every $\mathbf{x}$ in $\partial\Omega$, there exists an i such that $\mathbf{x}$ is on the part of the curve from $\mathbf{x}_i$ to $\mathbf{x}_{i+1}$ or from $\mathbf{x}_N$ to $\mathbf{x}_0$. By the triangle inequality, this implies $\overline{B}(\mathbf{x},y)\subset\overline{B}(\mathbf{x}_i,2y)$. Thus

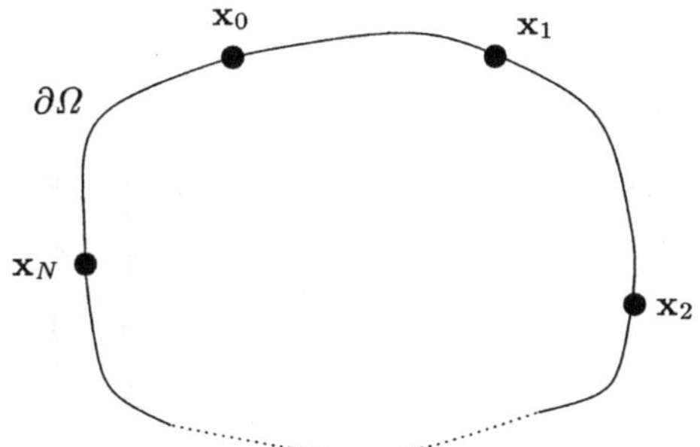

Fig. 3.2. Mesh on $\partial\Omega$ for $d = 2$

$$\{\mathbf{x} \in \mathbb{R}^d \; : \; d(x, \partial\Omega) \le y\} \subset \bigcup_{0 \le i \le N} \overline{B}(\mathbf{x}_i, 2y),$$

from which

$$\begin{aligned}\left|\{\mathbf{x} \in \mathbb{R}^d \; : \; d(\mathbf{x}, \partial\Omega) \le y\}\right| &\le \sum_{i=0}^{N} \left|\overline{B}(\mathbf{x}_i, 2y)\right| < \left(\frac{|\partial\Omega|}{y} + 1\right) 4\pi y^2 \\ &= 4\pi |\partial\Omega| y + 4\pi y^2\end{aligned}$$

follows.

In the case $d = 3$, $\partial\Omega$ is a compact manifold. Then, for every $\mathbf{x} \in \partial\Omega$, there exists a neighbourhood $U_{\mathbf{x}} \subset \partial\Omega$ such that its closure $\overline{U}_{\mathbf{x}}$ is homeomorphic to a closed square $\overline{V}_{\mathbf{x}} \subset \mathbb{R}^2$ through the homeomorphism $\phi_{\mathbf{x}} : \overline{V}_{\mathbf{x}} \to \overline{U}_{\mathbf{x}}$. The homeomorphism is Lipschitz continuous with a constant L. We cover the manifold by

$$\partial\Omega = \bigcup_{\mathbf{x} \in \partial\Omega} U_{\mathbf{x}}$$

and, because $\partial\Omega$ is compact, we can choose a finite cover $(U_{\mathbf{x}_i})_{0 \le i \le N}$ which will be fixed. Let the length of the sides of $\overline{V}_{\mathbf{x}_i}$ be equal to a_i. We create a mesh over on $\overline{V}_{\mathbf{x}_i}$ of cells of size y/L (or smaller). On this mesh, there are less than $(a_i L/y + 2)^2$ vertices and we denote them by $(\mathbf{z}_j)_{0 \le j \le P_i}$ where $P_i < (a_i L/y + 2)^2$. The order of the vertices is not important. Next, we consider $\mathbf{z} \in U_{\mathbf{x}_i}$. Then, we find the closest vertex on the mesh to $\phi^{-1}(\mathbf{z})$ and denote it by $\mathbf{z}_k$. It is easy to see that

$$\left\|\mathbf{z}_k - \phi^{-1}(\mathbf{z})\right\|_2 \le \frac{y}{L}$$

and the Lipschitz continuity of ϕ gives

$$\|\phi(\mathbf{z}_k) - \mathbf{z}\|_2 \le L \left\|\mathbf{z}_k - \phi^{-1}(\mathbf{z})\right\|_2 \le y.$$

By the triangle inequality follows now

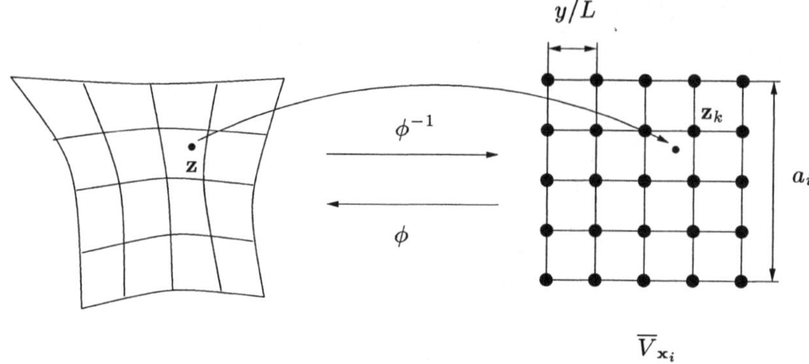

Fig. 3.3. Homeomorphic map to the square $\overline{V}_{\mathbf{x}_i}$, $d = 3$

$$B(\mathbf{z}, y) \subset B(\phi(\mathbf{z}_k), 2y). \tag{3.35}$$

Because $\mathbf{z} \in U_{\mathbf{x}_i}$ was chosen arbitrary, for every $\mathbf{z} \in U_{\mathbf{x}_i}$ there exists $\mathbf{z}_k \in \overline{V}_{\mathbf{x}_i}$ such that (3.35) holds. Combining (3.35) for $U_{\mathbf{x}_i}$, $0 \le i \le N$, gives

$$\{\mathbf{x} \in \mathbb{R}^3 \,:\, d(\mathbf{x}, \partial\Omega) \le y\} \subset \bigcup_{0 \le i \le N} \bigcup_{0 \le k \le P_i} \overline{B}(\phi(\mathbf{z}_k), 2y).$$

By the sub-additivity and monotonicity of Lebesgue measure, we obtain

$$\begin{aligned} \left|\{\mathbf{x} \in \mathbb{R}^3 \,:\, d(\mathbf{x}, \partial\Omega) \le y\}\right| &\le \sum_{i=0}^{N} \sum_{k=0}^{P_i} \left|\overline{B}(\phi(z_k), 2y)\right| \\ &\le \sum_{i=0}^{N} \left(\frac{a_i L}{y} + 2\right)^2 \frac{4}{3}\pi y^3 \le C\left(y^3 + y\right) \end{aligned}$$

for an appropriately chosen positive constant C. Note, the quadratic term in y can be absorbed into the linear term for $y \le 1$ and into the cubic term for $y > 1$. □

Theorem 3.13. *Let Ω be a bounded domain in $\mathbb{R}^d$ with Lipschitz boundary $\partial\Omega$, $\psi \in L^p(\partial\Omega)$ for some $p > 1$ and $p^{-1} + q^{-1} = 1$. Then for every $\theta \in (0,1)$ and $k \in (0, \infty)$ there exist constants $C > 0$ and $\epsilon > 0$ such that*

$$\int_{\mathbb{R}^d} \left| \int_{\partial\Omega} g_\delta(\mathbf{x} - \mathbf{s}) \psi(\mathbf{s})\, d\mathbf{s} \right|^k d\mathbf{x} \le C\delta^{1 + k\left(\frac{(d-1)\theta}{q} - d\right)} \|\psi\|^k_{L^p(\partial\Omega)} \tag{3.36}$$

for every $\delta \in (0, \epsilon)$ where C and ϵ depend on θ, k and $|\partial\Omega|$.

Proof. We fix a $\theta \in (0,1)$. From Hölder's inequality, we obtain

$$\int_{\mathbb{R}^d} \left| \int_{\partial\Omega} g_\delta(\mathbf{x} - \mathbf{s}) \psi(\mathbf{s})\, d\mathbf{s} \right|^k d\mathbf{x} \le \int_{\mathbb{R}^d} \left(\int_{\partial\Omega} g_\delta^q(\mathbf{x} - \mathbf{s})\, d\mathbf{s} \right)^{k/q} d\mathbf{x}\ \|\psi\|^k_{L^p(\partial\Omega)}.$$

Let $B\left(\mathbf{x},\delta^\theta\right)$ be the ball centred at $\mathbf{x}\in\mathbb{R}^d$ and with radius δ^θ. Then, the term containing the Gaussian filter function can be estimated by the triangle inequality

$$\begin{aligned}
&\int_{\mathbb{R}^d}\left(\int_{\partial\Omega} g_\delta^q(\mathbf{x}-\mathbf{s})\,ds\right)^{k/q} d\mathbf{x}\\
&=\int_{\mathbb{R}^d}\left(\int_{\partial\Omega\cap B(\mathbf{x},\delta^\theta)} g_\delta^q(\mathbf{x}-\mathbf{s})\,ds+\int_{\partial\Omega\setminus B(\mathbf{x},\delta^\theta)} g_\delta^q(\mathbf{x}-\mathbf{s})\,ds\right)^{k/q} d\mathbf{x}\\
&\le\int_{\mathbb{R}^d}\left[\left(\int_{\partial\Omega\cap B(\mathbf{x},\delta^\theta)} g_\delta^q(\mathbf{x}-\mathbf{s})\,ds\right)^{1/q}\right.\\
&\qquad\left.+\left(\int_{\partial\Omega\setminus B(\mathbf{x},\delta^\theta)} g_\delta^q(\mathbf{x}-\mathbf{s})\,ds\right)^{1/q}\right]^k d\mathbf{x}\\
&=:\int_{\mathbb{R}^d}\left(B_\delta(\mathbf{x})+C_\delta(\mathbf{x})\right)^k d\mathbf{x}\\
&\le C(k)\left(\int_{\mathbb{R}^d} B_\delta^k(\mathbf{x})\,d\mathbf{x}+\int_{\mathbb{R}^d} C_\delta^k(\mathbf{x})\,d\mathbf{x}\right). \qquad (3.37)
\end{aligned}$$

The terms in the last line are estimated separately.

We start with the estimate for $C_\delta(\mathbf{x})$. Let $\mathbf{x}$ such that $d(\mathbf{x},\partial\Omega)<\delta^\theta$. Then,

$$d\left(\mathbf{x},\partial\Omega\setminus B\left(\mathbf{x},\delta^\theta\right)\right)=\delta^\theta$$

and from the monotonicity of the Gaussian filter function follows $g_\delta(\mathbf{x}-\mathbf{s})\le g_\delta\left(\delta^\theta\right)$ for $\mathbf{s}\in\partial\Omega\setminus B\left(\mathbf{x},\delta^\theta\right)$. Similarly, $g_\delta(\mathbf{x}-\mathbf{s})$ can estimated from above by $g_\delta(d(\mathbf{x},\partial\Omega))$ if $d(\mathbf{x},\partial\Omega)\ge\delta^\theta$. One obtains the following estimate for $C_\delta^k(\mathbf{x})$

$$C_\delta^k(\mathbf{x})\le C\begin{cases} g_\delta^k\left(\delta^\theta\right) & \text{if } d(\mathbf{x},\partial\Omega)<\delta^\theta,\\ g_\delta^k(d(\mathbf{x},\partial\Omega)) & \text{if } d(\mathbf{x},\partial\Omega)\ge\delta^\theta,\end{cases}$$

where $C=C(|\partial\Omega|)$. We refer to the function behind the brace as bounding function, see Figure 3.4 for a sketch in a special situation. Thus, $C_\delta^k(\mathbf{x})$ is less or equal than the constant value $Cg_\delta^k\left(\delta^\theta\right)$ in a δ^θ-neighbourhood of $\partial\Omega$ and it decays exponentially fast as the distance from $\mathbf{x}$ to $\partial\Omega$ approaches infinity.

Let $C(t)=\{(\mathbf{z},t)\ :\ d(\mathbf{z},\partial\Omega)\le y, t=g_\delta^k(y),\delta^\theta\le y<\infty\}$ be the cross section of the bounding function at the function value t and $A(t)=|C(t)|$ the area of the cross section. Then

$$\int_{\mathbb{R}^d} C_\delta^k(\mathbf{x})\,d\mathbf{x}\le C\int_0^{g_\delta^k(\delta^\theta)} A(t)\,dt.$$

From Lemma 3.12, we know $A(t)\le C\left(y^d+y\right)$. Using $g_\delta^k(y)=t$, changing variables and integrating by parts yield

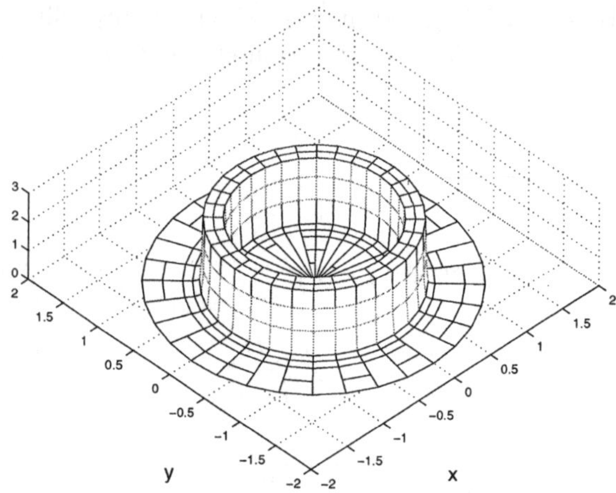

Fig. 3.4. Bounding function of $C_\delta^k(\mathbf{x})$, $d = 2$, $\partial\Omega = B(\mathbf{0}, 1)$, $\delta = 0.1$, $\theta = 0.99$, $k = 1$, $C = 2\pi$

$$
\begin{aligned}
&\int_0^{g_\delta^k(\delta^\theta)} A(t)\,dt \\
&\le C \int_0^{g_\delta^k(\delta^\theta)} (y^d + y)\,dt = C \int_\infty^{\delta^\theta} (y^d + y) \frac{d}{dy}\left(g_\delta^k(y)\right) dy \\
&= C\left((y^d + y)\, g_\delta^k(y) \Big|_\infty^{\delta^\theta} - \int_\infty^{\delta^\theta} \left(dy^{d-1} + 1\right) g_\delta^k(y)\,dy \right) \\
&= C\left(\left(\delta^{d\theta} + \delta^\theta\right) g_\delta^k\left(\delta^\theta\right) - d \int_\infty^{\delta^\theta} y^{d-1} g_\delta^k(y)\,dy - \int_\infty^{\delta^\theta} g_\delta^k(y)\,dy \right).
\end{aligned}
$$

With the change of variables $y = \delta/t$, we obtain

$$
\begin{aligned}
-\int_\infty^{\delta^\theta} y^{d-1} g_\delta^k(y)\,dy &= \int_0^{\delta^{1-\theta}} \frac{\delta^d}{t^{d+1}} g_\delta^k\left(\frac{\delta}{t}\right) dt \\
&= C \int_0^{\delta^{1-\theta}} \frac{1}{t^{d+1}\delta^{kd-d}} \exp\left(-\frac{6k}{t^2}\right) dt.
\end{aligned}
$$

The function $t^{-(d+1)} \exp\left(-6k/t^2\right)$, $k > 0$, is monotonically increasing for $t \in \left(0, \sqrt{12k/(d+1)}\right)$ since its derivative is positive in this interval. Choosing δ small enough, such that $\delta^{1-\theta} \le \sqrt{12k/(d+1)}$, we conclude

$$
C \int_0^{\delta^{1-\theta}} \frac{1}{t^{d+1}\delta^{kd-d}} \exp\left(-\frac{6k}{t^2}\right) dt \le C\delta^{(\theta-k)d} \exp\left(-\frac{6k}{\delta^{2(1-\theta)}}\right).
$$

Doing the same computations for the other integral, we obtain

$$-\int_{\infty}^{\delta^\theta} g_\delta^k(y)\,dy \le C\delta^{\theta-kd}\exp\left(-\frac{6k}{\delta^{2(1-\theta)}}\right).$$

This gives, for δ sufficiently small, the estimate for $C_\delta(\mathbf{x})$

$$\int_{\mathbb{R}^d} C_\delta^k(\mathbf{x})\,d\mathbf{x} \le C\left(\delta^{d(\theta-k)} + \delta^{\theta-kd}\right)\exp\left(-\frac{6k}{\delta^{2(1-\theta)}}\right),$$

from what follows, since $\theta < 1$,

$$\lim_{\delta\to 0}\int_{\mathbb{R}^d} C_\delta^k(\mathbf{x})\,d\mathbf{x} = 0.$$

Now we will bound the second term in (3.37). The function $B_\delta^k(\mathbf{x})$ can be estimated from above in the following way

$$B_\delta^k(\mathbf{x}) \le \begin{cases} \left|\partial\Omega\cap B\left(\mathbf{x},\delta^\theta\right)\right|^{\frac{k}{q}} g_\delta^k\left(d\left(\mathbf{x},\partial\Omega\right)\right) & \text{if } d\left(\mathbf{x},\partial\Omega\right) < \delta^\theta, \\ 0 & \text{if } d\left(\mathbf{x},\partial\Omega\right) \ge \delta^\theta, \end{cases}$$

see Figure 3.5 for an illustration of the bounding function in a special situation. The bounding function is discontinuous, having a jump from the value 0 to the value $Cg_\delta^k\left(\delta^\theta\right)$ at $\{\mathbf{x}\in\mathbb{R}^d\ :\ d\left(\mathbf{x},\partial\Omega\right) = \delta^\theta\}$.

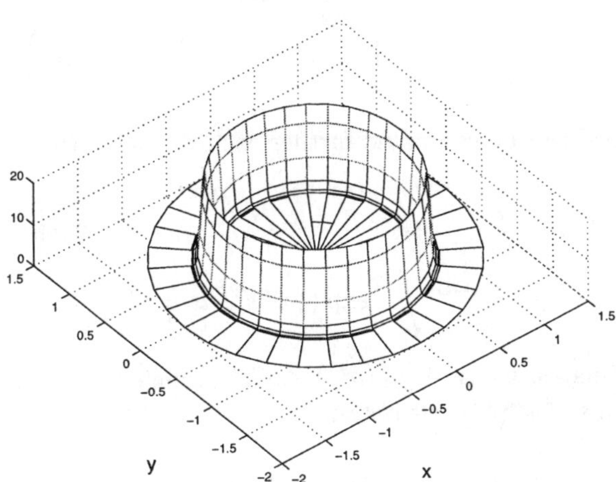

Fig. 3.5. Bounding function of $B_\delta^k(\mathbf{x})$, $d=2$, $\partial\Omega = B(\mathbf{0},1)$, $\delta = 0.1$, $\theta = 0.99$, $k=1$, $C=\delta^\theta$

Since $\partial\Omega$ is smooth, we have $\left|\partial\Omega\cap B\left(\mathbf{x},\delta^\theta\right)\right| \le C\delta^{(d-1)\theta}$ if δ is small enough. It follows

$$\int_{\mathbb{R}^d} B_\delta^k(\mathbf{x})\,d\mathbf{x} \le C\int_{\{d(\mathbf{x},\partial\Omega)<\delta^\theta\}} \delta^{\frac{(d-1)\theta k}{q}} g_\delta^k\left(d\left(\mathbf{x},\partial\Omega\right)\right)d\mathbf{x}.$$

We will estimate the integral by integrating over the cross sections of the function in the integral. For the function values t, $0 \le t \le g_\delta^k(\delta^\theta)$, all cross sections have the same form. For function values $t = g_\delta^k(y)$, $\delta^\theta > y \ge 0$, the cross section is $\{\mathbf{x} \in \mathbb{R}^d \ : \ d(\mathbf{x}, \partial\Omega) \le y\}$. We denote the area of the cross sections by $A(t)$. Integration of the areas gives

$$\begin{aligned}\int_{\{d(\mathbf{x},\partial\Omega)<\delta^\theta\}} g_\delta^k\left(d\left(\mathbf{x},\partial\Omega\right)\right) d\mathbf{x} &= \int_0^{g_\delta^k(\delta^\theta)} A(t)\, dt + \int_{g_\delta^k(\delta^\theta)}^{g_\delta^k(0)} A(t)\, dt \\ &= A\left(g_\delta^k\left(\delta^\theta\right)\right) g_\delta^k\left(\delta^\theta\right) + \int_{g_\delta^k(\delta^\theta)}^{g_\delta^k(0)} A(t)\, dt.\end{aligned}$$

We will use now the estimate of the areas of the cross sections given in Lemma 3.12. If y is small enough, the term y^d can be absorbed into the term y in this estimate. Thus, if δ is small enough, we have $\left|\{\mathbf{x} \in \mathbb{R}^d \ : \ d(\mathbf{x},\partial\Omega) \le y\}\right| \le Cy$, $0 \le y < \delta^\theta$. We obtain, changing variables and applying integration by parts

$$\begin{aligned}\int_{g_\delta^k(\delta^\theta)}^{g_\delta^k(0)} A(t)\, dt &\le -C \int_0^{\delta^\theta} y \frac{d}{dy} g_\delta^k(y)\, dy \\ &= -Cy g_\delta^k(y)\Big|_{y=0}^{y=\delta^\theta} + C \int_0^{\delta^\theta} g_\delta^k(y)\, dy \\ &= C\left(-\delta^\theta g_\delta^k\left(\delta^\theta\right) + \int_0^{\delta^\theta} g_\delta^k(y)\, dy\right).\end{aligned}$$

The last integral can be estimated further with the substitution $y = \delta s$

$$\begin{aligned}\int_0^{\delta^\theta} g_\delta^k(y)\, dy &= \int_0^{\delta^{\theta-1}} \delta g_\delta^k(\delta s)\, ds = C \int_0^{\delta^{\theta-1}} \delta^{1-kd} \exp\left(-6ks^2\right) ds \\ &\le C\delta^{1-kd} \int_0^\infty \exp\left(-6ks^2\right) ds = C\delta^{1-kd}.\end{aligned}$$

Collecting estimates, using $A\left(g_\delta\left(\delta^\theta\right)\right) \le C\delta^\theta$, which results in the cancellation of the terms $\delta^\theta g_\delta^k\left(\delta^\theta\right)$, we obtain

$$\int_{\mathbb{R}^d} B_\delta^k(\mathbf{x})\, d\mathbf{x} \le C\delta^{1-kd+\frac{(d-1)\theta k}{q}}.$$

The estimate for $C_\delta(\mathbf{x})$ converges exponentially for $\delta \to 0$. Thus, for δ sufficiently small, the estimate of $B_\delta(\mathbf{x})$ will dominate. This proves the theorem.
□

An inspection of the proof of Theorem 3.13 reveals that the commutation error $A_\delta(\mathbb{S}(\mathbf{u}, p))$ is largest at the boundary and decays rapidly as one moves away from the boundary.

3.6 Error Estimate of the Commutation Error Term in the $H^{-1}(\Omega)$ Norm

Variational methods, such as finite element, spectral or spectral element methods, discretise the weak form of the relevant equations. These methods are known to depend on the size of the $H^{-1}(\Omega)$ norm of any omitted term. The main result of this section is that the commutation error tends to zero in $H^{-1}(\Omega)$ as $\delta \to 0$, see Theorem 3.15. The order of convergence is at least $\mathcal{O}(\delta^{1/2})$. Thus, using variational methods leads to the expected asymptotic vanishing of the commutation error.

Lemma 3.14. *There exists a constant C, which depends only on d, such that*

$$\|\overline{v} - v\|_{H^{1/2}(\mathbb{R}^d)} \leq C\delta^{1/2} \|v\|_{H^1(\mathbb{R}^d)} \tag{3.38}$$

for any $v \in H^1(\mathbb{R}^d)$ and any $\delta > 0$.

Proof. By using the definition of $\|\cdot\|_{H^{1/2}(\mathbb{R}^d)}$, we have

$$\begin{aligned}
\|\overline{v} - v\|_{H^{1/2}(\mathbb{R}^d)}^2 &= \int_{\mathbb{R}^d} \left(1 + \|\mathbf{x}\|_2^2\right)^{1/2} |1 - \mathcal{F}(g_\delta)|^2 |\mathcal{F}(v)|^2 \, d\mathbf{x} \\
&= \int_{\{\|\mathbf{x}\|_2 > \pi/\delta\}} \left(1 + \|\mathbf{x}\|_2^2\right)^{1/2} |1 - \mathcal{F}(g_\delta)|^2 |\mathcal{F}(v)|^2 \, d\mathbf{x} \\
&\quad + \int_{\{\|\mathbf{x}\|_2 \leq \pi/\delta\}} \left(1 + \|\mathbf{x}\|_2^2\right)^{1/2} |1 - \mathcal{F}(g_\delta)|^2 |\mathcal{F}(v)|^2 \, d\mathbf{x}.
\end{aligned}$$

To begin with, we prove a bound for the first integral. There exists a constant $C > 0$, which does not depend on δ and v, such that

$$\left(1 + \|\mathbf{x}\|_2^2\right)^{-1/2} < C\delta$$

for $\|\mathbf{x}\|_2 > \pi/\delta$. From (3.27) follows the point-wise estimate

$$|1 - \mathcal{F}(g_\delta)(\mathbf{x})| \leq 1$$

for any $\mathbf{x} \in \mathbb{R}^d$. Thus, the first integral can be bounded by

$$\begin{aligned}
&\left| \int_{\{\|\mathbf{x}\|_2 > \pi/\delta\}} \left(1 + \|\mathbf{x}\|_2^2\right)^{1/2} |1 - \mathcal{F}(g_\delta)|^2 |\mathcal{F}(v)|^2 \, d\mathbf{x} \right| \\
&\leq \int_{\{\|\mathbf{x}\|_2 > \pi/\delta\}} \left(1 + \|\mathbf{x}\|_2^2\right) \left(1 + \|\mathbf{x}\|_2^2\right)^{-1/2} |\mathcal{F}(v)|^2 \, d\mathbf{x} \\
&\leq C\delta \int_{\{\|\mathbf{x}\|_2 > \pi/\delta\}} \left(1 + \|\mathbf{x}\|_2^2\right) |\mathcal{F}(v)|^2 \, d\mathbf{x}.
\end{aligned} \tag{3.39}$$

Now, the bound for the second integral is proved. A Taylor series expansion of (3.27) at $\|\mathbf{x}\|_2 = 0$ and for fixed δ, compare (4.16), gives

$$\mathcal{F}(g_\delta)(\mathbf{x}) = 1 - \frac{\delta^2 \|\mathbf{x}\|_2^2}{24} + \mathcal{O}\left(\delta^4 \|\mathbf{x}\|_2^4\right),$$

such that we have the point-wise bound

$$|1 - \mathcal{F}(g_\delta)(\mathbf{x})|^2 \leq C\delta \|\mathbf{x}\|_2$$

for any $\|\mathbf{x}\|_2 \leq \pi/\delta$ where C does not depend on δ or $\mathbf{x}$. In addition, $\|\mathbf{x}\|_2 \leq \left(1 + \|\mathbf{x}\|_2^2\right)^{1/2}$ and consequently the second integral can be bounded as follows

$$\left| \int_{\{\|\mathbf{x}\|_2 \leq \pi/\delta\}} \left(1 + \|\mathbf{x}\|_2^2\right)^{1/2} |1 - \mathcal{F}(g_\delta)|^2 |\mathcal{F}(v)|^2 \, d\mathbf{x} \right| \tag{3.40}$$
$$\leq C\delta \int_{\{\|\mathbf{x}\|_2 \leq \pi/\delta\}} \left(1 + \|\mathbf{x}\|_2^2\right) |\mathcal{F}(v)|^2 \, d\mathbf{x}.$$

Combining (3.39) and (3.40) gives

$$\|\overline{v} - v\|_{H^{1/2}(\mathbb{R}^d)}^2 \leq C\delta \int_{\mathbb{R}^d} \left(1 + \|\mathbf{x}\|_2^2\right) |\mathcal{F}(v)|^2 \, d\mathbf{x} = C\delta \|v\|_{H^1(\mathbb{R}^d)} .$$

□

The next theorem contains the estimate of the commutation error term in the $H^{-1}(\Omega)$ norm.

Theorem 3.15. *Let $\psi \in L^2(\partial\Omega)$, then there exists a constant $C > 0$ which depends only on Ω such that*

$$\left\| \int_{\partial\Omega} g_\delta(\mathbf{x} - \mathbf{s}) \psi(\mathbf{s}) \, d\mathbf{s} \right\|_{H^{-1}(\Omega)} \leq C\delta^{1/2} \|\psi\|_{L^2(\partial\Omega)}$$

for every $\delta > 0$.

Proof. Let $v \in H_0^1(\Omega)$. Extending v by zero outside Ω, applying Fubini's theorem, using that v vanishes on $\partial\Omega$, applying the Cauchy-Schwarz inequality, the trace theorem and Lemma 3.14 give

$$\begin{aligned}
\int_\Omega \left(\int_{\partial\Omega} g_\delta(\mathbf{x} - \mathbf{s}) \psi(\mathbf{s}) \, d\mathbf{s} \right) v(\mathbf{x}) \, d\mathbf{x} &= \int_{\partial\Omega} \psi(\mathbf{s}) \, \overline{v}(\mathbf{s}) \, d\mathbf{s} \\
&= \int_{\partial\Omega} \psi(\mathbf{s}) \left(\overline{v}(\mathbf{s}) - v(\mathbf{s}) \right) d\mathbf{s} \\
&\leq \|\overline{v} - v\|_{L^2(\partial\Omega)} \|\psi\|_{L^2(\partial\Omega)} \\
&\leq C \|\overline{v} - v\|_{H^{1/2}(\Omega)} \|\psi\|_{L^2(\partial\Omega)} \\
&\leq C\delta^{1/2} \|v\|_{H^1(\Omega)} \|\psi\|_{L^2(\partial\Omega)} .
\end{aligned}$$

Division by $\|v\|_{H^1(\Omega)}$ and using the definition of the $H^{-1}(\Omega)$ norm, (2.3), give the desired result. □

Let

$$\mathcal{H} = \left\{ v \in H^1\left(\mathbb{R}^d\right) \;:\; v|_{\partial\Omega} = 0 \right\}$$

and let the assumption of Theorem 3.15 be fulfilled. An inspection of the proof shows that only $v|_{\partial\Omega} = 0$ was used and not that v vanishes on $\mathbb{R}^d \setminus \Omega$. Consequently, for the functions from $\mathcal{H}$ also holds

$$\left\| \int_{\partial\Omega} g_\delta\left(\mathbf{x}-\mathbf{s}\right)\psi\left(\mathbf{s}\right) ds \right\|_{H_{\mathcal{H}}^{-1}(\mathbb{R}^d)} := \sup_{v\in\mathcal{H}} \frac{\displaystyle\int_{\mathbb{R}^d} v(\mathbf{x}) \int_{\partial\Omega} g_\delta\left(\mathbf{x}-\mathbf{s}\right)\psi\left(\mathbf{s}\right) ds d\mathbf{x}}{\|v\|_{H^1(\mathbb{R}^d)}} \le C\delta^{1/2}\left\|\psi\right\|_{L^2(\partial\Omega)}.$$

3.7 Error Estimate for a Weak Form of the Commutation Error

In this section, we consider a weak form of the commutation error (3.30), multiplied with a suitable test function $\overline{v}\left(\mathbf{x}\right)$ and integrated on $\mathbb{R}^d$. The following theorem shows that this weak form converges to zero as δ tends to zero for fixed $\overline{v}\left(\mathbf{x}\right)$. For $d = 2$, Corollary 3.18 and Remark 3.19 show that the convergence is (at least) almost of order one if ψ is sufficiently smooth.

Lemma 3.16. *Let $v \in H^1\left(\mathbb{R}^d\right)$ such that $v|_\Omega \in H_0^1\left(\Omega\right) \cap H^2\left(\Omega\right)$ and $v\left(\mathbf{x}\right) = 0$ if $\mathbf{x} \notin \overline{\Omega}$ and let $\psi \in L^p\left(\partial\Omega\right)$, $1 \le p \le \infty$. Then*

$$\lim_{\delta\to 0} \int_{\mathbb{R}^d} \overline{v}\left(\mathbf{x}\right) \left(\int_{\partial\Omega} g_\delta\left(\mathbf{x}-\mathbf{s}\right)\psi\left(\mathbf{s}\right) ds \right) d\mathbf{x} = 0,$$

where $\overline{v}\left(\mathbf{x}\right) = \left(g_\delta * v\right)\left(\mathbf{x}\right)$.

Proof. By Fubini's theorem and the symmetry of g_δ, we obtain

$$\begin{aligned} &\lim_{\delta\to 0} \int_{\mathbb{R}^d} \overline{v}\left(\mathbf{x}\right) \left(\int_{\partial\Omega} g_\delta\left(\mathbf{x}-\mathbf{s}\right)\psi\left(\mathbf{s}\right) ds \right) d\mathbf{x} \\ &= \lim_{\delta\to 0} \int_{\partial\Omega} \psi\left(\mathbf{s}\right) \left(\int_{\mathbb{R}^d} g_\delta\left(\mathbf{s}-\mathbf{x}\right)\overline{v}\left(\mathbf{x}\right) d\mathbf{x} \right) ds. \end{aligned}$$

By the Sobolev embedding theorem $H^2\left(\Omega\right) \to L^\infty\left(\Omega\right)$, we get $v \in L^\infty\left(\Omega\right)$ from what follows by the construction of v that $v \in L^\infty\left(\mathbb{R}^d\right)$. Young's inequality for convolutions (2.18) gives $\overline{v} \in L^\infty\left(\mathbb{R}^d\right)$. In addition, $\overline{v}$ is uniformly continuous on the compact set $\partial\Omega$. The same holds for v since $v \in C^0\left(\overline{\Omega}\right)$ by the Sobolev embedding $H^2\left(\Omega\right) \to C^0\left(\overline{\Omega}\right)$. Applying twice Lemma 3.9 gives

$$\lim_{\delta\to 0} \int_{\mathbb{R}^d} \overline{v}\left(\mathbf{x}\right) \left(\int_{\partial\Omega} g_\delta\left(\mathbf{x}-\mathbf{s}\right)\psi\left(\mathbf{s}\right) ds \right) d\mathbf{x} = \int_{\partial\Omega} \psi\left(\mathbf{s}\right) v\left(\mathbf{s}\right) ds = 0,$$

since v vanishes on $\partial\Omega$. □

With the result of Theorem 3.13, we want to study the order of convergence with respect to δ of the weak form of the commutation error.

Theorem 3.17. *Let v and ψ be defined as in Theorem 3.16 and let the assumption of Theorem 3.13 be fulfilled. Then, there exists an $\epsilon > 0$ such that for $\delta \in (0, \epsilon)$*

$$\int_{\mathbb{R}^d} \left| \overline{v}(\mathbf{x}) \int_{\partial\Omega} g_\delta(\mathbf{x}-\mathbf{s}) \psi(\mathbf{s})\, d\mathbf{s} \right|^k d\mathbf{x} \\ \leq C\delta^{1+\left(-d+\frac{(d-1)\theta}{q}+\beta\theta\right)k} \|\psi\|^k_{L^p(\partial\Omega)} \|v\|^k_{H^2(\Omega)},$$

where $k \in [1, \infty)$, $\beta \in (0,1)$ if $d = 2$ and $\beta = 1/2$ if $d = 3$, $p^{-1} + q^{-1} = 1$, $p > 1$, and C and ϵ depend on θ, k and $|\partial\Omega|$.

Proof. Analogously to the begin of the proof of Theorem 3.13, one obtains

$$\int_{\mathbb{R}^d} \left| \overline{v}(\mathbf{x}) \int_{\partial\Omega} g_\delta(\mathbf{x}-\mathbf{s}) \psi(\mathbf{s})\, d\mathbf{s} \right|^k d\mathbf{x} \\ \leq C(k) \left[\int_{\mathbb{R}^d} \left| \overline{v}(\mathbf{x}) B_\delta(\mathbf{x}) \right|^k d\mathbf{x} + \int_{\mathbb{R}^d} \left| \overline{v}(\mathbf{x}) C_\delta(\mathbf{x}) \right|^k d\mathbf{x} \right] \|\psi\|^k_{L^p(\partial\Omega)},$$

where $B_\delta(\mathbf{x})$ and $C_\delta(\mathbf{x})$ are defined in the proof of Theorem 3.13. The terms on the right hand side are treated separately.

In Theorem 3.13, it is proved that $C_\delta^k \in L^1(\mathbb{R}^d)$ for every $k \in (0, \infty)$. This implies

$$\left(C_\delta^k\right)^p = C_\delta^{kp} = C_\delta^{k'} \in L^1(\mathbb{R}^d),$$

since $k' \in (0, \infty)$. That means $C_\delta^k \in L^p(\mathbb{R}^d)$ for $p \in [1, \infty)$. From the bounding function of C_δ^k it is obvious that $C_\delta^k \in L^\infty(\mathbb{R}^d)$, too. Using Young's inequality for convolutions (2.18) and $\|g_\delta\|_{L^1(\mathbb{R}^d)} = 1$, it follows

$$\|\overline{v}\|_{L^q(\mathbb{R}^d)} \leq \|g_\delta\|_{L^1(\mathbb{R}^d)} \|v\|_{L^q(\mathbb{R}^d)} = \|v\|_{L^q(\mathbb{R}^d)}$$

where $1 \leq q < \infty$. With the same argument, we get for $qk \geq 1$

$$\left\|\overline{v}^k\right\|_{L^q(\mathbb{R}^d)} = \|\overline{v}\|^k_{L^{qk}(\mathbb{R}^d)} \leq \|v\|^k_{L^{qk}(\mathbb{R}^d)}.$$

By the regularity assumptions on v, it follows $v \in C^0(\mathbb{R}^d)$. This implies, together with $v = 0$ outside Ω, that $v \in L^p(\mathbb{R}^d)$ for every $1 \leq p \leq \infty$. Consequently, $\|v\|_{L^{qk}(\mathbb{R}^d)} < \infty$. Applying Hölder's inequality, we obtain

$$\int_{\mathbb{R}^d} \left| \overline{v}(\mathbf{x}) C_\delta(\mathbf{x}) \right|^k d\mathbf{x} \leq \|v\|^k_{L^{qk}(\mathbb{R}^d)} \left\| C_\delta^k(\mathbf{x}) \right\|_{L^p(\mathbb{R}^d)}.$$

For the second factor, we can use the bound obtained in the proof of Theorem 3.13, replacing k by kp. Thus if δ is small enough, it follows

$$\begin{aligned}&\int_{\mathbb{R}^d} |\overline{v}(\mathbf{x}) C_\delta(\mathbf{x})|^k \, d\mathbf{x} \\ &\le C\delta^{-kd} \left(\delta^{d\theta} + \delta^\theta\right)^{1/p} \exp\left(-\frac{6k}{\delta^{2(1-\theta)}}\right) \|v\|^k_{L^{qk}(\mathbb{R}^d)} \end{aligned} \tag{3.41}$$

for every test function v which satisfies the regularity assumptions stated in Lemma 3.16.

The estimate of the second term starts by noting that the domain of integration can be restricted to a small neighbourhood of $\partial\Omega$

$$\begin{aligned}\int_{\mathbb{R}^d} |\overline{v}(\mathbf{x}) B_\delta(\mathbf{x})|^k \, d\mathbf{x} &= \int_{\{d(\mathbf{x},\partial\Omega)\le\delta^\theta\}} |\overline{v}(\mathbf{x}) B_\delta(\mathbf{x})|^k \, d\mathbf{x} \\ &\le \|\overline{v}\|^k_{L^\infty(\{d(\mathbf{x},\partial\Omega)\le\delta^\theta\})} \int_{\{d(\mathbf{x},\partial\Omega)\le\delta^\theta\}} B^k_\delta(\mathbf{x}) \, d\mathbf{x} \\ &\le \|\overline{v}\|^k_{L^\infty(\{d(\mathbf{x},\partial\Omega)\le\delta^\theta\})} \delta^{1+\left(-d+\frac{(d-1)\theta}{q}\right)k}, \end{aligned} \tag{3.42}$$

where $\theta \in (0,1)$ and $p^{-1} + q^{-1} = 1$. The last estimate is taken from the proof of Theorem 3.13. It remains to estimate the norm of $\overline{v}$. By the triangle inequality, we obtain

$$\begin{aligned}&\|\overline{v}\|_{L^\infty(\{d(\mathbf{x},\partial\Omega)\le\delta^\theta\})} \\ &\le \|\overline{v} - v\|_{L^\infty(\{d(\mathbf{x},\partial\Omega)\le\delta^\theta\})} + \|v\|_{L^\infty(\{d(\mathbf{x},\partial\Omega)\le\delta^\theta\})} . \end{aligned} \tag{3.43}$$

Since $v \in H^2(\Omega)$, we have by the Sobolev embedding $H^2(\Omega) \to C^{0,\beta}(\overline{\Omega})$ that $v \in C^{0,\beta}(\overline{\Omega})$ with $\beta \in (0,1)$ if $d = 2$ and $\beta = 1/2$ if $d = 3$. That means, there exists a constant $C_H \ge 0$ such that

$$|v(\mathbf{x}) - v(\mathbf{y})| \le C_H \|\mathbf{x} - \mathbf{y}\|_2^\beta \text{ for all } \mathbf{x}, \mathbf{y} \in \overline{\Omega}.$$

By the definition of the norm in $C^{0,\beta}(\overline{\Omega})$ and the Sobolev embedding theorem, this constant can be estimated by $C_H \le C(\Omega) \|v\|_{H^2(\Omega)}$. We fix an arbitrary $\mathbf{x} \in \{d(\mathbf{x}, \partial\Omega) \le \delta^\theta\}$ and we take $\mathbf{y} \in \partial\Omega$ with $\|\mathbf{x} - \mathbf{y}\|_2 = d(\mathbf{x}, \mathbf{y})$. Since v vanishes on $\partial\Omega$, we obtain $|v(\mathbf{x})| \le C_H d(\mathbf{x}, \partial\Omega)^\beta$. It follows

$$\|v\|_{L^\infty(\{d(\mathbf{x},\partial\Omega)\le\delta^\theta\})} \le C_H \delta^{\theta\beta}.$$

The first term on the right hand side of (3.43) is, using that the $L^1(\mathbb{R}^d)$ norm of the Gaussian filter is equal to one,

$$\|\overline{v} - v\|_{L^\infty(\{d(\mathbf{x},\partial\Omega)\le\delta^\theta\})} = \operatorname*{esssup}_{\mathbf{x}\in\{d(\mathbf{x},\partial\Omega)\le\delta^\theta\}} \left| \int_{\mathbb{R}^d} g_\delta(\mathbf{x} - \mathbf{y}) (v(\mathbf{x}) - v(\mathbf{y})) \, d\mathbf{y} \right| .$$

Since v vanishes outside Ω, it can be easily proved that

$$|v(\mathbf{x}) - v(\mathbf{y})| \le C_H \|\mathbf{x} - \mathbf{y}\|_2^\beta$$

holds for all $\mathbf{x}, \mathbf{y} \in \mathbb{R}^d$. We obtain, using the symmetry of the Gaussian filter,

$$\begin{aligned}\left|\int_{\mathbb{R}^d} g_\delta(\mathbf{x}-\mathbf{y})(v(\mathbf{x})-v(\mathbf{y}))\,d\mathbf{y}\right| &\le C_H \int_{\mathbb{R}^d} g_\delta(\mathbf{x}-\mathbf{y})\,\|\mathbf{x}-\mathbf{y}\|_2^\beta\,d\mathbf{y}\\ &= C_H \int_{\mathbb{R}^d} g_\delta(\delta\mathbf{z})\,\delta^{\beta+d}\,\|\mathbf{z}\|_2^\beta\,d\mathbf{z}\\ &= C C_H \delta^\beta \int_{\mathbb{R}^d} \exp\left(-6\,\|\mathbf{z}\|_2^2\right)\|\mathbf{z}\|_2^\beta\,d\mathbf{z}.\end{aligned}$$

The last integral is finite. Thus, we can conclude

$$\|\overline{v}-v\|_{L^\infty(\{d(x,\partial\Omega)\le\delta^\theta\})} \le C C_H \delta^\beta.$$

Since δ^β decays faster for small δ than $\delta^{\beta\theta}$, we obtain the estimate

$$\|\overline{v}\|^k_{L^\infty(\{d(\mathbf{x},\partial\Omega)\le\delta^\theta\})} \le C_H^k \delta^{\beta\theta k}.$$

Combining this estimate with (3.42) and using the estimate for C_H, we get

$$\int_{\mathbb{R}^d} |\overline{v}(\mathbf{x})\,B_\delta(\mathbf{x})|^k\,d\mathbf{x} \le C\delta^{1+\left(-d+\frac{(d-1)\theta}{q}+\beta\theta\right)k}\,\|v\|^k_{H^2(\Omega)}.$$

This dominates estimate (3.41) for small δ. Collecting terms, gives the final result. □

An easy consequence of Theorem 3.17 is the following

Corollary 3.18. *Let the assumptions of Theorem 3.17 be fulfilled. Then, the weak form of the commutation error is bounded:*

$$\begin{aligned}&\left|\int_{\mathbb{R}^d} \overline{v}(\mathbf{x}) \int_{\partial\Omega} g_\delta(\mathbf{x}-\mathbf{s})\,\psi(\mathbf{s})\,d\mathbf{s}\right| d\mathbf{x}\\ &\le C\delta^{1-d+\frac{(d-1)\theta}{q}+\beta\theta}\,\|\psi\|_{L^p(\partial\Omega)}\,\|v\|_{H^2(\Omega)}.\end{aligned} \tag{3.44}$$

Remark 3.19. Order of convergence given by (3.44). Let $d = 2$ and $p < \infty$ arbitrary large, i.e. ψ is sufficiently smooth. Then q is arbitrary close to one. Choosing θ and β also arbitrary close to one leads to the following power of δ in (3.44)

$$1 + (-2 + (1-\epsilon_1) + (1-\epsilon_2)) = 1 - (\epsilon_1+\epsilon_2) = 1-\epsilon_3$$

for arbitrary small $\epsilon_1, \epsilon_2, \epsilon_3 > 0$. In this case, the convergence is almost of first order.

The result of Theorem 3.17 does not provide an order of convergence for $d = 3$. Lemma 3.4 suggests choosing $p = 4$, i.e. $q = 4/3$. Then, the power of δ in (3.44) becomes $2(\theta-1)$, which is negative for $\theta < 1$. □

4

LES Models Which are Based on Approximations in Wave Number Space

In this chapter, models for the tensor $\overline{\mathbf{u}\mathbf{u}^T}$ in the space averaged Navier-Stokes equations (3.11) are derived, where the aim is to model $\overline{\mathbf{u}\mathbf{u}^T}$ in terms of $(\overline{\mathbf{u}}, \overline{p})$. As mentioned in Remark 3.1, the entries of this tensor are a priori not related to $(\overline{\mathbf{u}}, \overline{p})$.

There exists a lot of proposals in the literature for modelling $\overline{\mathbf{u}\mathbf{u}^T}$. This monograph will concentrate on models which are based mainly on mathematical considerations, where we consider throughout this chapter the case $\Omega = \mathbb{R}^d$ and the filter width δ to be constant. The approach to obtain these models is the following:

- *Apply the Fourier transform to* $\overline{\mathbf{u}\mathbf{u}^T}$. Using the decomposition $\mathbf{u} = \overline{\mathbf{u}} + \mathbf{u}'$, it is possible to represent the Fourier transform of the subgrid scale component $\mathbf{u}'$ by the Fourier transform of the large scale component $\overline{\mathbf{u}}$, see (4.14). In this way, one gets rid of $\mathbf{u}'$.
- *Approximate the Fourier transform of the filter function such that the inverse Fourier transform can be computed explicitly.* The application of the inverse Fourier transform gives a partial model for $\overline{\mathbf{u}\mathbf{u}^T}$.

The most common, second order approximations of the filter function (with respect to δ) approximate the subgrid scale tensor $\overline{\mathbf{u}'\mathbf{u}'^T}$ by zero. This approximation turns out to be insufficient in computations, see, e.g., Chapter 10 or Iliescu et al. [IJL+03]. We consider two alternative proposals for modelling $\overline{\mathbf{u}'\mathbf{u}'^T}$:

- the Smagorinsky model [Sma63], which is the simpliest LES model,
- models by Iliescu and Layton [IL98].

The Smagorinsky model belongs to the class of eddy viscosity models. Since, we will present analytical investigations of this model and a dynamic version of this model is currently one of the most successful LES models, we will introduce also the class of eddy viscosity models.

4.1 Eddy Viscosity Models

In the eddy viscosity models, the momentum equation of the space averaged Navier-Stokes equations (3.11) is rewritten in the form

$$\overline{\mathbf{u}}_t - 2\nu\nabla\cdot\mathbb{D}(\,\overline{\mathbf{u}}\,) + \nabla\cdot\left(\overline{\mathbf{u}}\,\overline{\mathbf{u}}^T\right) + \nabla\cdot\mathbb{T} + \nabla\overline{p} = \overline{\mathbf{f}} \quad \text{in } (0,T]\times\Omega. \tag{4.1}$$

The tensor

$$\mathbb{T} = \overline{\mathbf{u}\mathbf{u}^T} - \overline{\mathbf{u}}\,\overline{\mathbf{u}}^T$$

is called Reynolds stress tensor. This tensor has to be modelled.

The simplest form of the eddy viscosity model is

$$\mathbb{T} - \frac{\operatorname{tr}(\mathbb{T})}{3}\mathbb{I} = -\nu_T\mathbb{D}(\,\overline{\mathbf{u}}\,), \tag{4.2}$$

where ν_T is called turbulent viscosity and $\operatorname{tr}(\mathbb{T})$ is the trace of $\mathbb{T}$. Extensions of this form, possessing additional terms on the right hand side, can be found, e.g., in Kosović [Kos97]. Model (4.2) is based on the Reynolds closure hypothesis or the Boussinesq hypothesis, see, e.g., Mohammadi and Pironneau [MP94].

Remark 4.1. Definition of a new pressure. The trace of $\mathbb{T}$ is usually added to the filtered pressure which defines a new pressure

$$\tilde{p} = \overline{p} + \frac{\operatorname{tr}(\mathbb{T})}{3}.$$

For convenience of notation, we will denote this new pressure in the following also by $\overline{p}$. □

4.1.1 The Smagorinsky Model

The Smagorinsky model [Sma63] is the simplest LES model and because of its simplicity still one of the most popular ones. The turbulent viscosity parameter is given in the Smagorinsky model by

$$\nu_T = \nu_S\,\|\mathbb{D}(\,\overline{\mathbf{u}}\,)\|_F = c_S\delta^2\,\|\mathbb{D}(\,\overline{\mathbf{u}}\,)\|_F\,. \tag{4.3}$$

Remark 4.2. The artificial viscosity interpretation. The Smagorinsky model introduces an artificial, non-linear viscosity into the space averaged Navier-Stokes equations (3.11):

$$-2\nabla\cdot(\nu\mathbb{D}(\,\overline{\mathbf{u}}\,)) - \nabla\cdot(\nu_T\mathbb{D}(\,\overline{\mathbf{u}}\,)) = -\nabla\cdot((2\nu+\nu_T)\,\mathbb{D}(\,\overline{\mathbf{u}}\,))\,. \tag{4.4}$$

Thus, one can think of ν_T acting as an artificial viscosity. It follows with (3.7)

$$\begin{aligned}\nabla\cdot((2\nu+\nu_T)\,(\mathbb{D}(\,\overline{\mathbf{u}}\,))) &= \mathbb{D}(\,\overline{\mathbf{u}}\,)\,\nabla\,(2\nu+\nu_T) + (2\nu+\nu_T)\,\nabla\cdot\mathbb{D}(\,\overline{\mathbf{u}}\,)\\ &= \mathbb{D}(\,\overline{\mathbf{u}}\,)\,\nabla\,(2\nu+\nu_T) + \nabla\cdot\left(\left(\nu+\frac{\nu_T}{2}\right)\nabla\,\overline{\mathbf{u}}\right).\end{aligned}$$

Since ν is constant, the first term vanishes only if ν_T is also a constant. In this case, (4.4) simplifies to

$$-\nabla \cdot \left(\left(\nu + \frac{\nu_T}{2} \right) \nabla \overline{\mathbf{u}} \right). \tag{4.5}$$

But if ν_T is constant, we are back to solving a laminar problem. However, in the literature one can find models with term (4.5). This form has some advantages in the implementation of the discretisation of the viscous term, see Remark 7.2. □

Remark 4.3. Drawbacks of the Smagorinsky model. The drawbacks of the Smagorinsky model are well known and documented, e.g., see Zhang et al. [ZSK93]. Some of them are

- The Smagorinsky model constant c_S is an a priori input. This single constant is incapable to represent correctly various turbulent flows.
- The eddy viscosity does not vanish for a laminar flow.
- The backscatter of energy is prevented completely since $c_S\delta^2 \left\|\mathbb{D}(\overline{\mathbf{u}})\right\|_F \geq 0$.
- The Smagorinsky model introduces, in general, too much diffusion into the flow.

□

Remark 4.4. Size of the Smagorinsky constant. Numerical tests with the Smagorinsky model which can be found in the literature use typically a Smagorinsky constant of size $c_S \in [0.01, 0.1]$, e.g., see Sagaut [Sag01, p. 95] or Piomelli [Pio99]. □

4.1.2 The Dynamic Subgrid Scale Model

The Smagorinsky model (4.3) contains the parameter c_S. As pointed out, a good choice of c_S depends on the concrete flow problem and is in general a priori hardly to achieve. It may even be advantageous to choose c_S different in different flow regions. On the other hand, an unfavourable choice of c_S may lead to poor numerical results. Germano et al. [GPMC91] proposed a dynamic eddy viscosity model which computes the parameter c_S as a function of space and time. The proposal in [GPMC91] was modified by Lilly [Lil92] to the form presented below.

The dynamic subgrid scale model starts with introducing a second filter, a so-called test filter denoted by a hat, with $\widehat{\delta} > \delta$. Then, the space averaged Navier-Stokes equations (4.1) are filtered once more with the test filter. Assuming that differentiation and filtering commute yields

$$\begin{aligned}
\widehat{\overline{\mathbf{u}}}_t - 2\nu\nabla \cdot \mathbb{D}\left(\widehat{\overline{\mathbf{u}}}\right) + \nabla \cdot \left(\widehat{\overline{\mathbf{u}}\,\overline{\mathbf{u}}^T}\right) + \nabla \cdot \widehat{\mathbb{T}} + \nabla \widehat{\overline{p}} &= \widehat{\overline{\mathbf{f}}} \quad \text{in } (0,T] \times \Omega, \\
\nabla \cdot \widehat{\overline{\mathbf{u}}} &= 0 \quad \text{in } [0,T] \times \Omega.
\end{aligned}$$

A direct calculation gives

$$\mathbb{K} - \widehat{\mathbb{T}} = \widehat{\overline{\mathbf{u}}\,\overline{\mathbf{u}}^T} - \widehat{\overline{\mathbf{u}}}\ \widehat{\overline{\mathbf{u}}}^{\,T} \tag{4.6}$$

with

$$\mathbb{K} = \widehat{\mathbf{u}\mathbf{u}^T} - \widehat{\overline{\mathbf{u}}}\ \widehat{\overline{\mathbf{u}}}^{\,T}.$$

With the ansatz

$$\begin{aligned}
\mathbb{T}(t,\mathbf{x}) - \frac{\operatorname{tr}(\mathbb{T})}{3}\mathbb{I} &= -c_S(t,\mathbf{x})\,\delta^2 \left\|\mathbb{D}(\overline{\mathbf{u}})\right\|_F \mathbb{D}(\overline{\mathbf{u}}), \\
\mathbb{K}(t,\mathbf{x}) - \frac{\operatorname{tr}(\mathbb{K})}{3}\mathbb{I} &= -c_S(t,\mathbf{x})\,\widehat{\delta}^{\,2} \left\|\mathbb{D}\left(\widehat{\overline{\mathbf{u}}}\right)\right\|_F \mathbb{D}\left(\widehat{\overline{\mathbf{u}}}\right)
\end{aligned}$$

into (4.6), one obtains

$$\begin{aligned}
\mathbf{0} = &- \widehat{\overline{\mathbf{u}}\,\overline{\mathbf{u}}^T} + \widehat{\overline{\mathbf{u}}}\ \widehat{\overline{\mathbf{u}}}^{\,T} + \frac{1}{3}\left(\operatorname{tr}(\mathbb{K}) - \widehat{\operatorname{tr}(\mathbb{T})}\right)\mathbb{I} \\
&+ \widehat{\left(c_S(t,\mathbf{x})\,\delta^2 \left\|\mathbb{D}(\overline{\mathbf{u}})\right\|_F \mathbb{D}(\overline{\mathbf{u}})\right)} - c_S(t,\mathbf{x})\,\widehat{\delta}^{\,2} \left\|\mathbb{D}\left(\widehat{\overline{\mathbf{u}}}\right)\right\|_F \mathbb{D}\left(\widehat{\overline{\mathbf{u}}}\right).
\end{aligned} \tag{4.7}$$

From the linearity of the filter (3.9), the linearity of the trace operator and (4.6) follow

$$\begin{aligned}
\operatorname{tr}(\mathbb{K}) - \widehat{\operatorname{tr}(\mathbb{T})} &= \operatorname{tr}(\mathbb{K}) - \operatorname{tr}\left(\widehat{\mathbb{T}}\right) = \operatorname{tr}\left(\mathbb{K} - \widehat{\mathbb{T}}\right) \\
&= \operatorname{tr}\left(\widehat{\overline{\mathbf{u}}\,\overline{\mathbf{u}}^T} - \widehat{\overline{\mathbf{u}}}\ \widehat{\overline{\mathbf{u}}}^{\,T}\right).
\end{aligned} \tag{4.8}$$

In order to obtain an equation for $c_S(t,\mathbf{x})$, one approximates

$$\widehat{\left(c_S(t,\mathbf{x})\,\delta^2 \left\|\mathbb{D}(\overline{\mathbf{u}})\right\|_F \mathbb{D}(\overline{\mathbf{u}})\right)} \approx c_S(t,\mathbf{x})\,\delta^2 \left(\widehat{\left\|\mathbb{D}(\overline{\mathbf{u}})\right\|_F \mathbb{D}(\overline{\mathbf{u}})}\right). \tag{4.9}$$

If c_S depends only on t but not on $\mathbf{x}$, one has an equality instead of an approximation. Inserting (4.9) and (4.8) into (4.7) gives

$$\begin{aligned}
\mathbf{0} \approx &- \widehat{\overline{\mathbf{u}}\,\overline{\mathbf{u}}^T} + \widehat{\overline{\mathbf{u}}}\ \widehat{\overline{\mathbf{u}}}^{\,T} + \frac{1}{3}\operatorname{tr}\left(\widehat{\overline{\mathbf{u}}\,\overline{\mathbf{u}}^T} - \widehat{\overline{\mathbf{u}}}\ \widehat{\overline{\mathbf{u}}}^{\,T}\right)\mathbb{I} \\
&+ c_S(t,\mathbf{x})\left(\delta^2\left(\widehat{\left\|\mathbb{D}(\overline{\mathbf{u}})\right\|_F \mathbb{D}(\overline{\mathbf{u}})}\right) - \widehat{\delta}^{\,2}\left\|\mathbb{D}\left(\widehat{\overline{\mathbf{u}}}\right)\right\|_F \mathbb{D}\left(\widehat{\overline{\mathbf{u}}}\right)\right) \\
=: &\ \mathbb{L} + c_S\mathbb{M}.
\end{aligned} \tag{4.10}$$

Equations for $c_S(t,\mathbf{x})$ are obtained by replacing the approximation sign in (4.10) with the equal sign. Then, there are $d(d+1)/2$ equations to determine a single constant for given t and $\mathbf{x}$. Because of the divergence constraint, the traces of the deformation tensors vanish, such that only $d(d+1)/2-1$ equations are linearly independent. Lilly [Lil92] proposed to determine the parameter $c_S(t, \mathbf{x})$ by the least squares method, i.e. to find $c_S(t, \mathbf{x})$ such that $\|\mathbb{L} + c_S(t, \mathbf{x})\, \mathbb{M}\|_F^2$ is minimised. A straightforward calculation gives

$$c_S(t, \mathbf{x}) = \frac{\mathbb{L} : \mathbb{M}}{\mathbb{M} : \mathbb{M}}(t, \mathbf{x}). \tag{4.11}$$

In practical computations, the test filter can be applied by solving the space averaged Navier-Stokes equations on a coarse grid. If the next coarser grid of the current grid is used, then $\widehat{\delta} = 2\delta$.

Remark 4.5. Numerical instabilities of the dynamic subgrid scale model. The dynamic subgrid scale model can predict negative values for $c_S(t, \mathbf{x})$. This is an advantage since the model allows thus backscatter of energy, in contrast to the Smagorinsky model. However, numerical tests show that $c_S(t, \mathbf{x})$ can vary strongly in space and may contain negative values with a very large amplitude. These two properties may strongly destabilize the numerical solution process. In practice, the nominator and denominator of (4.11) are averaged, often in time, to compute a smoother function $c_S(t, \mathbf{x})$, e.g., see Lesieur [Les97, p. 405], Breuer [Bre98] or Sagaut [Sag01, Sect. 4.3.3]. □

Remark 4.6. Interpretation of approximation (4.9). Applying the Fourier transform to the left hand side of (4.9) gives

$$\delta^2 \mathcal{F}\left(g_{\hat{\delta}}\right)(t, \mathbf{y}) \left[\mathcal{F}\left(c_S\right)(t, \mathbf{y}) * \mathcal{F}\left(\|\mathbb{D}(\overline{\mathbf{u}})\|_F\, \mathbb{D}(\overline{\mathbf{u}})\right)(t, \mathbf{y})\right].$$

This expression will be small for $\mathbf{y}$ with $\|\mathbf{y}\|_2 \geq y_c$, where y_c is the cut-off wave number of the test filter, since $\mathcal{F}\left(g_{\hat{\delta}}\right)$ is almost identical to zero for these $\mathbf{y}$. Thus, approximation (4.9) assumes essentially that $\mathcal{F}\left(c_S\right)(t, \mathbf{y})$ is constant for all $\mathbf{y}$ whose Euclidean norm is less than the cut-off wave number of the test filter. □

4.2 Modelling of the Large Scale and Cross Terms

The starting point of the derivation of the models based on approximations in wave number space is the decomposition (3.12)

$$\overline{\mathbf{u}\mathbf{u}^T} = \overline{\overline{\mathbf{u}}\,\overline{\mathbf{u}}^T} + \overline{\overline{\mathbf{u}}\mathbf{u}'^T} + \overline{\mathbf{u}'\overline{\mathbf{u}}^T} + \overline{\mathbf{u}'\mathbf{u}'^T}.$$

The modelling of all terms is done in a similar way. However, the model obtained for the subgrid scale tensor $\overline{\mathbf{u}'\mathbf{u}'^T}$ turns out to be unsatisfactorily such that an extra treatment of this term is necessary.

Let the averaging be done with the Gaussian filter

$$\overline{\mathbf{u}}(\mathbf{x}, t) = g_\delta * \mathbf{u}(\mathbf{x}, t)$$

where g_δ is defined in (3.28) and δ is a constant.

The model of the large scale and cross terms is obtained in five steps:

1. computing the Fourier transform,
2. replacing $\mathcal{F}(\mathbf{u}')$ by a function of $\mathcal{F}(\overline{\mathbf{u}})$ if necessary,
3. approximating the Fourier transform of the Gaussian filter by an appropriate simpler function,
4. neglecting all terms which are in a certain sense of higher order in δ,
5. computing the inverse Fourier transform.

There are two approaches in the literature which differ in the third point. The first approach, Leonard [Leo74], Clark et al. [CFR79], Aldama [Ald90], approximates $\mathcal{F}(g_\delta)$ by a Taylor polynomial whereas the second approach, Galdi and Layton [GL00], uses a rational approximation.

The Fourier transform of the large scale term is, using the convolution theorem,

$$\mathcal{F}\left(\overline{\overline{\mathbf{u}}\,\overline{\mathbf{u}}^T}\right) = \mathcal{F}(g_\delta)\,\mathcal{F}\left(\overline{\mathbf{u}}\,\overline{\mathbf{u}}^T\right), \tag{4.12}$$

and the Fourier transforms of the cross terms are

$$\begin{aligned}\mathcal{F}\left(\overline{\overline{\mathbf{u}}\mathbf{u}'^T}\right) &= \mathcal{F}(g_\delta)\left(\mathcal{F}(\overline{\mathbf{u}}) * \mathcal{F}(\mathbf{u}')^T\right),\\ \mathcal{F}\left(\overline{\mathbf{u}'\overline{\mathbf{u}}^T}\right) &= \mathcal{F}(g_\delta)\left(\mathcal{F}(\mathbf{u}') * \mathcal{F}(\overline{\mathbf{u}})^T\right).\end{aligned} \tag{4.13}$$

Since $\mathcal{F}(g_\delta) \neq 0$, we have

$$\mathcal{F}(\mathbf{u}) = \frac{\mathcal{F}(g_\delta)\,\mathcal{F}(\mathbf{u})}{\mathcal{F}(g_\delta)} = \frac{\mathcal{F}(\overline{\mathbf{u}})}{\mathcal{F}(g_\delta)}.$$

With the decomposition $\mathbf{u} = \overline{\mathbf{u}} + \mathbf{u}'$, it follows

$$\mathcal{F}(\mathbf{u}') = \left(\frac{1}{\mathcal{F}(g_\delta)} - 1\right)\mathcal{F}(\overline{\mathbf{u}}). \tag{4.14}$$

Inserting this into (4.13) gives

$$\begin{aligned}\mathcal{F}\left(\overline{\overline{\mathbf{u}}\mathbf{u}'^T}\right) &= \mathcal{F}(g_\delta)\left(\mathcal{F}(\overline{\mathbf{u}}) * \left(\frac{1}{\mathcal{F}(g_\delta)} - 1\right)\mathcal{F}(\overline{\mathbf{u}})^T\right),\\ \mathcal{F}\left(\overline{\mathbf{u}'\overline{\mathbf{u}}^T}\right) &= \mathcal{F}(g_\delta)\left(\left(\frac{1}{\mathcal{F}(g_\delta)} - 1\right)\mathcal{F}(\overline{\mathbf{u}}) * \mathcal{F}(\overline{\mathbf{u}})^T\right).\end{aligned} \tag{4.15}$$

4.2.1 The Taylor LES Model

Taylor series expansions of $\mathcal{F}(g_\delta)$ and of $1/\mathcal{F}(g_\delta)$ with respect to δ and for fixed $\mathbf{y}$ give

$$\mathcal{F}(g_\delta)(\delta, \mathbf{y}) = 1 - \frac{\|\mathbf{y}\|_2^2}{4\gamma}\delta^2 + \mathcal{O}\left(\delta^4\right) \tag{4.16}$$

and

$$\frac{1}{\mathcal{F}(g_\delta)}(\delta, \mathbf{y}) = 1 + \frac{\|\mathbf{y}\|_2^2}{4\gamma}\delta^2 + \mathcal{O}\left(\delta^4\right). \tag{4.17}$$

Now, $\mathcal{F}(g_\delta)$ and $1/\mathcal{F}(g_\delta)$ are approximated in (4.12) and (4.15) by quadratic polynomials which are obtained by neglecting the terms of $\mathcal{O}(\delta^4)$ in (4.16) and (4.17), see Figure 4.1 for the one dimensional situation. It can be seen that the polynomial approximation of $\mathcal{F}(g_\delta)$ is a good approximation only for small wave numbers and it is completely wrong for high wave numbers. That means, the most important property of the Gaussian filter function, the damping of the high wave number components of $(\mathbf{u}, p)$, is not preserved by its Taylor polynomial approximation !

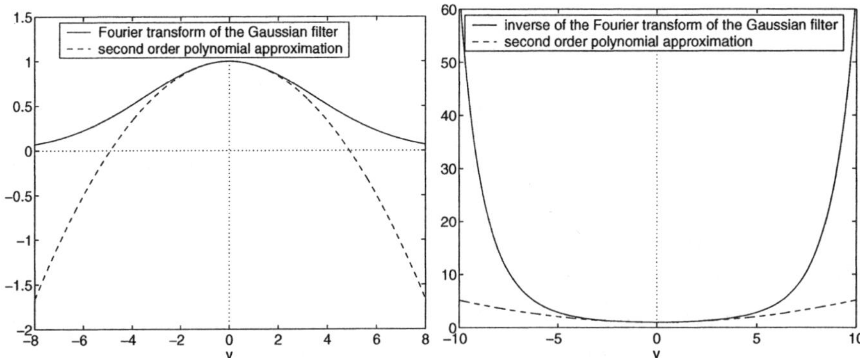

Fig. 4.1. $\mathcal{F}(g_\delta)$ and $1/\mathcal{F}(g_\delta)$ with their polynomial approximations, $\gamma = 6, \delta = 1$

Inserting (4.16) and (4.17) into (4.12) and (4.15) gives

$$\mathcal{F}\left(\overline{\overline{\mathbf{u}}\,\overline{\mathbf{u}}^T}\right) = \left(1 - \frac{\|\mathbf{y}\|_2^2}{4\gamma}\delta^2 + \mathcal{O}(\delta^4)\right)\mathcal{F}\left(\overline{\mathbf{u}}\,\overline{\mathbf{u}}^T\right),$$

$$\mathcal{F}\left(\overline{\overline{\mathbf{u}}\mathbf{u}'^T}\right) = \left(1 - \frac{\|\mathbf{y}\|_2^2}{4\gamma}\delta^2 + \mathcal{O}(\delta^4)\right) \times \left[\mathcal{F}(\overline{\mathbf{u}}) * \left(\frac{\|\mathbf{y}\|_2^2}{4\gamma}\delta^2 + \mathcal{O}(\delta^4)\right)\mathcal{F}(\overline{\mathbf{u}})^T\right],$$

$$\mathcal{F}\left(\overline{\mathbf{u}'\overline{\mathbf{u}}^T}\right) = \left(1 - \frac{\|\mathbf{y}\|_2^2}{4\gamma}\delta^2 + \mathcal{O}(\delta^4)\right) \times \left[\left(\frac{\|\mathbf{y}\|_2^2}{4\gamma}\delta^2 + \mathcal{O}(\delta^4)\right)\mathcal{F}(\overline{\mathbf{u}}) * \mathcal{F}(\overline{\mathbf{u}})^T\right].$$

Now, these expressions are simplified using properties of the Fourier transform, see Section 2.3. All terms which has as factor the fourth or a higher power of δ are neglected. However, the other factors of these terms depend on $\overline{\mathbf{u}}$ which in turn depends on δ in some unknown way. That means, the neglected terms are only formally of fourth order in δ. We get

$$
\begin{aligned}
\mathcal{F}\left(\overline{\overline{\mathbf{u}}\,\overline{\mathbf{u}}^T}\right) &= \mathcal{F}\left(\overline{\mathbf{u}}\,\overline{\mathbf{u}}^T\right) + \frac{\delta^2}{4\gamma}\mathcal{F}\left(\Delta\left(\overline{\mathbf{u}}\,\overline{\mathbf{u}}^T\right)\right) + \mathcal{O}^{formal}\left(\delta^4\right),\\
\mathcal{F}\left(\overline{\overline{\mathbf{u}}\,\mathbf{u}'^T}\right) &= -\frac{\delta^2}{4\gamma}\mathcal{F}\left(\overline{\mathbf{u}}\,\Delta\left(\overline{\mathbf{u}}\right)^T\right) + \mathcal{O}^{formal}\left(\delta^4\right),\\
\mathcal{F}\left(\overline{\mathbf{u}'\,\overline{\mathbf{u}}^T}\right) &= -\frac{\delta^2}{4\gamma}\mathcal{F}\left(\Delta\left(\overline{\mathbf{u}}\right)\overline{\mathbf{u}}^T\right) + \mathcal{O}^{formal}\left(\delta^4\right).
\end{aligned}
$$

The final approximation of the individual terms is computed by applying the inverse Fourier transform

$$
\begin{aligned}
\overline{\overline{\mathbf{u}}\,\overline{\mathbf{u}}^T} &= \overline{\mathbf{u}}\,\overline{\mathbf{u}}^T + \frac{\delta^2}{4\gamma}\Delta\left(\overline{\mathbf{u}}\,\overline{\mathbf{u}}^T\right) + \mathcal{O}^{formal}\left(\delta^4\right),\\
\overline{\overline{\mathbf{u}}\,\mathbf{u}'^T} &= -\frac{\delta^2}{4\gamma}\overline{\mathbf{u}}\,\Delta\left(\overline{\mathbf{u}}\right)^T + \mathcal{O}^{formal}\left(\delta^4\right),\\
\overline{\mathbf{u}'\,\overline{\mathbf{u}}^T} &= -\frac{\delta^2}{4\gamma}\Delta\left(\overline{\mathbf{u}}\right)\overline{\mathbf{u}}^T + \mathcal{O}^{formal}\left(\delta^4\right).
\end{aligned}
$$

This gives the following model obtained with the Taylor polynomial approximation of the Gaussian filter

$$
\begin{aligned}
&\overline{\overline{\mathbf{u}}\,\overline{\mathbf{u}}^T} + \overline{\overline{\mathbf{u}}\,\mathbf{u}'^T} + \overline{\mathbf{u}'\,\overline{\mathbf{u}}^T}\\
&\approx \overline{\mathbf{u}}\,\overline{\mathbf{u}}^T + \frac{\delta^2}{4\gamma}\left(\Delta\left(\overline{\mathbf{u}}\,\overline{\mathbf{u}}^T\right) - \overline{\mathbf{u}}\,\Delta\left(\overline{\mathbf{u}}\right)^T - \Delta\left(\overline{\mathbf{u}}\right)\overline{\mathbf{u}}^T\right)\\
&= \overline{\mathbf{u}}\,\overline{\mathbf{u}}^T + \frac{\delta^2}{2\gamma}\nabla\overline{\mathbf{u}}\nabla\overline{\mathbf{u}}^T,
\end{aligned}
\tag{4.18}
$$

where

$$
\Delta\left(\overline{\mathbf{u}}\,\overline{\mathbf{u}}^T\right) - \overline{\mathbf{u}}\,\Delta\left(\overline{\mathbf{u}}\right)^T - \Delta\left(\overline{\mathbf{u}}\right)\overline{\mathbf{u}}^T = 2\nabla\overline{\mathbf{u}}\nabla\overline{\mathbf{u}}^T \tag{4.19}
$$

has been used.

4.2.2 The Second Order Rational LES Model

The second order rational LES model will be called in the following chapters only *rational LES model.*

Based on the observation that the Fourier transform of the Gaussian filter is approximated very badly with the Taylor polynomial approximation for large wave numbers, Galdi and Layton proposed in [GL00] to use an rational approximation of the exponential

$$
e^{ax} = \frac{1}{1+ax} + \mathcal{O}\left(a^2x^2\right).
$$

Applying this subdiagonal Padé approximation to $\mathcal{F}\left(g_\delta\right)$ gives

$$
\mathcal{F}\left(g_\delta\right)\left(\delta,\mathbf{y}\right) = \frac{1}{1+\dfrac{\|\mathbf{y}\|_2^2}{4\gamma}\delta^2} + \mathcal{O}\left(\delta^4\right) \tag{4.20}
$$

and transforming this formula to $1/\mathcal{F}(g_\delta)$ yields

$$\frac{1}{\mathcal{F}(g_\delta)}(\delta, \mathbf{y}) = 1 + \frac{\|\mathbf{y}\|_2^2}{4\gamma}\delta^2 + \mathcal{O}^{formal}\left(\delta^4\right). \tag{4.21}$$

The last term in (4.21) is actually $\mathcal{O}\left(\delta^4\right)/\mathcal{F}(g_\delta)$ such that it is only formally of fourth order. The rational approximations of $\mathcal{F}(g_\delta)$ and $1/\mathcal{F}(g_\delta)$ are obtained by neglecting all (formal) fourth order terms in (4.20) and (4.21). The behaviour of $\mathcal{F}(g_\delta)$ for high wave numbers is much better approximated than with the Taylor polynomial, see Figure 4.2 for a one dimensional sketch. The approximation of $1/\mathcal{F}(g_\delta)$ is the same as in the polynomial case.

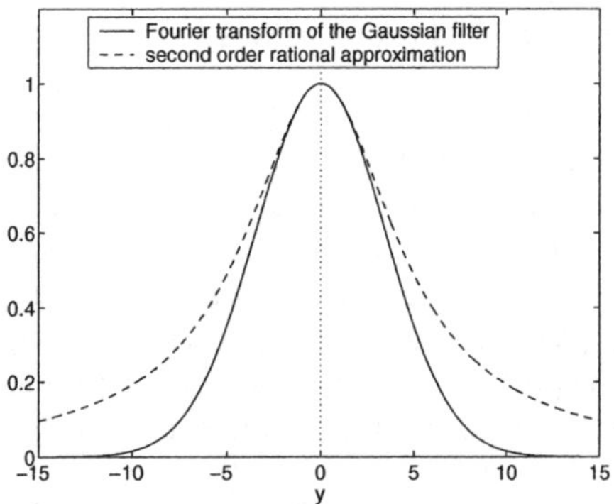

Fig. 4.2. $\mathcal{F}(g_\delta)$ with its second order rational approximations, $\gamma = 6, \delta = 1$

The derivation of the model continues now in the same way as in the polynomial case. Inserting (4.20) and (4.21) into (4.12) and (4.15), simplifying the arising terms using properties of the Fourier transform, see Section 2.3, neglecting all terms which are formally of fourth order with respect to δ and applying the inverse Fourier transform give

$$\begin{aligned}
\overline{\overline{\mathbf{u}}\,\overline{\mathbf{u}}^T} &= \left(I - \frac{\delta^2}{4\gamma}\Delta\right)^{-1}\left(\overline{\mathbf{u}}\,\overline{\mathbf{u}}^T\right) + \mathcal{O}^{formal}\left(\delta^4\right), \\
\overline{\overline{\mathbf{u}}\,\mathbf{u}'^T} &= -\frac{\delta^2}{4\gamma}\left(I - \frac{\delta^2}{4\gamma}\Delta\right)^{-1}\left(\overline{\mathbf{u}}\,\Delta\left(\overline{\mathbf{u}}\right)^T\right) + \mathcal{O}^{formal}\left(\delta^4\right), \\
\overline{\mathbf{u}'\,\overline{\mathbf{u}}^T} &= -\frac{\delta^2}{4\gamma}\left(I - \frac{\delta^2}{4\gamma}\Delta\right)^{-1}\left(\Delta\left(\overline{\mathbf{u}}\right)\overline{\mathbf{u}}^T\right) + \mathcal{O}^{formal}\left(\delta^4\right).
\end{aligned}$$

Finally, one obtains

$$\begin{aligned}
&\overline{\overline{\mathbf{u}}\,\overline{\mathbf{u}}^T} + \overline{\overline{\mathbf{u}}\,\mathbf{u}'^T} + \overline{\mathbf{u}'\,\overline{\mathbf{u}}^T} \\
&\approx \left(I - \frac{\delta^2}{4\gamma}\Delta\right)^{-1} \left[\overline{\mathbf{u}}\,\overline{\mathbf{u}}^T - \frac{\delta^2}{4\gamma}\left(\overline{\mathbf{u}}\,\Delta\left(\overline{\mathbf{u}}\right)^T + \Delta\left(\overline{\mathbf{u}}\right)\overline{\mathbf{u}}^T\right)\right] \\
&= \left(I - \frac{\delta^2}{4\gamma}\Delta\right)^{-1} \left[\overline{\mathbf{u}}\,\overline{\mathbf{u}}^T - \frac{\delta^2}{4\gamma}\Delta\left(\overline{\mathbf{u}}\,\overline{\mathbf{u}}^T\right) + \frac{\delta^2}{2\gamma}\nabla\overline{\mathbf{u}}\nabla\overline{\mathbf{u}}^T\right] \\
&= \overline{\mathbf{u}}\,\overline{\mathbf{u}}^T + \frac{\delta^2}{2\gamma}\left(I - \frac{\delta^2}{4\gamma}\Delta\right)^{-1} \nabla\overline{\mathbf{u}}\nabla\overline{\mathbf{u}}^T. \qquad (4.22)
\end{aligned}$$

In the corresponding formula (2.10) of the paper of Galdi and Layton [GL00], there is a misprint (minus sign instead of plus sign). The operator $\left(I - \frac{\delta^2}{4\gamma}\Delta\right)^{-1}$ describes an elliptic, second order problem which has to be solved. In this monograph, this problem will be called auxiliary problem.

Remark 4.7. The auxiliary problem in a bounded domain. If Ω is a bounded domain, which is usually the case in computations, the auxiliary problem has to be equipped with boundary conditions on $\partial\Omega$. Galdi and Layton [GL00] proposed to use homogeneous Neumann boundary conditions. These boundary conditions will be applied in the computations presented in this monograph. The only exception is the case that periodic boundary conditions are prescribed for the flow problem at some part of the boundary. Then, the auxiliary problem is equipped also with periodic boundary conditions at those parts of the boundary.

There are neither investigations if the use of homogeneous Neumann boundary conditions is optimal nor numerical comparisons to other boundary conditions. The numerical studies presented in Chapters 10 and 11 show that the use of homogeneous Neumann boundary conditions gives often reasonable or good results. Nevertheless, the question of mathematically supported boundary conditions for the auxiliary problem has to be studied. □

Remark 4.8. The auxiliary problem is an approximation of the convolution. A simple computation, using the rational approximation (4.20) of $\mathcal{F}(g_\delta)$ and properties of the Fourier transform from Section 2.3, gives

$$\mathcal{F}(g_\delta * \mathbf{u}) = \mathcal{F}(g_\delta)\,\mathcal{F}(\mathbf{u}) \approx \frac{1}{1 + \frac{\|\mathbf{y}\|_2^2}{4\gamma}\delta^2}\mathcal{F}(\mathbf{u}) = \mathcal{F}\left(\left(I - \frac{\delta^2}{4\gamma}\Delta\right)^{-1}\mathbf{u}\right),$$

from what follows

$$g_\delta * \mathbf{u} \approx \left(I - \frac{\delta^2}{4\gamma}\Delta\right)^{-1}\mathbf{u}.$$

Thus, the auxiliary problem is an approximation of the convolution operator. □

This remark suggests that the rational LES model can be defined with a convolution instead of the auxiliary problem

$$\overline{\overline{\mathbf{u}}\,\overline{\mathbf{u}}^T} + \overline{\overline{\mathbf{u}}\mathbf{u}'^T} + \overline{\mathbf{u}'\overline{\mathbf{u}}^T} \approx \overline{\mathbf{u}}\,\overline{\mathbf{u}}^T + \frac{\delta^2}{2\gamma} g_\delta * \left(\nabla \overline{\mathbf{u}} \nabla \overline{\mathbf{u}}^T\right).$$

This model is called rational LES model with convolution.

Remark 4.9. The computation of the convolution of a function given in a bounded domain. In the numerical tests, Ω is usually a bounded domain. If a function defined on Ω has to be convolved, we start by extending this function to $\mathbb{R}^d$ by the trivial extension. Then, the convolution is well defined and can be computed. In the last step, the convolved function is restricted to Ω. □

Remark 4.10. Interpretation of the auxiliary problem as backward Euler scheme. We consider the time dependent Poisson problem

$$\begin{aligned} u_t - \Delta u &= \frac{4\gamma}{\delta^2} f \text{ in } (0,T] \times \Omega, \\ \frac{\partial u}{\partial \mathbf{n}_{\partial\Omega}} &= 0 \quad \text{ in } [0,T] \times \partial\Omega, \\ u(0,\cdot) &= 0 \quad \text{ in } \Omega \end{aligned} \tag{4.24}$$

for given right hand side f and $T = \delta^2/(4\gamma)$. The discretisation of (4.24) with a backward Euler scheme with time step $\Delta t_1 = \delta^2/(4\gamma)$ gives the following system at t_1

$$\begin{aligned} \left(I - \frac{\delta^2}{4\gamma}\Delta\right) u(t_1) &= f(t_1) \text{ in } \Omega, \\ \frac{\partial u(t_1, \mathbf{x})}{\partial \mathbf{n}_{\partial\Omega}} &= 0 \qquad \text{on } \partial\Omega. \end{aligned}$$

Thus, the auxiliary problem can be interpreted as approximation of (4.24) with one step of a backward Euler scheme. □

4.2.3 The Fourth Order Rational LES Model

As we shall see in Section 4.3.1, the second order polynomial and rational approximations model the subgrid scale term $\overline{\mathbf{u}'\mathbf{u}'^T}$ by $\mathbf{0}$. However, this model has proved to be insufficient in numerical tests. Thus, a model of the subgrid scale term for the second order approximations has to be obtained by different considerations than the approximation of $\mathcal{F}(g_\delta)$, see Section 4.3. Here, we will derive a fourth order rational approximation which yields a non-vanishing model for the subgrid scale term.

The way of derivation of the model is the same as for the second order rational model. The second order Padé approximation of the exponential is

$$e^{ax} = \frac{1}{1 + ax + \frac{a}{2}x^2} + \mathcal{O}\left(a^3 x^3\right).$$

This gives

$$\mathcal{F}(g_\delta)(\delta,\mathbf{y}) = \frac{1}{1+\frac{\|\mathbf{y}\|_2^2}{4\gamma}\delta^2+\frac{\|\mathbf{y}\|_2^4}{32\gamma^2}\delta^4} + \mathcal{O}\left(\delta^6\right), \tag{4.25}$$

see Figure 4.3 for a graphical representation of the fourth order rational approximation of $\mathcal{F}(g_\delta)$. To approximate $1/\mathcal{F}(g_\delta)(\delta,\mathbf{y})$, the Taylor series expansion can be used. This yields

$$\frac{1}{\mathcal{F}(g_\delta)}(\delta,\mathbf{y}) = 1+\frac{\|\mathbf{y}\|_2^2}{4\gamma}\delta^2+\frac{\|\mathbf{y}\|_2^4}{32\gamma^2}\delta^4+\mathcal{O}\left(\delta^6\right). \tag{4.26}$$

Like in the second order rational approximation, one can obtain (4.26) also from (4.25) with a formal sixth order term.

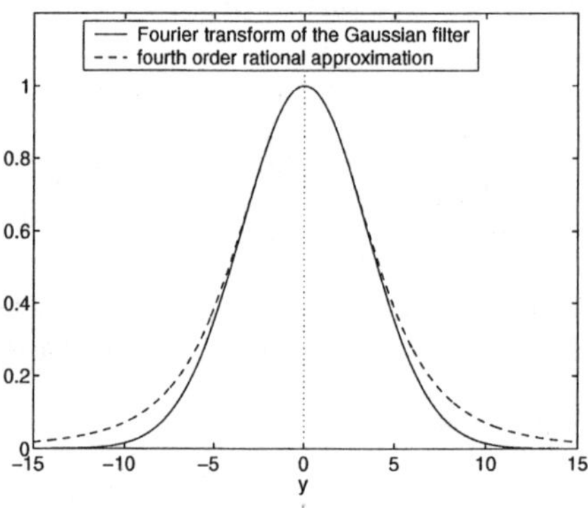

Fig. 4.3. $\mathcal{F}(g_\delta)$ with its fourth order rational approximations, $\gamma=6, \delta=1$

The derivation of the rational fourth order approximation is continued in the standard way. One obtains

$$\begin{aligned}
\mathcal{F}\left(\overline{\overline{\mathbf{u}}\,\overline{\mathbf{u}}^T}\right) &= \mathcal{F}(g_\delta)\,\mathcal{F}\left(\overline{\mathbf{u}}\,\overline{\mathbf{u}}^T\right),\\
\mathcal{F}\left(\overline{\overline{\mathbf{u}}\,\mathbf{u}'^T}\right) &= \mathcal{F}(g_\delta)\,\mathcal{F}\left(-\frac{\delta^2}{4\gamma}\overline{\mathbf{u}}\,\Delta\overline{\mathbf{u}}^T+\frac{\delta^4}{32\gamma^2}\overline{\mathbf{u}}\left(\Delta^2\overline{\mathbf{u}}\right)^T+\mathcal{O}^{formal}\left(\delta^6\right)\right),\\
\mathcal{F}\left(\overline{\mathbf{u}'\,\overline{\mathbf{u}}^T}\right) &= \mathcal{F}(g_\delta)\,\mathcal{F}\left(-\frac{\delta^2}{4\gamma}\Delta\overline{\mathbf{u}}\,\overline{\mathbf{u}}^T+\frac{\delta^4}{32\gamma^2}\left(\Delta^2\overline{\mathbf{u}}\right)\overline{\mathbf{u}}^T+\mathcal{O}^{formal}\left(\delta^6\right)\right).
\end{aligned}$$

Applying (4.25) and using properties of the Fourier transform give the model

$$\begin{aligned}
\overline{\overline{\mathbf{u}}\,\overline{\mathbf{u}}^T}+\overline{\overline{\mathbf{u}}\,\mathbf{u}'^T}+\overline{\mathbf{u}'\,\overline{\mathbf{u}}^T} \approx g_\delta * \Bigg[&\overline{\mathbf{u}}\,\overline{\mathbf{u}}^T-\frac{\delta^2}{4\gamma}\left(\overline{\mathbf{u}}\,\Delta\overline{\mathbf{u}}^T+\Delta\overline{\mathbf{u}}\,\overline{\mathbf{u}}^T\right)\\
&+\frac{\delta^4}{32\gamma^2}\left(\overline{\mathbf{u}}\left(\Delta^2\overline{\mathbf{u}}\right)^T+\left(\Delta^2\overline{\mathbf{u}}\right)\overline{\mathbf{u}}^T\right)\Bigg].
\end{aligned}$$

This can be rewritten, using (4.19) and

$$\Delta^2\left(\overline{\mathbf{u}}\,\overline{\mathbf{u}}^T\right) = 2\Delta\overline{\mathbf{u}}\left(\Delta\overline{\mathbf{u}}\right)^T + \overline{\mathbf{u}}\left(\Delta^2\overline{\mathbf{u}}\right)^T + \left(\Delta^2\overline{\mathbf{u}}\right)\overline{\mathbf{u}}^T + 2\nabla\left(\Delta\overline{\mathbf{u}}\right)\nabla\overline{\mathbf{u}}^T \\ +2\nabla\overline{\mathbf{u}}\nabla\left(\Delta\overline{\mathbf{u}}\right)^T + 2\Delta\left(\nabla\overline{\mathbf{u}}\nabla\overline{\mathbf{u}}^T\right),$$

in the form

$$\overline{\overline{\mathbf{u}}\,\overline{\mathbf{u}}^T} + \overline{\overline{\mathbf{u}}\mathbf{u}'^T} + \overline{\mathbf{u}'\overline{\mathbf{u}}^T} \\ \approx g_\delta * \left[\left(I - \frac{\delta^2}{4\gamma}\Delta + \frac{\delta^4}{32\gamma^2}\Delta^2\right)\overline{\mathbf{u}}\,\overline{\mathbf{u}}^T + \frac{\delta^2}{2\gamma}\nabla\overline{\mathbf{u}}\nabla\overline{\mathbf{u}}^T \right. \\ \left. -\frac{\delta^4}{16\gamma^2}\left(\Delta\overline{\mathbf{u}}\Delta\overline{\mathbf{u}}^T + \nabla\left(\Delta\overline{\mathbf{u}}\right)\nabla\overline{\mathbf{u}}^T + \nabla\overline{\mathbf{u}}\nabla\left(\Delta\overline{\mathbf{u}}\right)^T + \Delta\left(\nabla\overline{\mathbf{u}}\nabla\overline{\mathbf{u}}^T\right)\right)\right].$$

The approximation (4.25) of $\mathcal{F}(g_\delta)$ now yields

$$\mathcal{F}(g_\delta * \mathbf{u}) = \mathcal{F}(g_\delta)\mathcal{F}(\mathbf{u}) \approx \frac{1}{1 + \frac{\|\mathbf{y}\|_2^2}{4\gamma}\delta^2 + \frac{\|\mathbf{y}\|_2^4}{32\gamma^2}\delta^4}\mathcal{F}(\mathbf{u}) \\ = \mathcal{F}\left(\left(I - \frac{\delta^2}{4\gamma}\Delta + \frac{\delta^4}{32\gamma^2}\Delta^2\right)^{-1}\mathbf{u}\right)$$

from which follows

$$g_\delta * \mathbf{u} \approx \left(I - \frac{\delta^2}{4\gamma}\Delta + \frac{\delta^4}{32\gamma^2}\Delta^2\right)^{-1}\mathbf{u}.$$

This gives the final fourth order rational model of the large scale and the cross terms with a fourth order partial differential equation as auxiliary problem

$$\overline{\overline{\mathbf{u}}\,\overline{\mathbf{u}}^T} + \overline{\overline{\mathbf{u}}\mathbf{u}'^T} + \overline{\mathbf{u}'\overline{\mathbf{u}}^T} \\ \approx \overline{\mathbf{u}}\,\overline{\mathbf{u}}^T + \left(I - \frac{\delta^2}{4\gamma}\Delta + \frac{\delta^4}{32\gamma^2}\Delta^2\right)^{-1}\left[\frac{\delta^2}{2\gamma}\nabla\overline{\mathbf{u}}\nabla\overline{\mathbf{u}}^T \right. \\ \left. -\frac{\delta^4}{16\gamma^2}\left(\Delta\overline{\mathbf{u}}\Delta\overline{\mathbf{u}}^T + \nabla\left(\Delta\overline{\mathbf{u}}\right)\nabla\overline{\mathbf{u}}^T + \nabla\overline{\mathbf{u}}\nabla\left(\Delta\overline{\mathbf{u}}\right)^T + \Delta\left(\nabla\overline{\mathbf{u}}\nabla\overline{\mathbf{u}}^T\right)\right)\right]. \quad (4.27)$$

4.3 Models for the Subgrid Scale Term

The subgrid scale term $\overline{\mathbf{u}'\mathbf{u}'^T}$ is considered to possess a great influence on the formation of turbulence. Thus, its modelling is of great importance.

4.3.1 The Second Order Fourier Transform Approach

If the subgrid scale term is modelled with the second order approaches which were used for the large scale term and the cross terms, Section 4.2, one obtains with the second order polynomial approximation of the Fourier transform of the Gaussian filter

$$\overline{\mathbf{u}'\mathbf{u}'^T} = \frac{\delta^4}{16\gamma^2}\left(\Delta\overline{\mathbf{u}}\,\Delta\overline{\mathbf{u}}^T\right) + \mathcal{O}^{formal}\left(\delta^6\right)$$

and with the second order rational approximation

$$\overline{\mathbf{u}'\mathbf{u}'^T} = \frac{\delta^4}{16\gamma^2}\left(I - \frac{\delta^2}{4\gamma}\Delta\right)^{-1}\left(\Delta\overline{\mathbf{u}}\,\Delta\overline{\mathbf{u}}^T\right) + \mathcal{O}^{formal}\left(\delta^6\right).$$

Both approximations are formally of fourth order in δ and therefore will be neglected in these approaches. That means, one obtains the approximation

$$\overline{\mathbf{u}'\mathbf{u}'^T} \approx \mathbf{0},$$

which proves to be unsatisfactorily, see Section 10.3 for a numerical example.

4.3.2 The Fourth Order Rational LES Model

The fourth order rational approximation, Section 4.2.3, yields

$$\mathcal{F}\left(\overline{\mathbf{u}'\mathbf{u}'^T}\right) = \mathcal{F}\left(g_\delta\right)\left(\frac{\delta^4}{16\gamma^2}\mathcal{F}\left(\Delta\overline{\mathbf{u}}\,\Delta\overline{\mathbf{u}}^T\right) + \mathcal{O}^{formal}\left(\delta^6\right)\right).$$

Thus,

$$\overline{\mathbf{u}'\mathbf{u}'^T} = g_\delta * \frac{\delta^4}{16\gamma^2}\Delta\overline{\mathbf{u}}\,\Delta\overline{\mathbf{u}}^T + \mathcal{O}^{formal}\left(\delta^6\right).$$

Neglecting the sixth order term gives with (4.27) the fourth order rational LES model

$$\begin{aligned} &\overline{\overline{\mathbf{u}}\,\overline{\mathbf{u}}^T} + \overline{\overline{\mathbf{u}}\mathbf{u}'^T} + \overline{\mathbf{u}'\overline{\mathbf{u}}^T} + \overline{\mathbf{u}'\mathbf{u}'^T} \\ &\approx \overline{\mathbf{u}}\,\overline{\mathbf{u}}^T + \left(I - \frac{\delta^2}{4\gamma}\Delta + \frac{\delta^4}{32\gamma^2}\Delta^2\right)^{-1}\left[\frac{\delta^2}{2\gamma}\nabla\overline{\mathbf{u}}\nabla\overline{\mathbf{u}}^T\right. \\ &\left.-\frac{\delta^4}{16\gamma^2}\left(\nabla\left(\Delta\overline{\mathbf{u}}\right)\nabla\overline{\mathbf{u}}^T + \nabla\overline{\mathbf{u}}\nabla\left(\Delta\overline{\mathbf{u}}\right)^T + \Delta\left(\nabla\overline{\mathbf{u}}\nabla\overline{\mathbf{u}}^T\right)\right)\right]. \end{aligned} \tag{4.28}$$

Remark 4.11. Usability of the fourth order rational LES model. In order to use the fourth order rational LES model in numerical simulations of turbulent flows, one has to replace the function $\overline{\mathbf{u}}$ in (4.28) by a discrete function. In finite element methods, the discrete functions belong usually only to $C^0\left(\overline{\Omega}\right)$. An approximation of the third order derivatives in the fourth order rational LES model can be achieved in the following way:

- compute the first order derivatives,
- project the discontinuous first order derivatives to a continuous function by an averaging process,

- compute the second order derivatives using the averaged first order derivatives,
- project the discontinuous second order derivatives to a continuous function by an averaging process,
- compute the third order derivatives using the averaged second order derivatives.

Thus, it seems not to be impossible to use the fourth order rational LES model in the context of finite element methods. However, the accuracy of the computed third order derivatives has to be investigated, as well analytically as in numerical studies.

If the domain Ω is bounded, the fourth order auxiliary problem in (4.28) has to be equipped with boundary conditions. The derivation of mathematically supported boundary conditions will be fundamental for the applicability of the fourth order rational LES model.

We will not use the fourth order rational LES model in the numerical studies presented in this monograph. □

4.3.3 The Smagorinsky Model

A popular approach consists in using the Smagorinsky model (4.3) for approximating the subgrid scale term. However, with the Fourier space approach, Section 4.3.1, one obtains that the subgrid scale term is formally of fourth order in δ whereas the Smagorinsky model is formally of second order in δ. That is a contradiction, at least formally. We want to point out again that in the formal higher order terms there is some dependence on δ which is not known. Numerical studies show that the Smagorinsky model introduces often too much viscosity. This is a hint that the power of δ is indeed too small.

Remark 4.12. Choice of the Smagorinsky constant in the computations. If the Smagorinsky model is used as LES model, the constant c_S is chosen accordingly to Remark 4.4. In the Taylor and the rational LES models, the Smagorinsky model is used only as a model of the subgrid scale tensor. Thus, its influence should be kept smaller than in computations with the pure Smagorinsky model. To achieve this, we choose c_S small in the computations with the Taylor and the rational LES models, e.g., $c_S \leq 0.01$. □

4.3.4 Models Proposed by Iliescu and Layton

Iliescu and Layton [IL98] proposed, based on physical arguments, that the turbulent diffusion parameter should depend on the (mean) kinetic energy of the small eddies, i.e.

$$\nu_T = \nu_T\left(\frac{1}{2}\rho \left\|\mathbf{u}'\right\|_2^2\right) \quad \text{or } \nu_T = \nu_T\left(\overline{\frac{1}{2}\rho \left\|\mathbf{u}'\right\|_2^2}\right).$$

The so-called Kolmogorov-Prandtl expression suggests that the correct form of ν_T is given by

$$\nu_T = c l_m \sqrt{\frac{1}{2}\rho \left\| \mathbf{u}' \right\|_2^2}, \tag{4.29}$$

where l_m is the mixing length which is $l_m = \delta$ in LES.

From (4.14) and (4.21) one gets

$$\mathbf{u}' \approx -\frac{\delta^2}{4\gamma} \Delta \overline{\mathbf{u}}.$$

Terms, which are formally of fourth order in δ are neglected. Inserting this approximation into (4.29), Iliescu and Layton proposed to use the turbulent diffusion parameter

$$\nu_T = c_S \frac{\delta^3}{\gamma} \left\| \Delta \overline{\mathbf{u}} \right\|_2.$$

With the average of the kinetic energy of the small eddies and the approximation $g_\delta \approx g_\delta^2$, they obtained in the same way another proposal:

$$\nu_T = c_S \frac{\delta^3}{\gamma} \left\| g_\delta * \Delta \overline{\mathbf{u}} \right\|_2.$$

With

$$\begin{aligned} \mathcal{F}\left(\overline{\mathbf{u}} - g_\delta * \overline{\mathbf{u}} \right) &= \mathcal{F}\left(g_\delta\right) \left(\frac{1}{\mathcal{F}\left(g_\delta\right)} - 1 \right) \mathcal{F}\left(\overline{\mathbf{u}} \right) \\ &\approx \mathcal{F}\left(g_\delta\right) \left(\frac{\left\| \mathbf{y} \right\|_2^2}{4\gamma} \delta^2 \right) \mathcal{F}\left(\overline{\mathbf{u}} \right) = -\frac{\delta^2}{4\gamma} \mathcal{F}\left(g_\delta * \Delta \overline{\mathbf{u}} \right) \end{aligned}$$

a third model is obtained:

$$\nu_T = c_S \delta \left\| \overline{\mathbf{u}} - g_\delta * \overline{\mathbf{u}} \right\|_2. \tag{4.30}$$

These models are all formally of third order in δ.

Model (4.30) will be applied in the computations where the convolution operator is approximated by a second order partial differential operator, see Remark 4.8,

$$\nu_T = c_S \delta \left\| \overline{\mathbf{u}} - \left(I - \frac{\delta^2}{4\gamma} \Delta \right)^{-1} \overline{\mathbf{u}} \right\|_2. \tag{4.31}$$

5

The Variational Formulation of the LES Models

We have seen in the previous chapters that it was not possible to derive equations for $(\overline{\mathbf{u}}, \overline{p})$ only from the Navier-Stokes equations since a modelling process was also necessary. Thus the quantities which will be computed using these models will not be $(\overline{\mathbf{u}}, \overline{p})$ but, hopefully good, approximations to $(\overline{\mathbf{u}}, \overline{p})$. To have a clear distinction between the large scale quantities $(\overline{\mathbf{u}}, \overline{p})$ and their approximations, we will denote the solution obtained by the LES models by $(\mathbf{w}, r)$.

In this chapter, we will consider the case that Ω is a bounded domain. In this case, the space averaged Navier-Stokes equations and the LES models are simply restricted to Ω. This is the way usually used in practice. We like to stress that this way leads especially near the boundary of Ω to additional errors.

The chapter starts by deriving a weak or variational form of the LES models. This form will be used in the analysis of the existence and uniqueness of solutions in Chapter 6. Moreover, we have already seen that one of the errors committed by the simple restriction to Ω, the commutation error, tends to zero as the filter width $\delta \to 0$ in a variational formulation of the space averaged Navier-Stokes equations, Section 3.7, whereas this is not the case for the strong formulation, Section 3.5. From this point of view, the variational formulation is preferable to the strong formulation. Last, a variational formulation is also the basis of finite element methods, see Chapter 7.

The formulation of equations for $(\mathbf{w}, r)$ in a bounded domain requires to equip them with boundary conditions. In Section 5.2, a number of boundary conditions for incompressible flow problems are presented and their application to the LES models is discussed.

The final section of this chapter introduces some function spaces for the LES models which will be used in the analysis presented in subsequent chapters.

5.1 The Weak Formulation of the Equations

The strong formulation of the LES models derived in the previous chapters has the form

$$
\begin{aligned}
\mathbf{w}_t - \nabla\cdot\left((2\nu+\nu_T)\,\mathbb{D}(\mathbf{w})\right) + (\mathbf{w}\cdot\nabla)\,\mathbf{w}& \\
+\nabla r + \nabla\cdot\frac{\delta^2}{2\gamma}\left(A\left(\nabla\mathbf{w}\nabla\mathbf{w}^T\right)\right) &= \overline{\mathbf{f}} \ \text{ in } (0,T]\times\Omega, \\
\nabla\cdot\mathbf{w} &= 0 \quad \text{in } [0,T]\times\Omega, \\
\mathbf{w}(0,\cdot) &= \mathbf{w}_0 \text{ in } \Omega,
\end{aligned}
\tag{5.1}
$$

where the operator A depends on the approximation of the Fourier transform of the Gaussian filter. In Chapter 4, the following choices for A have been proposed:

- $A = I$ for the Taylor LES model (4.18),
- $A = \left(I - \delta^2/(4\gamma)\,\Delta\right)^{-1}$ for the second order rational LES model with auxiliary problem (4.22),
- $A = g_\delta *$ for the second order rational LES model with convolution (4.23).

Possible choices for the turbulent viscosity ν_T are given in Section 4.3. In addition, (5.1) has to be equipped with boundary conditions, see Section 5.2 for a discussion of this topic. Depending on the boundary condition, an additional condition on the pressure has to be prescribed to achieve uniqueness of r, see Section 5.3.

The weak or variational formulation of (5.1) is obtained in the standard way by multiplying (5.1) with suitable test functions, integrating over $(0,T)\times\Omega$ and using integration by parts. Let (V_a, Q_a) be a suitable pair of ansatz spaces and (V_t, Q_t) a pair of test spaces. The weak formulation of (5.1) is:

Find $(\mathbf{w}, r) \in (V_a, Q_a)$ such that for all $(\mathbf{v}, q) \in (V_t, Q_t)$

$$
\begin{aligned}
&\int_0^T \Big[(\mathbf{w}_t,\mathbf{v}) + \left((2\nu+\nu_T)\,\mathbb{D}(\mathbf{w}),\mathbb{D}(\mathbf{v})\right) + b(\mathbf{w},\mathbf{w},\mathbf{v}) - (r,\nabla\cdot\mathbf{v}) \\
&\qquad - \left(\frac{\delta^2}{2\gamma}\left(A\left(\nabla\mathbf{w}\nabla\mathbf{w}^T\right)\right),\nabla\mathbf{v}\right)\Big]\,dt \\
&= \int_0^T \Big[(\overline{\mathbf{f}},\mathbf{v}) \\
&\qquad + \int_{\partial\Omega}\left[(2\nu+\nu_T)\,\mathbb{D}(\mathbf{w}) - r\mathbb{I} - \frac{\delta^2}{2\gamma}\left(A\left(\nabla\mathbf{w}\nabla\mathbf{w}^T\right)\right)\right]\mathbf{n}_{\partial\Omega}\cdot\mathbf{v}ds\Big]\,dt, \\
&0 = (\nabla\cdot\mathbf{w}, q) \quad \text{for almost all } t\in[0,T],
\end{aligned}
\tag{5.2}
$$

where

$$
b(\mathbf{u},\mathbf{w},\mathbf{v}) = \left((\mathbf{u}\cdot\nabla)\,\mathbf{w},\mathbf{v}\right). \tag{5.3}
$$

In the derivation of this formulation, we have used

$$
\begin{aligned}
&((2\nu + \nu_T)\,\mathbb{D}(\mathbf{w})\,,\nabla\mathbf{v}) \\
&= \left(\left(\nu + \frac{\nu_T}{2}\right)\mathbb{D}(\mathbf{w})\,,\nabla\mathbf{v}\right) + \left(\left(\nu + \frac{\nu_T}{2}\right)\mathbb{D}(\mathbf{w})^T\,,\nabla\mathbf{v}^T\right) \\
&= \left(\left(\nu + \frac{\nu_T}{2}\right)\mathbb{D}(\mathbf{w})\,,\nabla\mathbf{v}\right) + \left(\left(\nu + \frac{\nu_T}{2}\right)\mathbb{D}(\mathbf{w})\,,\nabla\mathbf{v}^T\right) \\
&= ((2\nu + \nu_T)\,\mathbb{D}(\mathbf{w})\,,\mathbb{D}(\mathbf{v}))\,.
\end{aligned}
$$

If ν_T is constant in space and if $\mathbf{v}$ vanishes on $\partial\Omega$, the viscous term can be transformed into

$$
\begin{aligned}
((2\nu + \nu_T)\,\mathbb{D}(\mathbf{w})\,,\mathbb{D}(\mathbf{v})) &= \left(\nu + \frac{\nu_T}{2}\right)\left((\nabla\mathbf{w}, \nabla\mathbf{v}) + (\nabla\mathbf{w}^T, \nabla\mathbf{v})\right) \\
&= \left(\left(\nu + \frac{\nu_T}{2}\right)\nabla\mathbf{w}, \nabla\mathbf{v}\right),
\end{aligned}
$$

where in the last step integration by parts and $\nabla\cdot\mathbf{w} = 0$ have been used. This gradient formulation has some computational advantages in comparison to the deformation tensor formulation, see Remark 7.2, and is therefore sometimes used in computations also for non-constant ν_T.

The stress tensor of the LES models has the form

$$
\mathbb{S}\,(\mathbf{w}, r, A) = (2\nu + \nu_T)\,\mathbb{D}(\mathbf{w}) - r\mathbb{I} - \frac{\delta^2}{2\gamma}\left(A\left(\nabla\mathbf{w}\nabla\mathbf{w}^T\right)\right).
$$

The vector $\mathbb{S}\,(\mathbf{w}, r, A)\,\mathbf{n}_{\partial\Omega}$ is sometimes called Cauchy stress vector. If A is a symmetric operator, which is assumed in the following, then $\mathbb{S}\,(\mathbf{w}, r, A)$ is symmetric, too.

The trilinear form $b\,(\mathbf{u}, \mathbf{v}, \mathbf{w})$ defined in (5.3) is called convective form of the convective term. If $\mathbf{u}, \mathbf{v}, \mathbf{w} \in H^1\,(\Omega)$, integration by parts yields

$$
b\,(\mathbf{u}, \mathbf{v}, \mathbf{w}) + b\,(\mathbf{u}, \mathbf{w}, \mathbf{v}) = -\,(\nabla\cdot\mathbf{u}, \mathbf{v}\cdot\mathbf{w}) + \int_{\partial\Omega}(\mathbf{v}\cdot\mathbf{w})\,(\mathbf{u}\cdot\mathbf{n}_{\partial\Omega})\,ds. \tag{5.4}
$$

This formula simplifies if one of the functions vanishes on $\partial\Omega$ or if $\mathbf{u}$ is divergence free. The so-called skew-symmetric form of the convective term is defined by

$$
b_s\,(\mathbf{u}, \mathbf{v}, \mathbf{w}) = \frac{1}{2}\,(b\,(\mathbf{u}, \mathbf{v}, \mathbf{w}) - b\,(\mathbf{u}, \mathbf{w}, \mathbf{v}))\,. \tag{5.5}
$$

From (5.4) follows that the convective form and the skew-symmetric form of the convective term are equivalent if $\mathbf{u}$ is divergence free and the boundary integral in (5.4) vanishes.

5.2 Boundary Conditions for the LES Models

An overview of boundary conditions used for the incompressible Navier-Stokes equations can be found in Gresho and Sani [GS00, Section 3.8].

5.2.1 Dirichlet Boundary Condition

Dirichlet boundary conditions prescribe the value of the velocity at the boundary. If the values are non-homogeneous, they prescribe an inflow or an outflow. Homogeneous Dirichlet boundary conditions, so-called no slip boundary conditions, describe that the fluid is fixed at the wall, i.e. the fluid does not penetrate the wall and it does not slip along the wall.

If Dirichlet boundary conditions for $\mathbf{w}$ are prescribed on a part $\Gamma_{\text{diri}} \subset \partial\Omega$, then the test function $\mathbf{v}$ in (5.2) can be chosen such that $\mathbf{v}|_{\Gamma_{\text{diri}}} = \mathbf{0}$. In this case, the boundary integral on Γ_{diri} in the right hand side of (5.2) vanishes which simplifies the variational formulation.

If the velocity $\mathbf{u}$ obeys the no slip boundary condition $\mathbf{u} = \mathbf{0}$ on $\Gamma_{\text{diri,hom}} \subset \Gamma_{\text{diri}}$, then the no slip boundary condition $\mathbf{w} = \mathbf{0}$ on $\Gamma_{\text{diri,hom}}$ is certainly incorrect, see Figure 5.1. The large scale velocity $\mathbf{w}(\mathbf{x})$, $\mathbf{x} \in \Gamma_{\text{diri,hom}}$, should be an approximation of $\overline{\mathbf{u}}(\mathbf{x})$ which in turn is an average of the unknown velocity $\mathbf{u}$ in a δ-neighbourhood of $\mathbf{x}$, where $\mathbf{u}$ is continued outside Ω by $\mathbf{0}$. This average will in general not vanish on $\Gamma_{\text{diri,hom}}$. Setting $\mathbf{w} = \mathbf{0}$ on $\Gamma_{\text{diri,hom}}$ will result in incorrect solutions in a neighbourhood of $\Gamma_{\text{diri,hom}}$. It has been proposed by Galdi and Layton [GL00] better to use slip with friction boundary conditions, see Sections 5.2.4 and 5.2.5.

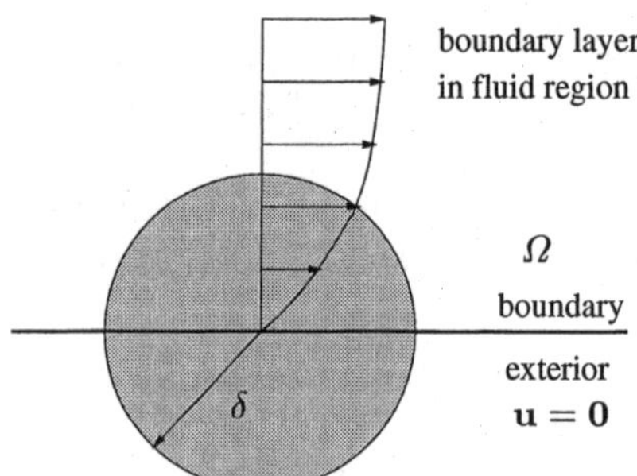

Fig. 5.1. Averaging the velocity (with no slip boundary conditions) at the boundary

The usual way of treating given inflow and outflow Dirichlet boundary conditions for $\mathbf{u}$ is to apply the same boundary conditions for $\mathbf{w}$. In general, these are also wrong boundary conditions. The changes of the flow within a δ-neighbourhood of such an inflow or outflow boundary are not considered. Even, under the assumption that the flow stays unchanged in a δ-neighbourhood of the inflow boundary, the averaging of a non-constant (e.g., parabolic) inflow of $\mathbf{u}$ will lead to a different inflow for $\mathbf{w}$.

5.2.2 Outflow or Do-Nothing Boundary Condition

Outflow or do-nothing boundary conditions are given by setting the Cauchy stress vector equal to zero, i.e.

$$\mathbb{S}\left(\mathbf{w}, r, A\right) \mathbf{n}_{\partial\Omega} = \mathbf{0} \text{ on } \Gamma_{\text{out}} \subset \partial\Omega. \tag{5.6}$$

Since $\mathbb{S}\left(\mathbf{w}, r, A\right)$ depends on A, (5.6) gives different boundary conditions for different operators A.

With (5.6), the boundary integral on the right hand side of (5.2) vanishes on Γ_{out}. This kind of boundary condition is used in flow problems where no other outflow boundary conditions are available. It can be shown already at the Poiseuille flow that the do-nothing boundary condition leads to incorrect results if the deformation tensor formulation of the viscous term is used or if the skew-symmetric form of the convective term is applied, see Heywood et al. [HRT96]. If the domain is large enough, the incorrect outflow boundary condition does not influence the computed solution in the main part of the domain. That means for practical applications that one has to use a domain which is sufficiently large such that the do-nothing boundary condition is applied far enough from regions of interest.

5.2.3 Free Slip Boundary Condition

The free slip boundary condition must be applied on boundaries without friction. It has the form

$$\begin{aligned} \mathbf{w} \cdot \mathbf{n}_{\partial\Omega} &= g \quad && \text{on } \Gamma_{\text{slip}} \subset \partial\Omega, \\ \mathbf{n}_{\partial\Omega}^T \mathbb{S}\left(\mathbf{w}, r, A\right) \boldsymbol{\tau}_k &= 0 \quad && \text{on } \Gamma_{\text{slip}}, \quad 1 \le k \le d-1, \end{aligned} \tag{5.7}$$

where $\{\boldsymbol{\tau}_k\}_{k=1}^{d-1}$ is an orthonormal system of tangential vectors on Γ_{slip}. There is no penetration through the wall if $g = 0$ on Γ_{slip}. For different operators A, different boundary conditions are prescribed by (5.7).

5.2.4 Slip With Linear Friction and No Penetration Boundary Condition

The slip with linear friction and no penetration boundary condition has the form

$$\begin{aligned} \mathbf{w} \cdot \mathbf{n}_{\partial\Omega} &= 0 \quad && \text{on } \Gamma_{\text{slfr}} \subset \partial\Omega, \\ \mathbf{w} \cdot \boldsymbol{\tau}_k + \beta^{-1}\left(\delta, \nu\right) \mathbf{n}_{\partial\Omega}^T \mathbb{S}\left(\mathbf{w}, r, A\right) \boldsymbol{\tau}_k &= 0 \quad && \text{on } \Gamma_{\text{slfr}}, \quad 1 \le k \le d-1. \end{aligned} \tag{5.8}$$

This boundary condition states that the fluid does not penetrate the wall and it slips along the wall whereas it loses energy. The loss of energy is given by the friction parameter $\beta\left(\delta, \nu\right)$. In the limit case $\beta^{-1}\left(\delta, \nu\right) \to 0$, the no slip condition is recovered and in the limit case $\beta^{-1}\left(\delta, \nu\right) \to \infty$ the free slip condition. Different operators A prescribe again different boundary conditions. Such boundary conditions were studied already by Maxwell [Max79] and Navier [Nav23].

In the case $d = 3$, the boundary integral on the right hand side of (5.2) can be split into three terms, using

$$\mathbf{v} = (\mathbf{v} \cdot \mathbf{n}_{\partial\Omega})\, \mathbf{n}_{\partial\Omega} + (\mathbf{v} \cdot \boldsymbol{\tau}_1)\, \boldsymbol{\tau}_1 + (\mathbf{v} \cdot \boldsymbol{\tau}_2)\, \boldsymbol{\tau}_2$$

and the symmetry of $\mathbb{S}(\mathbf{w}, r, A)$,

$$\begin{aligned} \int_{\Gamma_{\mathrm{slfr}}} \mathbb{S}(\mathbf{w}, r, A)\, \mathbf{n}_{\partial\Omega} \cdot \mathbf{v} ds = & \int_{\Gamma_{\mathrm{slfr}}} \mathbf{n}_{\partial\Omega}^T \mathbb{S}(\mathbf{w}, r, A)\, \mathbf{n}_{\partial\Omega}\ \mathbf{v} \cdot \mathbf{n}_{\partial\Omega} ds \\ & + \int_{\Gamma_{\mathrm{slfr}}} \mathbf{n}_{\partial\Omega}^T \mathbb{S}(\mathbf{w}, r, A)\, \boldsymbol{\tau}_1\ \mathbf{v} \cdot \boldsymbol{\tau}_1 ds \\ & + \int_{\Gamma_{\mathrm{slfr}}} \mathbf{n}_{\partial\Omega}^T \mathbb{S}(\mathbf{w}, r, A)\, \boldsymbol{\tau}_2\ \mathbf{v} \cdot \boldsymbol{\tau}_2 ds. \end{aligned}$$

The last two terms can be simplified by using the boundary conditions (5.8)

$$\begin{aligned} & \int_{\Gamma_{\mathrm{slfr}}} \mathbb{S}(\mathbf{w}, r, A)\, \mathbf{n}_{\partial\Omega} \cdot \mathbf{v} ds \\ & = \int_{\Gamma_{\mathrm{slfr}}} \mathbf{n}_{\partial\Omega}^T \mathbb{S}(\mathbf{w}, r, A)\, \mathbf{n}_{\partial\Omega}\ \mathbf{v} \cdot \mathbf{n}_{\partial\Omega} ds \\ & \quad - \int_{\Gamma_{\mathrm{slfr}}} \beta(\delta, \nu) \left[(\mathbf{w} \cdot \boldsymbol{\tau}_1)(\mathbf{v} \cdot \boldsymbol{\tau}_1) + (\mathbf{w} \cdot \boldsymbol{\tau}_2)(\mathbf{v} \cdot \boldsymbol{\tau}_2) \right] ds. \end{aligned}$$

The usual velocity test space in this case consists of functions whose normal component vanishes at Γ_{slfr}. Thus, the no penetration boundary condition will be imposed strongly whereas the slip with friction boundary condition results in additional integrals on Γ_{slfr}.

Galdi and Layton [GL00] proposed to apply slip with friction and no penetration boundary conditions for the large eddies. Such boundary conditions are more suitable than Dirichlet boundary conditions to describe phenomena which can be observed in nature. E.g., the main vortices of a hurricane do not stick at the boundary such that homogeneous Dirichlet boundary conditions are not satisfied. These vortices move on the boundary (slip), loosing energy while moving (friction) and do not penetrate the boundary.

Remark 5.1. The choice of the friction parameter $\beta(\delta, \nu)$. An open problem in using the slip with friction boundary condition is the choice of the friction parameter $\beta(\delta, \nu)$. A correct choice of $\beta(\delta, \nu)$ depends, besides on δ and ν, e.g., on the material and the roughness of the wall. Of course, $\beta(\delta, \nu)$ may be also a function of space and time on Γ_{slfr}.

The friction parameters for some model situations are computed in John et al. [JLS03]. This paper contains also first steps towards non-linear friction parameters which depend on the local tangential velocity.

A numerical study of the slip with friction boundary condition for the stationary Navier-Stokes equations can be found in John [Joh02b]. The numerical tests on two and three dimensional channel flows across a step study the influence of the friction parameter on the position of the reattachment point and the reattachment line, respectively, of the recirculating vortex. □

5.2.5 Slip With Linear Friction and Penetration With Resistance Boundary Condition

This boundary condition has the form

$$\begin{aligned} \mathbf{w} \cdot \mathbf{n}_{\partial\Omega} + \alpha \mathbf{n}_{\partial\Omega}^T \mathbb{S}(\mathbf{w}, r, A)\, \mathbf{n}_{\partial\Omega} &= 0 \quad \text{on } \Gamma_{\text{sfpr}} \subset \partial\Omega, \\ \mathbf{w} \cdot \boldsymbol{\tau}_k + \beta^{-1}(\delta, \nu)\, \mathbf{n}_{\partial\Omega}^T \mathbb{S}(\mathbf{w}, r, A)\, \boldsymbol{\tau}_k &= 0 \quad \text{on } \Gamma_{\text{sfpr}}, \quad 1 \le k \le d-1. \end{aligned} \tag{5.9}$$

In addition to the slip with linear friction boundary condition, the fluid is allowed to penetrate the walls at Γ_{sfpr}. For $\alpha = 0$, the no penetration boundary condition is recovered. Once more, different operators A give different boundary conditions.

Analogously to the slip with linear friction boundary condition, the test function $\mathbf{v}$ in (5.2) is decomposed into three orthogonal components and the boundary integral in (5.2) becomes, using the boundary condition (5.9),

$$\begin{aligned} &\int_{\Gamma_{\text{sfpr}}} \mathbb{S}(\mathbf{w}, r, A)\, \mathbf{n}_{\partial\Omega} \cdot \mathbf{v} ds \\ &= \int_{\Gamma_{\text{sfpr}}} \mathbf{n}_{\partial\Omega}^T \mathbb{S}(\mathbf{w}, r, A)\, \mathbf{n}_{\partial\Omega}\ \mathbf{v} \cdot \mathbf{n}_{\partial\Omega} ds \\ &\quad + \int_{\Gamma_{\text{sfpr}}} \mathbf{n}_{\partial\Omega}^T \mathbb{S}(\mathbf{w}, r, A)\, \boldsymbol{\tau}_1\ \mathbf{v} \cdot \boldsymbol{\tau}_1 ds + \int_{\Gamma_{\text{sfpr}}} \mathbf{n}_{\partial\Omega}^T \mathbb{S}(\mathbf{w}, r, A)\, \boldsymbol{\tau}_2\ \mathbf{v} \cdot \boldsymbol{\tau}_2 ds \\ &= -\int_{\Gamma_{\text{sfpr}}} \alpha^{-1} (\mathbf{w} \cdot \mathbf{n}_{\partial\Omega})(\mathbf{v} \cdot \mathbf{n}_{\partial\Omega})\, ds \\ &\quad - \int_{\Gamma_{\text{sfpr}}} \beta(\delta, \nu) \left[(\mathbf{w} \cdot \boldsymbol{\tau}_1)(\mathbf{v} \cdot \boldsymbol{\tau}_1) + (\mathbf{w} \cdot \boldsymbol{\tau}_2)(\mathbf{v} \cdot \boldsymbol{\tau}_2) \right] ds. \end{aligned} \tag{5.10}$$

The first term can be considered as a weak imposition of the no penetration constraint as penalty formulation with the penalty parameter α^{-1}.

The implementation of this boundary condition is described in Section 7.6. We use it also in the case of slip and no penetration, imposing the no penetration condition weakly by choosing an appropriate parameter α, see Remark 7.10.

5.2.6 Periodic Boundary Condition

Periodic boundary conditions do not posses any physical meaning. They are used to simulate an infinite extension of Ω in one or more directions. Let, e.g., this direction be $\mathbf{e}_i$, where $\mathbf{e}_i$ is a vector of the canonical basis of $\mathbb{R}^d$. It is assumed that the flow is periodic in this direction with the length l of the period. In computations, a finite domain is used which possesses two boundaries perpendicular to the periodic direction with distance l. The periodic boundary conditions are given by

$$\mathbf{w}(t, \mathbf{x} + l\mathbf{e}_i) = \mathbf{w}(t, \mathbf{x}) \quad \forall\, \mathbf{x} \in \mathbb{R}^d,\ \forall\, t > 0.$$

From the point of view of the finite computational domain, all appearing functions have to be extended periodically in the periodic direction to return to the original problem. One has to pay attention to this point in the computation of a convolution for a periodic function. This function has to be extended off the computational domain of course periodically in the periodic direction and not trivially as in the non-periodic directions, see Remark 4.9.

The use of space periodic boundary conditions may also facilitate analytical investigations, see Temam [Tem95, p. 4].

5.3 Function Spaces for the LES Models

We will discuss in this section the choice of appropriate function spaces for the LES models which are investigated in the following chapters. An introduction of the basic spaces can be found in Section 2.1.

We assume that the functions in the velocity ansatz space V_a and test space V_t have the same regularity. These spaces might differ only in the incorporated boundary conditions. The choice of V_a, V_t depends on the kind of turbulent viscosity ν_T and on the boundary conditions which are used.

Considering the Navier-Stokes equations, i.e. $\nu_T = 0$ and $A = 0$, the choice $V_a, V_t \subset L^2\left(0, T; H^1(\Omega)\right)$ guarantees that all terms are well defined in (5.2) which are computed with functions from V_a, V_t only. In the case of the Smagorinsky model , $\nu_T = c_S\delta^2 \left\|\mathbb{D}(\mathbf{w})\right\|_F$, the nonlinear viscous term has the form

$$\int_0^T \int_\Omega c_S \delta^2 \left\|\mathbb{D}(\mathbf{w})\right\|_F \mathbb{D}(\mathbf{w}) : \mathbb{D}(\mathbf{v})\, d\mathbf{x} dt. \tag{5.11}$$

Since c_S and δ^2 are constants and

$$\left\|\mathbb{D}(\mathbf{w})\right\|_F \le C(d) \max_{1\le i,j\le d} \left|\left(\mathbb{D}(\mathbf{w})\right)_{ij}\right| \quad \text{in } \Omega,$$

where $C(d)$ depends only on the dimension d, (5.11) is well defined if $V_a, V_t \subset L^3\left(0, T; W^{1,3}(\Omega)\right)$. In the case of the Taylor LES model and the rational LES model, the choice $V_a, V_t \subset L^3\left(0, T; W^{1,3}(\Omega)\right)$ guarantees that the nonlinear term with the operator A in (5.2) is well defined. Boundary conditions which are incorporated into V_a and V_t are Dirichlet boundary conditions and prescribed penetration boundary conditions, e.g., no penetration conditions.

The terms involving the pressure ansatz and test spaces in (5.2) are well defined for the above chosen velocity spaces if $Q_a, Q_t \subset L^2\left(0,T;L^2\left(\Omega\right)\right)$. If the normal component of the functions of the velocity test space V_t vanish on $\partial\Omega$, i.e. $\mathbf{v}\cdot\mathbf{n}_{\partial\Omega} = 0$ on $\partial\Omega$, the pressure r is defined only up to an additive constant. Let c be a constant pressure, then integration by parts gives

$$(c, \nabla\cdot\mathbf{v}) = 0$$

and the orthogonality of the normal and tangential vectors on $\partial\Omega$ yields

$$\begin{aligned}\int_{\partial\Omega} c\mathbb{I}\mathbf{n}_{\partial\Omega} ds &= c\int_{\partial\Omega}\Big[\left(\mathbf{n}_{\partial\Omega}\cdot\mathbf{n}_{\partial\Omega}\right)\left(\mathbf{v}\cdot\mathbf{n}_{\partial\Omega}\right) + \left(\mathbf{n}_{\partial\Omega}\cdot\boldsymbol{\tau}_1\right)\left(\mathbf{v}\cdot\boldsymbol{\tau}_1\right)\\ &\quad + \left(\mathbf{n}_{\partial\Omega}\cdot\boldsymbol{\tau}_2\right)\left(\mathbf{v}\cdot\boldsymbol{\tau}_2\right)\Big] ds\\ &= 0.\end{aligned}$$

This constant can be fixed by an additional condition. We require that the integral mean value of r vanishes,

$$\int_\Omega r d\mathbf{x} = 0 \text{ in } (0,T],$$

such that $Q_a = L^2\left(0,T;L_0^2\left(\Omega\right)\right)$. If the functions of the velocity ansatz space vanish on $\partial\Omega$, the pressure test space can be restricted in the same way as the pressure ansatz space. Integration by parts of the divergence constraint in (5.2) shows that an additive constant in q does not lead to a new condition. In particular, the correct pressure space is $L^2\left(0,T;L^2\left(\Omega\right)\right)$ for the slip with friction and penetration with resistance boundary condition (5.9) for positive values of the resistance parameter α.

In Chapter 7, system (5.1) will be discretised first in time and then a variational problem for the arising system in space will be formulated. Appropriate ansatz and test spaces for the spatial systems are given by the spatial components of the spaces introduced above.

6

Existence and Uniqueness of Solutions of the LES Models

This chapter presents analytical investigations of the existence and uniqueness of solutions of the LES models (5.1) in a bounded domain Ω. Since Ω is bounded, (5.1) has to be equipped with boundary conditions. The analysis presented in this chapter uses homogeneous Dirichlet boundary conditions. This is just for simplicity of presentation, extensions to other boundary conditions are possible, see Remark 6.16.

The main part of this chapter, Section 6.1, consists in the proof of the existence and uniqueness of the solution of the Smagorinsky model following a classical paper by Ladyzhenskaya [Lad67]. It turns out that the existence and uniqueness of a weak solution can be proved in 2d and 3d. The proof uses the Galerkin method and relies upon a priori error estimates (Lemmata 6.1, 6.2, 6.3) and the monotonicity of the non-linear viscous operator coming from the Smagorinsky model (Lemma 6.9). Difficulties in the analysis arise form the non-linearity of this viscous term.

The proof of Ladyzhenskaya for the Smagorinsky model could be extended by Coletti [Col98] to the Taylor LES model under the assumption that the Smagorinsky subgrid scale term dominates the Taylor LES term, see Section 6.2. However, the correct situation is the other way around since the Taylor LES model represents terms of formally second order in δ and the Smagorinsky subgrid scale term stands for terms of formally fourth order. But there is a clear numerical evidence, Section 10.2, that for the correct situation a similar result like for the case considered by Coletti cannot be expected.

The current state of art is that the proof from Ladyzhenskaya has not been extended to the rational LES model. Section 6.3 demonstrates that the arguments which are used to prove the monotonicity of an operator in the Smagorinsky model and the Taylor LES model fail for the rational LES model. On the other hand, a counter example which states that the operator might be non-monotone for the rational LES model does not exist, too.

6.1 The Smagorinsky Model

The results presented in this section were obtained by Ladyzhenskaya [Lad67], for an overview see also Ladyzhenskaya [Lad69].

We consider the Smagorinsky model with homogeneous Dirichlet boundary conditions (no slip boundary conditions)

$$\begin{aligned} \mathbf{w}_t - \nabla\cdot\left(\left(\nu+\nu_S\left\|\nabla\mathbf{w}\right\|_F\right)\nabla\mathbf{w}\right)+(\mathbf{w}\cdot\nabla)\,\mathbf{w}+\nabla r &= \mathbf{f} \quad \text{in } (0,T]\times\Omega,\\ \nabla\cdot\mathbf{w} &= 0 \quad \text{in } [0,T]\times\Omega,\\ \mathbf{w} &= \mathbf{0} \quad \text{on } [0,T]\times\Gamma,\\ \mathbf{w}\,(0,\cdot) &= \mathbf{w}_0 \text{ in } \Omega,\\ \int_\Omega r d\mathbf{x} &= 0 \quad \text{in } (0,T], \end{aligned} \tag{6.1}$$

with $\nu_S > 0$, $\mathbf{f}\in L^2\left(0,T;L^2\left(\Omega\right)\right)$ and finite final time $T<\infty$.

The Banach space

$$W^{1,3}_{0,div}\left(\Omega\right)=\left\{\mathbf{v}\in W^{1,3}\left(\Omega\right)\ :\ \mathbf{v}|_{\partial\Omega}=\mathbf{0}\ ,\ \nabla\cdot\mathbf{v}=0 \text{ in } \Omega\right\} \tag{6.2}$$

is equipped with the same norm as $W^{1,3}_0\left(\Omega\right)$. In addition, let

$$V=H^1\left(0,T;L^2\left(\Omega\right)\right)\cap L^3\left(0,T;W^{1,3}_{0,div}\left(\Omega\right)\right) \tag{6.3}$$

equipped with

$$\left\|\mathbf{v}\right\|_V=\left\|\nabla\mathbf{v}\right\|_{L^3(0,T;L^3(\Omega))}+\left\|\mathbf{v}_t\right\|_{L^2(0,T;L^2(\Omega))}\,.$$

The weak formulation of (6.1) is as follows:

Find $\mathbf{w}\in V$ such that $\mathbf{w}\,(0,\mathbf{x})=\mathbf{w}_0\in W^{1,3}_{0,div}\left(\Omega\right)$ and for all $\mathbf{v}\in V$

$$\int_0^T\left(\mathbf{w}_t+(\mathbf{w}\cdot\nabla)\,\mathbf{w},\mathbf{v}\right)+\left(\left(\nu+\nu_S\left\|\nabla\mathbf{w}\right\|_F\right)\nabla\mathbf{w},\nabla\mathbf{v}\right)dt=\int_0^T\left(\mathbf{f},\mathbf{v}\right)dt. \tag{6.4}$$

The function $\mathbf{w}\in V$ is called weak solution of the Smagorinsky model.

The solvability of (6.4) is proved with the Galerkin method, i.e. a sequence $\{\mathbf{w}^n\}\in V$ of functions is constructed which solve finite dimensional approximations of (6.4). Then it is shown that a subsequence of $\{\mathbf{w}^n\}$ converges to a solution of (6.4).

6.1.1 A priori error estimates

The convergence proof is based on three a priori error estimates which will be given in the following lemmata.

Lemma 6.1. *Assume that (6.1) has a sufficiently smooth solution* $(\mathbf{w}, r)$. *Then, this solution satisfies*

$$\|\mathbf{w}(T,\mathbf{x})\|_{L^2(\Omega)} \le \|\mathbf{w}_0(\mathbf{x})\|_{L^2(\Omega)} + \int_0^T \|\mathbf{f}(t,\mathbf{x})\|_{L^2(\Omega)}\,dt, \quad T > 0. \tag{6.5}$$

Proof. Since $(\mathbf{w}, r)$ is a solution of (6.1), one obtains by testing the momentum equation of (6.1) with $\mathbf{w}$ and integration by parts

$$(\mathbf{w}_t, \mathbf{w}) + ((\nu + \nu_S \|\nabla \mathbf{w}\|_F)\nabla \mathbf{w}, \nabla \mathbf{w}) + b(\mathbf{w},\mathbf{w},\mathbf{w}) - (r, \nabla\cdot\mathbf{w}) = (\mathbf{f}, \mathbf{w}).$$

Because $\nabla\cdot\mathbf{w} = 0$, the convective term and the term with the pressure r vanish, such that

$$\frac{1}{2}\frac{d}{dt}(\mathbf{w},\mathbf{w}) + ((\nu + \nu_S \|\nabla \mathbf{w}\|_F)\nabla \mathbf{w}, \nabla \mathbf{w}) = (\mathbf{f}, \mathbf{w}). \tag{6.6}$$

Neglecting the non-negative second term on the left hand side, applying the chain rule on the first term of the left hand side and the Cauchy-Schwarz inequality for the right hand side give

$$\|\mathbf{w}\|_{L^2(\Omega)} \frac{d}{dt}\|\mathbf{w}\|_{L^2(\Omega)} \le \|\mathbf{f}\|_{L^2(\Omega)}\|\mathbf{w}\|_{L^2(\Omega)}.$$

Cancellation of $\|\mathbf{w}\|_{L^2(\Omega)}$ and integration on $(0,T)$ completes the proof. □

Lemma 6.2. *With the same assumptions as in Lemma 6.1, it holds for* $T > 0$

$$\begin{aligned}
&\|\mathbf{w}(T,\mathbf{x})\|_{L^2(\Omega)}^2 + 2\int_0^T ((\nu + \nu_S \|\nabla \mathbf{w}\|_F)\nabla \mathbf{w}, \nabla \mathbf{w})\,dt \\
&\le 2\|\mathbf{w}_0(\mathbf{x})\|_{L^2(\Omega)}^2 + 3\left(\int_0^T \|\mathbf{f}(t,\mathbf{x})\|_{L^2(\Omega)}\,dt\right)^2 = c_1(T).
\end{aligned} \tag{6.7}$$

Proof. Starting with (6.6), one gets by integration on $(0,T)$

$$\begin{aligned}
&\|\mathbf{w}(T,\mathbf{x})\|_{L^2(\Omega)}^2 + 2\int_0^T ((\nu + \nu_S \|\nabla \mathbf{w}\|_F)\nabla \mathbf{w}, \nabla \mathbf{w})\,dt \\
&\le \|\mathbf{w}_0(\mathbf{x})\|_{L^2(\Omega)}^2 + 2\int_0^T (\mathbf{f}, \mathbf{w})\,dt.
\end{aligned} \tag{6.8}$$

By the Cauchy-Schwarz inequality and inequality (6.5) (which is valid for all times t) follow for the second term on the right hand side

$$\begin{aligned}
\int_0^T (\mathbf{f}(t,\mathbf{x}), \mathbf{w}(t,\mathbf{x}))\,dt &\le \|\mathbf{w}_0(\mathbf{x})\|_{L^2(\Omega)} \int_0^T \|\mathbf{f}(t,\mathbf{x})\|_{L^2(\Omega)}\,dt \\
&\quad + \int_0^T \|\mathbf{f}(t,\mathbf{x})\|_{L^2(\Omega)} \left(\int_0^t \|\mathbf{f}(t',\mathbf{x})\|_{L^2(\Omega)}\,dt'\right) dt.
\end{aligned}$$

Using inequality (2.6) and the non-negativeness of $\|\mathbf{f}(t',\mathbf{x})\|_{L^2(\Omega)}$ yield

$$\begin{aligned}\int_0^T (\mathbf{f}(t,\mathbf{x}),\mathbf{w}(t,\mathbf{x}))\,dt &\le \frac{\|\mathbf{w}_0(\mathbf{x})\|^2_{L^2(\Omega)}}{2} + \frac{1}{2}\left(\int_0^T \|\mathbf{f}(t,\mathbf{x})\|_{L^2(\Omega)}\,dt\right)^2 \\ &\quad + \int_0^T \|\mathbf{f}(t,\mathbf{x})\|_{L^2(\Omega)}\left(\int_0^T \|\mathbf{f}(t',\mathbf{x})\|_{L^2(\Omega)}\,dt'\right)dt \\ &= \frac{\|\mathbf{w}_0(\mathbf{x})\|^2_{L^2(\Omega)}}{2} + \frac{3}{2}\left(\int_0^T \|\mathbf{f}(t,\mathbf{x})\|_{L^2(\Omega)}\,dt\right)^2.\end{aligned}$$

Inserting this estimate into (6.8) gives (6.7). □

Lemma 6.3. *If the assumptions of Lemma 6.1 are valid, then*

$$\|\nabla\mathbf{w}(T,\mathbf{x})\|^3_{L^3(\Omega)} + \frac{3}{2\nu_S}\int_0^T \|\mathbf{w}_t\|^2_{L^2(\Omega)}\,dt \le c_2(T). \tag{6.9}$$

Proof. To prove (6.9), one starts by testing the momentum equation of (6.1) with $\mathbf{w}_t$. This gives, after integration on $(0,T)$ and integration by parts

$$\begin{aligned}&\int_0^T \|\mathbf{w}_t\|^2_{L^2(\Omega)}\,dt + \frac{1}{2}(\nu\nabla\mathbf{w}(T,\mathbf{x}),\mathbf{w}(T,\mathbf{x})) + \frac{\nu_S}{3}\int_\Omega \|\nabla\mathbf{w}(T,\mathbf{x})\|^3_F\,d\mathbf{x} \quad (6.10)\\ &= \frac{1}{2}(\nu\nabla\mathbf{w}_0,\mathbf{w}_0) + \frac{\nu_S}{3}\int_\Omega \|\nabla\mathbf{w}_0\|^3_F\,d\mathbf{x} - \int_0^T b(\mathbf{w},\mathbf{w},\mathbf{w}_t)\,dt + \int_0^T (\mathbf{f},\mathbf{w}_t)\,dt.\end{aligned}$$

Here, we have used

$$(\|\nabla\mathbf{w}\|_F\nabla\mathbf{w},\nabla\mathbf{w}_t) = \frac{1}{3}\int_\Omega \frac{d}{dt}\|\nabla\mathbf{w}\|^3_F\,d\mathbf{x}.$$

First, the convective term will be estimated. Hölder's inequality (2.7) gives

$$\begin{aligned}\int_\Omega (\mathbf{w}^T\mathbf{w})(\nabla\mathbf{w}:\nabla\mathbf{w})\,d\mathbf{x} &= \int_\Omega \mathbf{w}^2(\nabla\mathbf{w})^2\,d\mathbf{x} \\ &\le \left\|\mathbf{w}^2\right\|_{L^3(\Omega)}\left\|(\nabla\mathbf{w})^2\right\|_{L^{3/2}} \\ &= \|\mathbf{w}\|^2_{L^6(\Omega)}\|\nabla\mathbf{w}\|^2_{L^3(\Omega)}.\end{aligned}$$

With the Sobolev embedding theorem (2.13) follows

$$\int_\Omega (\mathbf{w}^T\mathbf{w})(\nabla\mathbf{w}:\nabla\mathbf{w})\,d\mathbf{x} \le c\|\nabla\mathbf{w}\|^4_{L^3(\Omega)}. \tag{6.11}$$

The convective term is now estimated in the following form

$$
\begin{aligned}
\int_0^T b(\mathbf{w}, \mathbf{w}, \mathbf{w}_t)\, dt &= \int_0^T \int_\Omega (\mathbf{w} \cdot \nabla)\, \mathbf{w} \mathbf{w}_t d\mathbf{x} dt \\
&\leq \int_0^T \int_\Omega \left(\frac{\mathbf{w}_t^2}{4} + ((\mathbf{w} \cdot \nabla)\, \mathbf{w})^2 \right) d\mathbf{x} dt \\
&= \int_0^T \int_\Omega \left(\frac{\mathbf{w}_t^2}{4} + \left(\mathbf{w}^T \mathbf{w}\right) (\nabla \mathbf{w} : \nabla \mathbf{w}) \right) d\mathbf{x} dt \\
&\leq \frac{1}{4} \int_0^T \|\mathbf{w}_t\|_{L^2(\Omega)}\, dt + c \int_0^T \|\nabla \mathbf{w}\|_{L^3(\Omega)}^4\, dt.
\end{aligned}
$$

Applying also the estimate

$$
\int_0^T (\mathbf{f}, \mathbf{w}_t)\, dt \leq \int_0^T \int_\Omega \left(\mathbf{f}^2 + \frac{\mathbf{w}_t^2}{4} \right) d\mathbf{x} dt,
$$

one obtains from (6.10)

$$
\begin{aligned}
&\int_0^T \|\mathbf{w}_t\|_{L^2(\Omega)}^2\, dt + \nu \|\nabla \mathbf{w}(T, \mathbf{x})\|_{L^2(\Omega)}^2 + \frac{2\nu_S}{3} \|\nabla \mathbf{w}(T, \mathbf{x})\|_{L^3(\Omega)}^3 \\
&\leq \nu \|\nabla \mathbf{w}_0\|_{L^2(\Omega)}^2 + \frac{2\nu_S}{3} \|\nabla \mathbf{w}_0\|_{L^3(\Omega)}^3 + 2 \int_0^T \|\mathbf{f}\|_{L^2(\Omega)}^2\, dt \qquad (6.12) \\
&\quad + 2c \int_0^T \|\nabla \mathbf{w}\|_{L^3(\Omega)}^4\, dt.
\end{aligned}
$$

In particular,

$$
\begin{aligned}
\|\nabla \mathbf{w}(T, \mathbf{x})\|_{L^3(\Omega)}^3 \leq{}& \frac{3\nu}{2\nu_S} \|\nabla \mathbf{w}_0\|_{L^2(\Omega)}^2 + \|\nabla \mathbf{w}_0\|_{L^3(\Omega)}^3 + \frac{3}{\nu_S} \int_0^T \|\mathbf{f}\|_{L^2(\Omega)}^2\, dt \\
&+ \frac{3c}{\nu_S} \int_0^T \|\nabla \mathbf{w}\|_{L^3(\Omega)}^4\, dt.
\end{aligned}
$$

The application of Gronwall's lemma (Lemma 2.1) gives

$$
\begin{aligned}
&\|\nabla \mathbf{w}(T, \mathbf{x})\|_{L^3(\Omega)}^3 \\
&\leq \left(\frac{3\nu}{2\nu_S} \|\nabla \mathbf{w}_0\|_{L^2(\Omega)}^2 + \|\nabla \mathbf{w}_0\|_{L^3(\Omega)}^3 + \frac{3}{\nu_S} \int_0^T \|\mathbf{f}\|_{L^2(\Omega)}^2\, dt \right) \\
&\quad \times \exp\left(\frac{3c}{\nu_S} \int_0^T \|\nabla \mathbf{w}\|_{L^3(\Omega)}\, dt \right).
\end{aligned}
$$

We still have to estimate the term in the exponential. Using inequality (2.5), one obtains

$$\int_0^T \|\nabla \mathbf{w}\|_{L^3(\Omega)} \, dt = \int_0^T \left(\int_\Omega \sum_{i,j=1}^d \left| \frac{\partial \mathbf{w}_i}{\partial x_j} \right|^3 d\mathbf{x} \right)^{1/3} dt$$
$$\leq \int_0^T \left(\int_\Omega \left(\sum_{i,j=1}^d \left| \frac{\partial \mathbf{w}_i}{\partial x_j} \right|^2 \right)^{3/2} d\mathbf{x} \right)^{1/3} dt.$$

On the other hand, by the definition of the Frobenius norm

$$\int_\Omega \|\nabla \mathbf{w}\|_F (\nabla \mathbf{w} : \nabla \mathbf{w}) \, d\mathbf{x} = \int_\Omega (\nabla \mathbf{w} : \nabla \mathbf{w})^{3/2} \, d\mathbf{x}$$
$$= \int_\Omega \left(\sum_{i,j=1}^d \left| \frac{\partial \mathbf{w}_i}{\partial x_j} \right|^2 \right)^{3/2} d\mathbf{x}.$$

One obtains now by Hölder's inequality (2.7) and the a priori estimate (6.5)

$$\int_0^T \|\nabla \mathbf{w}\|_{L^3(\Omega)} \, dt \leq \int_0^T \left(\int_\Omega \|\nabla \mathbf{w}\|_F (\nabla \mathbf{w} : \nabla \mathbf{w}) \, d\mathbf{x} \right)^{1/3} dt$$
$$\leq \left(\int_0^T \int_\Omega \|\nabla \mathbf{w}\|_F (\nabla \mathbf{w} : \nabla \mathbf{w}) \, d\mathbf{x} dt \right)^{1/3} \left(\int_0^T dt \right)^{2/3}$$
$$\leq T^{2/3} \left(\frac{c_1 (T)}{2\nu_S} \right)^{1/3}.$$

Thus

$$\|\nabla \mathbf{w} (T, \mathbf{x})\|_{L^3(\Omega)}^3 \leq c_3 (T).$$

From (6.12) follows finally

$$\|\nabla \mathbf{w} (T, \mathbf{x})\|_{L^3(\Omega)}^3 + \frac{3}{2\nu_S} \int_0^T \|\mathbf{w}_t\|_{L^2(\Omega)}^2 \, dt \leq c_2 (T).$$

□

6.1.2 The Galerkin Method

We will now start with the construction of a sequence $\{\mathbf{w}^n\} \subset V$ such that a subsequence converges to a solution $\mathbf{w} \in V$ of (6.4).

Let $\{\mathbf{v}^l (\mathbf{x})\} \subset W_{0,div}^{1,3}$ be a sequence of linearly independent functions which are orthonormal with respect to the L^2-inner product in Ω and with $\mathbf{v}^1 = \mathbf{w}_0$. Then, we seek $\mathbf{w}^n$ in the form

$$\mathbf{w}^n (t, \mathbf{x}) = \sum_{l=1}^n \alpha_{ln} (t) \, \mathbf{v}^l (\mathbf{x}) \tag{6.13}$$

satisfying

$$\alpha_{ln}(0) = \begin{cases} 1 & l = 1 \\ 0 & l > 1 \end{cases}$$

and

$$\left(\mathbf{w}_t^n, \mathbf{v}^l\right) + \left(\left(\nu + \nu_S \left\|\nabla \mathbf{w}^n\right\|_F\right) \nabla \mathbf{w}^n, \nabla \mathbf{v}^l\right) + b\left(\mathbf{w}^n, \mathbf{w}^n, \mathbf{v}^l\right) = \left(\mathbf{f}, \mathbf{v}^l\right) \quad (6.14)$$

$l = 1, \ldots, n$. System (6.14) is an autonomous, quasi-linear system of ordinary differential equations with respect to the unknown functions $\alpha_{ln}(t)$.

Lemma 6.4. *System (6.14) admits a unique solution for all $T > 0$. Moreover, the estimates (6.5), (6.7) and (6.9) are valid for $\mathbf{w}^n$.*

Proof. The Picard-Lindelöf theorem is applied to prove the unique solvability of (6.14). That means, one has to prove a Lipschitz condition and an a priori bound for the right hand side of

$$\alpha'_{ln}(t) = F\left(\alpha_{ln}\right), \quad t \in (0, T], \quad (6.15)$$

see Zeidler [Zei86, Section 3.3, Corollary 3.9]. The functions α_{ln} appear only as linears or squared in the right hand side of (6.15). Thus, the Lipschitz condition is fulfilled. It remains to prove the a priori boundedness of

$$\max_{[0,T]} \sum_{l=1}^{n} \alpha_{ln}^2(t).$$

From the $L^2(\Omega)$-orthonormality of $\{\mathbf{v}^l(\mathbf{x})\}$ follows

$$\max_{[0,T]} \sum_{l=1}^{n} \alpha_{ln}^2(t) = \max_{[0,T]} \left\|\mathbf{w}^n(t, \mathbf{x})\right\|_{L^2(\Omega)}^2.$$

The linear combination of the equations of (6.14) gives

$$\left(\mathbf{w}_t^n, \mathbf{w}^n\right) + \left(\left(\nu + \nu_S \left\|\mathbf{w}^n\right\|_F\right) \nabla \mathbf{w}^n, \nabla \mathbf{w}^n\right) = \left(\mathbf{f}, \mathbf{w}^n\right), \quad (6.16)$$

where $b\left(\mathbf{w}^n, \mathbf{w}^n, \mathbf{w}^n\right) = \mathbf{0}$ has been used. This equation has the same form as (6.6). Since $\mathbf{w}^n$ is defined as a solution of (6.14), it solves also (6.16) such that the techniques used for proving Lemma 6.1 and Lemma 6.2 can be applied. Hence, we have the estimates

$$\left\|\mathbf{w}^n(t, \mathbf{x})\right\|_{L^2(\Omega)} \leq \left\|\mathbf{w}_0(\mathbf{x})\right\|_{L^2(\Omega)} + \int_0^t \left\|\mathbf{f}(t', \mathbf{x})\right\|_{L^2(\Omega)} dt', \quad 0 \leq t \leq T$$

and

$$\left\|\mathbf{w}^n(T, \mathbf{x})\right\|_{L^2(\Omega)}^2 + 2 \int_0^T \left(\left(\nu + \nu_S \left\|\nabla \mathbf{w}^n\right\|_F\right) \nabla \mathbf{w}^n, \nabla \mathbf{w}^n\right) dt \leq c_1(T). \quad (6.17)$$

The first estimate provides the a priori boundedness which suffices to prove the unique solvability of (6.14). System (6.14) can be brought also in form (6.10) by multiplication with $d\alpha_{ln}(t)/dt$ and summation, such that the estimate

$$\|\nabla \mathbf{w}^n(T,\mathbf{x})\|_{L^3(\Omega)}^3 + \frac{3}{2\nu_S}\int_0^T \|\mathbf{w}_t^n\|_{L^2(\Omega)}^2\, dt \le c_2(T) \tag{6.18}$$

is valid for $T > 0$. □

We will choose a subsequence of $\{\mathbf{w}^n\}$ which converges in a certain sense to a solution $\mathbf{w}$ of (6.4).

Lemma 6.5. *There is a* $\mathbf{w} \in V$ *such that a subsequence of* $\{\mathbf{w}^n\}$

i) converges weakly to $\mathbf{w}$ *in* V,
ii) converges strongly to $\mathbf{w}$ *in* $L^2\left(0,T;L^2(\Omega)\right)$,
iii) converges strongly to $\mathbf{w}$ *in* $L^q(0,T;L^q(\Omega))$ *for* $q<4$.
iv) A subsequence of $\{\mathbf{w}_t^n\}$ *converges weakly to* $\mathbf{w}_t$ *in* $L^2\left(0,T;L^2(\Omega)\right)$.
v) A subsequence of $\frac{\partial w_j^n}{\partial x_i}$, $i,j=1,\ldots,d$ *converges weakly to* $\frac{\partial w_j}{\partial x_i}$ *in* $L^3\left(0,T;L^3(\Omega)\right)$.

Proof. For brevity, we will speak of the convergence of $\{\mathbf{w}^n\}$ instead of the convergence of a subsequence. The convergence in the senses given in i) - v) follows easily from theorems given in Appendix 2.2. The point to prove is that all kinds of convergence have the same limit $\mathbf{w}$.

i) The weak convergence of $\{\mathbf{w}^n\}$ to $\mathbf{w} \in V$, i.e.

$$\lim_{n\to\infty}\int_0^T\int_\Omega \mathbf{w}^n\mathbf{v}d\mathbf{x}dt = \int_0^T\int_\Omega \mathbf{w}\mathbf{v}d\mathbf{x}dt \quad \forall \mathbf{v}\in V^*, \tag{6.19}$$

where V^* is the dual space of V, follows from the uniform boundedness of $\{\mathbf{w}^n\}$ in the V norm, which is a consequence of (6.18) and that every bounded sequence in a reflexive Banach space has a weakly convergent subsequence.

ii) Since $V \subset L^2\left(0,T;L^2(\Omega)\right) \subset V^*$, (6.19) holds also for all $\mathbf{v} \in L^2\left(0,T;L^2(\Omega)\right)$ such that $\{\mathbf{w}^n\}$ converges weakly to $\mathbf{w}$ in $L^2\left(0,T;L^2(\Omega)\right)$. From the uniform boundedness of $\mathbf{w}^n(t,\mathbf{x})$ in $W_0^{1,3}(\Omega)$ for every $t \ge 0$, estimate (6.18), and the compact embedding $W_0^{1,3}(\Omega)$ into $L^2(\Omega)$, (2.14), follow that there is a subsequence of $\{\mathbf{w}^n\}$ which converges strongly in $L^2(\Omega)$ to $\tilde{\mathbf{w}} \in L^2(\Omega)$. Since this holds for all $t \ge 0$, $\{\mathbf{w}^n\}$ converges strongly to $\tilde{\mathbf{w}} \in L^2\left(0,T;L^2(\Omega)\right)$:

$$\int_\Omega \mathbf{w}^n\mathbf{v}d\mathbf{x} \to \int_\Omega \tilde{\mathbf{w}}\mathbf{v}d\mathbf{x} \Longrightarrow \int_0^T\int_\Omega \mathbf{w}^n\mathbf{v}d\mathbf{x}dt \to \int_0^T\int_\Omega \tilde{\mathbf{w}}\mathbf{v}d\mathbf{x}dt.$$

Accordingly, this subsequence of $\{\mathbf{w}^n\}$ converges also weakly to $\tilde{\mathbf{w}}$. Since the sequence $\{\mathbf{w}^n\}$ converges weakly to $\mathbf{w}$ and the weak limit is unique, it follows $\mathbf{w} = \tilde{\mathbf{w}}$. Hence, (a subsequence of) $\{\mathbf{w}^n\}$ converges to $\mathbf{w}$ strongly in $L^2\left(0,T;L^2\left(\Omega\right)\right)$.

iii) We first establish that $\{\mathbf{w}^n\}$ is uniformly bounded (with respect to n) in $L^4\left(0,T;L^4\left(\Omega\right)\right)$. From the Cauchy-Schwarz inequality and the Sobolev embedding (2.13) follow

$$\begin{aligned}\|\mathbf{w}^n\|_{L^4(\Omega)}^4 &= \int_\Omega (\mathbf{w}^n)^4\, d\mathbf{x} \le \left(\int_\Omega (\mathbf{w}^n)^6\, d\mathbf{x}\right)^{1/2}\left(\int_\Omega (\mathbf{w}^n)^2\, d\mathbf{x}\right)^{1/2} \\ &\le c\left(\int_\Omega (\nabla\mathbf{w}^n)^2\, d\mathbf{x}\right)^{3/2}\left(\int_\Omega (\mathbf{w}^n)^2\, d\mathbf{x}\right)^{1/2},\end{aligned}$$

hence

$$\|\mathbf{w}^n\|_{L^4(\Omega)} \le c\,\|\nabla\mathbf{w}^n\|_{L^2(\Omega)}^{3/4}\,\|\mathbf{w}^n\|_{L^2(\Omega)}^{1/4}.$$

This gives, using (6.17), (6.18) and (2.9)

$$\|\mathbf{w}^n\|_{L^4(0,T;L^4(\Omega))}^4 \le \int_0^T \|\nabla\mathbf{w}^n\|_{L^2(\Omega)}^3\,\|\mathbf{w}^n\|_{L^2(\Omega)}\, dt \le \int_0^T c_2\left(t\right) c_1\left(t\right)^{1/2} dt.$$

Now, we can prove the strong convergence $\mathbf{w}^n \to \mathbf{w}$ in $L^q\left(0,T;L^q\left(\Omega\right)\right)$, $q = 4-\varepsilon, \varepsilon > 0$. Hölder's inequality (2.7) gives

$$\begin{aligned}&\|\mathbf{w}-\mathbf{w}^n\|_{L^q(0,T;L^q(\Omega))}^q \\ &= \int_0^T\int_\Omega |\mathbf{w}-\mathbf{w}^n|^{2-\varepsilon}\,|\mathbf{w}-\mathbf{w}^n|\,|\mathbf{w}-\mathbf{w}^n|\, d\mathbf{x}dt \\ &\le \left\|(\mathbf{w}-\mathbf{w}^n)^{2-\varepsilon}\right\|_{L^2(0,T;L^2(\Omega))}\|\mathbf{w}-\mathbf{w}^n\|_{L^4(0,T;L^4(\Omega))}\,\|\mathbf{w}-\mathbf{w}^n\|_{L^4(0,T;L^4(\Omega))}.\end{aligned}$$

The last two factors are bounded by the triangle inequality and the uniform boundedness of $\{\mathbf{w}^n\}$ in $L^4\left(0,T;L^4\left(\Omega\right)\right)$. The first term can be written in the form

$$\left\|(\mathbf{w}-\mathbf{w}^n)^{2-\varepsilon}\right\|_{L^2(0,T;L^2(\Omega))}^2 = \int_0^T\int_\Omega |\mathbf{w}-\mathbf{w}^n|^{2-2\varepsilon}\,|\mathbf{w}-\mathbf{w}^n|\,|\mathbf{w}-\mathbf{w}^n|\, d\mathbf{x}dt.$$

Now, the proof continues as in the first estimate. After a finite number of applications of Hölder's inequality, we have to estimate

$$\left\|(\mathbf{w}-\mathbf{w}^n)^{2-\varepsilon_0}\right\|_{L^2(0,T;L^2(\Omega))}$$

with $2-\varepsilon_0 \le 1$. If $2-\varepsilon_0 = 1$, the result of ii) can be applied directly to prove iii). In the case $2-\varepsilon_0 < 1$, first Theorem 2.9 has to be used to obtain

$$\left\|(\mathbf{w}-\mathbf{w}^n)^{2-\varepsilon_0}\right\|_{L^2(0,T;L^2(\Omega))} \le c\,\|\mathbf{w}-\mathbf{w}^n\|_{L^2(0,T;L^2(\Omega))}$$

before ii) can be applied.

iv) For $\phi \in C_0^\infty\left(0,T;L^2\left(\Omega\right)\right)$ follows, using ii),

$$\lim_{n\to\infty}\int_\Omega\int_0^T \mathbf{w}_t^n\phi dtd\mathbf{x} = -\lim_{n\to\infty}\int_\Omega\int_0^T \mathbf{w}^n\phi_t dtd\mathbf{x}$$
$$= -\int_\Omega\int_0^T \mathbf{w}\phi_t dtd\mathbf{x} = \int_\Omega\int_0^T \mathbf{w}_t\phi dtd\mathbf{x}.$$

Now, iv) follows from the density of $C_0^\infty\left(0,T;L^2\left(\Omega\right)\right)$ in $L^2\left(0,T;L^2\left(\Omega\right)\right)$, see Adams [Ada75, Theorem 2.19].

v) We have to prove

$$\lim_{n\to\infty}\int_0^T\int_\Omega \frac{\partial w_j^n}{\partial x_i}\phi d\mathbf{x}dt = \int_0^T\int_\Omega \frac{\partial w_j}{\partial x_i}\phi d\mathbf{x}dt \quad \forall\phi\in L^{3/2}\left(0,T;L^{3/2}\left(\Omega\right)\right).$$

It suffices to prove this relation for a dense set in $L^{3/2}\left(0,T;L^{3/2}\left(\Omega\right)\right)$. Let $\phi \in C_0^0\left(0,T;C_0^1\left(\Omega\right)\right)$. The space $C_0^0\left(0,T;C_0^1\left(\Omega\right)\right)$ is dense in $L^{3/2}\left(0,T;L^{3/2}\left(\Omega\right)\right)$, see Adams [Ada75, Theorem 2.19]. It follows, using ii) and twice integration by parts,

$$\lim_{n\to\infty}\int_0^T\int_\Omega \frac{\partial w_j^n}{\partial x_i}\phi d\mathbf{x}dt = -\lim_{n\to\infty}\int_0^T\int_\Omega w_j^n\frac{\partial\phi}{\partial x_i}d\mathbf{x}dt$$
$$= -\int_0^T\int_\Omega w_j\frac{\partial\phi}{\partial x_i}d\mathbf{x}dt = \int_0^T\int_\Omega \frac{\partial w_j}{\partial x_i}\phi d\mathbf{x}dt.$$

Boundary integrals vanish since ϕ vanishes on the boundary. Here, property ii) could be applied since we have $C_0^0\left(0,T;C_0^1\left(\Omega\right)\right) \subset L^2\left(0,T;L^2\left(\Omega\right)\right)$. □

We will use in the following the notation that '$\{\mathbf{w}^n\}$ converges' instead of that a subsequence converges.

Let

$$P^n = \left\{\mathbf{v} : \mathbf{v} = \sum_{l=1}^n \beta_l\left(t\right)\mathbf{v}^l\left(\mathbf{x}\right)\right\}$$

where $\beta_l\left(t\right)$ are absolutely continuous functions of $t\in[0,T]$ with $\beta_l\left(t\right) \in H^1\left(0,T\right)$. A function $\beta_l : [a,b] \to \mathbb{R}$ is said to be absolutely continuous if for any $\varepsilon > 0$ there exists a $\delta > 0$ such that for any finite system of pairwise non-intersecting intervals $(a_k,b_k) \subset (a,b)$, $k = 1,\ldots,n$ for which $\sum_{k=1}^n (b_k - a_k) < \delta$ the inequality $\sum_{k=1}^n |\beta_l(b_k) - \beta_l(a_k)| < \varepsilon$ holds.

We choose a fixed function $\varphi \in P^n$. From (6.14) follows by linear combination that φ fulfils

$$\int_0^T (\mathbf{w}_t^n,\varphi) + ((\nu+\nu_S\|\nabla\mathbf{w}^n\|_F)\nabla\mathbf{w}^n,\nabla\varphi) + b(\mathbf{w}^n,\mathbf{w}^n,\varphi)\,dt = \int_0^T (\mathbf{f},\varphi)\,dt. \tag{6.20}$$

Lemma 6.6. *For $\varphi \in P^n$ holds*

$$\lim_{n\to\infty} \int_0^T (\mathbf{w}_t^n, \varphi)\, dt = \int_0^T (\mathbf{w}_t, \varphi)\, dt, \tag{6.21}$$

$$\lim_{n\to\infty} \int_0^T b(\mathbf{w}^n, \mathbf{w}^n, \varphi)\, dt = \int_0^T b(\mathbf{w}, \mathbf{w}, \varphi)\, dt. \tag{6.22}$$

Proof. (6.21) : This follows immediately from Lemma 6.5, iv).
(6.22) : It is

$$\int_0^T b(\mathbf{w}^n, \mathbf{w}^n, \varphi)\, dt = \sum_{i,j=1}^d \int_0^T \int_\Omega \mathbf{w}_i^n \frac{\partial \mathbf{w}_j^n}{\partial \mathbf{x}_i} \varphi_j d\mathbf{x}dt.$$

Considering an arbitrary term of this sum yields

$$\int_0^T \int_\Omega \mathbf{w}_i^n \frac{\partial \mathbf{w}_j^n}{\partial \mathbf{x}_i} \varphi_j d\mathbf{x}dt = \int_0^T \int_\Omega \mathbf{w}_i \frac{\partial \mathbf{w}_j^n}{\partial \mathbf{x}_i} \varphi_j d\mathbf{x}dt + \int_0^T \int_\Omega (\mathbf{w}_i^n - \mathbf{w}_i) \frac{\partial \mathbf{w}_j^n}{\partial \mathbf{x}_i} \varphi_j d\mathbf{x}dt.$$

Since $\frac{\partial \mathbf{w}_j^n}{\partial \mathbf{x}_i} \in L^3\left(0,T; L^3(\Omega)\right)$ converges weakly to $\frac{\partial \mathbf{w}_j}{\partial \mathbf{x}_i}$ in this space and we have $\mathbf{w}_i\varphi_j \in L^{3/2}\left(0,T; L^{3/2}(\Omega)\right)$, one obtains for the first term

$$\lim_{n\to\infty} \int_0^T \int_\Omega \mathbf{w}_i \frac{\partial \mathbf{w}_j^n}{\partial \mathbf{x}_i} \varphi_j d\mathbf{x}dt = \int_0^T \int_\Omega \mathbf{w}_i \frac{\partial \mathbf{w}_j}{\partial \mathbf{x}_i} \varphi_j d\mathbf{x}dt.$$

Applying Hölder's inequality to the second term gives

$$\left| \int_0^T \int_\Omega (\mathbf{w}_i^n - \mathbf{w}_i) \frac{\partial \mathbf{w}_j^n}{\partial \mathbf{x}_i} \varphi_j d\mathbf{x}dt \right| \le \left\| (\mathbf{w}_i^n - \mathbf{w}_i) \varphi_j \right\|_{L^{3/2}(0,T;L^{3/2}(\Omega))} \left\| \frac{\partial \mathbf{w}_j^n}{\partial \mathbf{x}_i} \right\|_{L^3(0,T;L^3(\Omega))}.$$

By the Cauchy-Schwarz inequality follows

$$\left\| (\mathbf{w}_i^n - \mathbf{w}_i) \varphi_j \right\|_{L^{3/2}(0,T;L^{3/2}(\Omega))} \le \left\| \mathbf{w}_i^n - \mathbf{w}_i \right\|_{L^3(0,T;L^3(\Omega))}^{3/2} \left\| \varphi_j \right\|_{L^3(0,T;L^3(\Omega))}^{3/2}.$$

By Lemma 6.5 iii), $\mathbf{w}_i^n$ converges strongly to $\mathbf{w}_i$ in $L^3\left(0,T; L^3(\Omega)\right)$. The term $\left\| \frac{\partial \mathbf{w}_j^n}{\partial \mathbf{x}_i} \right\|_{L^3(0,T;L^3(\Omega))}$ is uniformly bounded and $\left\| \varphi_j \right\|_{L^3(0,T;L^3(\Omega))}^{3/2}$ is just a constant. Altogther,

$$\lim_{n\to\infty} \left| \int_0^T \int_\Omega (\mathbf{w}_i^n - \mathbf{w}_i) \frac{\partial \mathbf{w}_j^n}{\partial \mathbf{x}_i} \varphi_j d\mathbf{x}dt \right| = 0$$

and

$$\lim_{n\to\infty}\int_0^T\int_\Omega \mathbf{w}_i^n\frac{\partial \mathbf{w}_j^n}{\partial \mathbf{x}_i}\varphi_j d\mathbf{x}dt = \int_0^T\int_\Omega \mathbf{w}_i\frac{\partial \mathbf{w}_j}{\partial \mathbf{x}_i}\varphi_j d\mathbf{x}dt.$$

□

Lemma 6.7. *The limiting equation of (6.20) is*

$$\int_0^T (\mathbf{w}_t,\varphi) + (\mathbf{B},\nabla\varphi) + b(\mathbf{w},\mathbf{w},\varphi)\,dt = \int_0^T (\mathbf{f},\varphi)\,dt \tag{6.23}$$

with $\mathbf{B}\in L^{3/2}\left(0,T;L^{3/2}(\Omega)\right)$.

Proof. The limits of the first and the third term have been established in Lemma 6.6.

By construction, we have $\nabla\varphi \in L^3\left(0,T;L^3(\Omega)\right)$. Now, if the sequence $\left(\nu+\nu_S\|\nabla\mathbf{w}^n\|_F\right)\nabla\mathbf{w}^n$ is uniformly bounded in $L^{3/2}\left(0,T;L^{3/2}(\Omega)\right)$ then there is a subsequence which converges weakly to an operator $\mathbf{B}\in L^{3/2}\left(0,T;L^{3/2}(\Omega)\right)$, i.e.

$$\int_0^T \left(\left(\nu+\nu_S\|\nabla\mathbf{w}^n\|_F\right)\nabla\mathbf{w}^n,\nabla\varphi\right)dt \to \int_0^T (\mathbf{B},\nabla\varphi)\,dt.$$

One has

$$\begin{aligned}&\left\|\left(\nu+\nu_S\|\nabla\mathbf{w}^n\|_F\right)\nabla\mathbf{w}^n\right\|^{3/2}_{L^{3/2}(0,T;L^{3/2}(\Omega))} &(6.24)\\ &= \nu^{3/2}\int_0^T\int_\Omega |\nabla\mathbf{w}^n|^{3/2}\,d\mathbf{x}dt + \nu_S^{3/2}\int_0^T\int_\Omega \|\nabla\mathbf{w}^n\|_F^{3/2}\,|\nabla\mathbf{w}^n|^{3/2}\,d\mathbf{x}dt.\end{aligned}$$

Using inequality (2.6) gives for an arbitrary component of the first term

$$\int_0^T\int_\Omega\left|\frac{\partial\mathbf{w}_i^n}{\partial\mathbf{x}_j}\right|^{3/2}d\mathbf{x}dt \le \frac{3}{4}\int_0^T\int_\Omega\left|\frac{\partial\mathbf{w}_i^n}{\partial\mathbf{x}_j}\right|^{2}d\mathbf{x}dt + \frac{1}{4}\int_0^T\int_\Omega d\mathbf{x}dt.$$

This is bounded by (6.17) and since Ω is bounded. The second term of (6.24) is considered also component-wise. Using inequality (2.6) gives

$$\begin{aligned}&\int_0^T\int_\Omega \|\nabla\mathbf{w}^n\|_F^{3/2}\left|\frac{\partial\mathbf{w}_i^n}{\partial\mathbf{x}_j}\right|^{3/2}d\mathbf{x}dt\\ &= \int_0^T\int_\Omega \|\nabla\mathbf{w}^n\|_F\left(\sum_{k,l=1}^d\left|\frac{\partial\mathbf{w}_k^n}{\partial\mathbf{x}_l}\right|^2\right)^{1/4}\left|\frac{\partial\mathbf{w}_i^n}{\partial\mathbf{x}_j}\right|^{3/2}d\mathbf{x}dt\\ &\le \frac{1}{4}\int_0^T \|\nabla\mathbf{w}^n\|_F\left(\nabla\mathbf{w}^n,\nabla\mathbf{w}^n\right)dt + \frac{3}{4}\int_0^T\int_\Omega \|\nabla\mathbf{w}^n\|_F\left|\frac{\partial\mathbf{w}_i^n}{\partial\mathbf{x}_j}\right|^{2}d\mathbf{x}dt.\end{aligned}$$

Both terms in the last line of the inequality are bounded by (6.17). Therefore, a subsequence of it converges weakly to some $\mathbf{B}(t,\mathbf{x})$ from which the statement of the lemma follows. □

Identity (6.23) is valid for $\varphi \in P^n$ for arbitrary n and thus for $\varphi \in \cup_{n=1}^{\infty} P^n$.

Lemma 6.8. *Identity (6.23) is valid for $\varphi \in V$.*

Proof. We refer for the proof of this lemma to Ladyshenskaya [Lad69, p. 159 f.]. □

The existence proof will use the main theorem for monotone operators by Minty and Browder [Min62, Bro65]. Therefore, it will be shown that the non-linear viscous operator is monotone. This operator is defined by $\mathbf{A} : L^3(\Omega) \to L^{3/2}(\Omega)$ with

$$\mathbf{A}(\nabla \mathbf{w}^n) = (\nu + \nu_S \|\nabla \mathbf{w}^n\|_F) \nabla \mathbf{w}^n. \tag{6.25}$$

Lemma 6.9. *For arbitrary functions $\mathbf{w}', \mathbf{w}'' \in W^{1,3}(\Omega)$ holds the estimate*

$$\int_\Omega (\mathbf{A}(\nabla \mathbf{w}') - \mathbf{A}(\nabla \mathbf{w}'')) : (\nabla \mathbf{w}' - \nabla \mathbf{w}'') \, d\mathbf{x} \geq \nu \|\nabla \mathbf{w}' - \nabla \mathbf{w}''\|_{L^2(\Omega)}^2 \tag{6.26}$$

with the operator $\mathbf{A}$ defined in (6.25). Especially,

$$\int_\Omega (\|\nabla \mathbf{w}'\|_F (\nabla \mathbf{w}') - \|\nabla \mathbf{w}''\|_F (\nabla \mathbf{w}'')) : (\nabla \mathbf{w}' - \nabla \mathbf{w}'') \, d\mathbf{x} \geq 0,$$

i.e. the Smagorinsky term defines a monotone operator from $L^3(\Omega)$ into $L^{3/2}(\Omega)$.

Proof. The operator $\mathbf{A}$ is monotone iff the function $f : [0,1] \to \mathbb{R}$ with

$$f(\tau) = \int_0^1 \omega \mathbf{A}(\tau \nabla \mathbf{w}' + (1-\tau) \nabla \mathbf{w}'') : \nabla (\mathbf{w}' - \mathbf{w}'') \, d\mathbf{x}$$

is monotone increasing for all $\mathbf{w}', \mathbf{w}'' \in W^{1,3}(\Omega)$, see, e.g., Zeidler [Zei90, Proposition 25.6].

Let $\mathbf{w}', \mathbf{w}'' \in C^1(\bar{\Omega})$. We define $\mathbf{w}^\tau = \tau \mathbf{w}' + (1-\tau)\mathbf{w}''$. With this definition follows

$$\begin{aligned}
&(\mathbf{A}(\nabla \mathbf{w}') - \mathbf{A}(\nabla \mathbf{w}'')) : (\nabla \mathbf{w}' - \nabla \mathbf{w}'') \\
&= \sum_{i,j=1}^{d} \left(\int_0^1 \frac{d}{d\tau} \mathbf{A}_{ij}(\nabla \mathbf{w}^\tau) \, d\tau \right) \left(\frac{\partial \mathbf{w}_i'}{\partial x_j} - \frac{\partial \mathbf{w}_i''}{\partial x_j} \right).
\end{aligned}$$

We have

$$\begin{aligned}
\frac{d}{d\tau} \mathbf{A}_{ij}(\nabla \mathbf{w}^\tau) = {} & \left(\nu_S \frac{\partial}{\partial \tau} \|\nabla \mathbf{w}^\tau\|_F \right) \left(\tau \frac{\partial \mathbf{w}_i'}{\partial x_j} + (1-\tau) \frac{\partial \mathbf{w}_i''}{\partial x_j} \right) \\
& + (\nu + \nu_S \|\nabla \mathbf{w}^\tau\|_F) \left(\frac{\partial \mathbf{w}_i'}{\partial x_j} - \frac{\partial \mathbf{w}_i''}{\partial x_j} \right)
\end{aligned}$$

and

$$\frac{\partial}{\partial \tau}\|\nabla \mathbf{w}^\tau\|_F = \frac{\partial}{\partial \tau}\left(\sum_{k,l=1}^{d}\left(\tau\frac{\partial \mathbf{w}_k'}{\partial x_l} + (1-\tau)\frac{\partial \mathbf{w}_k''}{\partial x_l}\right)^2\right)^{1/2}$$
$$= \frac{1}{\|\nabla \mathbf{w}^\tau\|_F}\sum_{k,l=1}^{d}\frac{\partial \mathbf{w}_k^\tau}{\partial x_l}\left(\frac{\partial \mathbf{w}_k'}{\partial x_l} - \frac{\partial \mathbf{w}_k''}{\partial x_l}\right).$$

Thus,

$$\begin{aligned}
&(\mathbf{A}(\nabla \mathbf{w}') - \mathbf{A}(\nabla \mathbf{w}'')) : (\nabla \mathbf{w}' - \nabla \mathbf{w}'') \\
&= \int_0^1 \sum_{i,j=1}^{d} (\nu + \nu_S \|\nabla \mathbf{w}^\tau\|_F)\left(\frac{\partial \mathbf{w}_i'}{\partial x_j} - \frac{\partial \mathbf{w}_k''}{\partial x_l}\right)\left(\frac{\partial \mathbf{w}_i'}{\partial x_j} - \frac{\partial \mathbf{w}_i''}{\partial x_j}\right) d\tau \\
&+ \int_0^1 \nu_S \|\nabla \mathbf{w}^\tau\|_F^{-1} \sum_{i,j,k,l=1}^{d} \frac{\partial \mathbf{w}_k^\tau}{\partial x_l}\frac{\partial \mathbf{w}_i^\tau}{\partial x_j}\left(\frac{\partial \mathbf{w}_k'}{\partial x_l} - \frac{\partial \mathbf{w}_k''}{\partial x_l}\right)\left(\frac{\partial \mathbf{w}_i'}{\partial x_j} - \frac{\partial \mathbf{w}_i''}{\partial x_j}\right) d\tau.
\end{aligned}$$

The first term is estimated as follows

$$\begin{aligned}
&\int_0^1 \sum_{i,j,=1}^{d} (\nu + \nu_S \|\nabla \mathbf{w}^\tau\|_F)\left(\frac{\partial \mathbf{w}_i'}{\partial x_j} - \frac{\partial \mathbf{w}_i''}{\partial x_j}\right)^2 d\tau \\
&\geq \sum_{i,j,=1}^{d} \nu \left(\frac{\partial \mathbf{w}_i'}{\partial x_j} - \frac{\partial \mathbf{w}_i''}{\partial x_j}\right)^2 \int_0^1 d\tau \\
&= \nu (\nabla \mathbf{w}' - \nabla \mathbf{w}'') : (\nabla \mathbf{w}' - \nabla \mathbf{w}'').
\end{aligned}$$

The second term is non-negative since

$$\begin{aligned}
&\sum_{i,j,k,l=1}^{d} \frac{\partial \mathbf{w}_k^\tau}{\partial x_l}\frac{\partial \mathbf{w}_i^\tau}{\partial x_j}\left(\frac{\partial \mathbf{w}_k'}{\partial x_l} - \frac{\partial \mathbf{w}_k''}{\partial x_l}\right)\left(\frac{\partial \mathbf{w}_i'}{\partial x_j} - \frac{\partial \mathbf{w}_i''}{\partial x_j}\right) \\
&= \left(\sum_{i,j=1}^{d} \frac{\partial \mathbf{w}_i^\tau}{\partial x_j}\left(\frac{\partial \mathbf{w}_i'}{\partial x_j} - \frac{\partial \mathbf{w}_i''}{\partial x_j}\right)\right)^2.
\end{aligned}$$

Thus, this term can be estimated from below by zero. The proof is completed with integration on Ω and using the density of $C^1(\bar{\Omega})$ in $W^{1,3}(\Omega)$. □

Lemma 6.10. *Let $\tilde{\varphi} \in V$, then*

$$-\int_0^T \int_\Omega (\mathbf{w}_t + (\mathbf{w}\cdot\nabla)\mathbf{w} - \mathbf{f})\cdot(\mathbf{w} - \tilde{\varphi}) + \mathbf{A}(\nabla \varphi) : (\nabla \mathbf{w} - \nabla \tilde{\varphi})\, d\mathbf{x}dt \geq 0. \tag{6.27}$$

Proof. To begin with, let $\tilde{\varphi} \in P^n$ for an arbitrary n. From (6.26) follows

$$\int_0^T \int_\Omega (\mathbf{A}(\nabla \mathbf{w}^n) - \mathbf{A}(\nabla \tilde{\varphi})) : (\nabla \mathbf{w}^n - \nabla \tilde{\varphi})\, d\mathbf{x}dt \geq 0.$$

Since $\mathbf{w}^n \in P^n$ and $\tilde{\varphi} \in P^n$, we obtain with (6.20)

$$\int_0^T \int_\Omega \mathbf{A}(\nabla \mathbf{w}^n) : (\nabla \mathbf{w}^n - \nabla \tilde{\varphi})\, d\mathbf{x}dt$$
$$= \int_0^T (\mathbf{f} - \mathbf{w}_t^n - (\mathbf{w}^n \cdot \nabla)\mathbf{w}^n, \mathbf{w}^n - \tilde{\varphi})\, dt$$

such that

$$-\int_0^T \int_\Omega \Big[(\mathbf{f} - \mathbf{w}_t^n - (\mathbf{w}^n \cdot \nabla)\mathbf{w}^n) \cdot (\mathbf{w}^n - \tilde{\varphi}) + \mathbf{A}(\nabla \tilde{\varphi}) : (\nabla \mathbf{w}^n - \nabla \tilde{\varphi}) \Big] d\mathbf{x}dt \geq 0. \tag{6.28}$$

Now, one passes to the limit $n \to \infty$. This gives

$$\int_0^T (\mathbf{f}, \mathbf{w}^n - \tilde{\varphi})\, dt = \int_0^T (\mathbf{f}, \mathbf{w} - \tilde{\varphi})\, dt + \int_0^T (\mathbf{f}, \mathbf{w}^n - \mathbf{w})\, dt.$$

The second term converges to zero since $\mathbf{w}^n$ converges to $\mathbf{w}$ strongly in $L^2(0,T;L^2(\Omega))$ and

$$\left| \int_0^T (\mathbf{f}, \mathbf{w}^n - \mathbf{w})\, dt \right| \leq \|\mathbf{f}\|_{L^2(0,T;L^2(\Omega))} \|\mathbf{w}^n - \mathbf{w}\|_{L^2(0,T;L^2(\Omega))}.$$

The non-linear term is considered component-wise

$$\int_0^T \int_\Omega (\mathbf{w}^n \cdot \nabla)\mathbf{w}^n \cdot \mathbf{w}^n d\mathbf{x}dt = \sum_{k,l=1}^d \int_0^T \int_\Omega \mathbf{w}_k^n \frac{\partial \mathbf{w}_l^n}{\partial \mathbf{x}_k} \mathbf{w}_l^n d\mathbf{x}dt.$$

By construction, $\frac{\partial \mathbf{w}_l^n}{\partial \mathbf{x}_k} \in L^3(0,T;L^3(\Omega))$ and it was proved in Lemma 6.5 v) that $\frac{\partial \mathbf{w}_l^n}{\partial \mathbf{x}_k} \rightharpoonup \frac{\partial \mathbf{w}_l}{\partial \mathbf{x}_k}$ in $L^3(0,T;L^3(\Omega))$. This gives

$$\lim_{n\to\infty} \int_0^T \int_\Omega \mathbf{w}_k^n \frac{\partial \mathbf{w}_l^n}{\partial \mathbf{x}_k} \mathbf{w}_l^n d\mathbf{x}dt$$
$$= \lim_{n\to\infty} \int_0^T \int_\Omega \mathbf{w}_k \frac{\partial \mathbf{w}_l^n}{\partial \mathbf{x}_k} \mathbf{w}_l d\mathbf{x}dt + \lim_{n\to\infty} \int_0^T \int_\Omega \frac{\partial \mathbf{w}_l^n}{\partial \mathbf{x}_k} (\mathbf{w}_k^n \mathbf{w}_l^n - \mathbf{w}_k \mathbf{w}_l)\, d\mathbf{x}dt$$
$$= \int_0^T \int_\Omega \mathbf{w}_k \frac{\partial \mathbf{w}_l}{\partial \mathbf{x}_k} \mathbf{w}_l d\mathbf{x}dt + \lim_{n\to\infty} \int_0^T \int_\Omega \frac{\partial \mathbf{w}_l^n}{\partial \mathbf{x}_k} (\mathbf{w}_k^n \mathbf{w}_l^n - \mathbf{w}_k \mathbf{w}_l)\, d\mathbf{x}dt.$$

Since

$$\left| \int_0^T \int_\Omega \frac{\partial \mathbf{w}_l^n}{\partial \mathbf{x}_k} (\mathbf{w}_k^n \mathbf{w}_l^n - \mathbf{w}_k \mathbf{w}_l)\, d\mathbf{x}dt \right|$$
$$\leq \left\| \frac{\partial \mathbf{w}_l^n}{\partial \mathbf{x}_k} \right\|_{L^3(0,T;L^3(\Omega))} \|\mathbf{w}_k^n \mathbf{w}_l^n - \mathbf{w}_k \mathbf{w}_l\|_{L^{3/2}(0,T;L^{3/2}(\Omega))},$$

one has to show that $\mathbf{w}_k^n\mathbf{w}_l^n \to \mathbf{w}_k\mathbf{w}_l$ strongly in $L^{3/2}\left(0,T;L^{3/2}\left(\Omega\right)\right)$. Using the triangle and the Cauchy-Schwarz inequality give

$$
\begin{aligned}
&\left\|\mathbf{w}_k^n\mathbf{w}_l^n - \mathbf{w}_k\mathbf{w}_l\right\|_{L^{3/2}(0,T;L^{3/2}(\Omega))} \\
&\leq \left\|\left(\mathbf{w}_k^n - \mathbf{w}_k\right)\mathbf{w}_l^n\right\|_{L^{3/2}(0,T;L^{3/2}(\Omega))} + \left\|\mathbf{w}_k\left(\mathbf{w}_l^n - \mathbf{w}_l\right)\right\|_{L^{3/2}(0,T;L^{3/2}(\Omega))} \\
&\leq \left\|\mathbf{w}_k^n - \mathbf{w}_k\right\|_{L^3(0,T;L^3(\Omega))}\left\|\mathbf{w}_l^n\right\|_{L^3(0,T;L^3(\Omega))} \\
&\quad + \left\|\mathbf{w}_l^n - \mathbf{w}_l\right\|_{L^3(0,T;L^3(\Omega))}\left\|\mathbf{w}_l^k\right\|_{L^3(0,T;L^3(\Omega))}.
\end{aligned}
$$

Since $\mathbf{w}_k^n \to \mathbf{w}_k$ strongly in $L^3\left(0,T;L^3\left(\Omega\right)\right)$ (Lemma 6.5 iii)), it follows $\mathbf{w}_k^n\mathbf{w}_l^n \to \mathbf{w}_k\mathbf{w}_l$ strongly in $L^{3/2}\left(0,T;L^{3/2}\left(\Omega\right)\right)$ and

$$
\lim_{n\to\infty}\int_0^T\int_\Omega \mathbf{w}_k^n\frac{\partial \mathbf{w}_l^n}{\partial \mathbf{x}_k}\mathbf{w}_l^n d\mathbf{x}dt = \int_0^T\int_\Omega \mathbf{w}_k\frac{\partial \mathbf{w}_l}{\partial \mathbf{x}_k}\mathbf{w}_l d\mathbf{x}dt.
$$

It was shown in the proof of Lemma 6.7 that $\mathbf{A}\left(\nabla\tilde{\varphi}\right) \in L^{3/2}\left(0,T;L^{3/2}\left(\Omega\right)\right)$. Since $\nabla\mathbf{w}^n \in L^3\left(0,T;L^3\left(\Omega\right)\right)$ and $\nabla\mathbf{w}^n \rightharpoonup \nabla\mathbf{w}$ (Lemma 6.5 v)) in that space follows

$$
\lim_{n\to\infty}\int_0^T\int_\Omega \mathbf{A}\left(\nabla\tilde{\varphi}\right):\nabla\mathbf{w}^n d\mathbf{x}dt = \int_0^T\int_\Omega \mathbf{A}\left(\nabla\tilde{\varphi}\right):\nabla\mathbf{w}d\mathbf{x}dt.
$$

In addition, we have

$$
\begin{aligned}
\lim_{n\to\infty}\int_0^T\int_\Omega \mathbf{w}_t^n\mathbf{w}^n d\mathbf{x}dt &= \int_0^T\int_\Omega \mathbf{w}_t^n\left(\mathbf{w}^n - \mathbf{w}\right)d\mathbf{x}dt + \lim_{n\to\infty}\int_0^T\int_\Omega \mathbf{w}_t^n\mathbf{w}d\mathbf{x}dt \\
&= 0 + \int_0^T\int_\Omega \mathbf{w}_t\mathbf{w}d\mathbf{x}dt
\end{aligned}
$$

by Lemma 6.5 ii) and iv).

Thus, inequality (6.27) has been proved for any $\tilde{\varphi} \in P^n$. But then it is also valid for arbitrary $\tilde{\varphi} \in P$ and also for arbitrary $\tilde{\varphi} \in V$ (Lemma 6.8). □

Lemma 6.11. *For all $\varphi \in V$ holds*

$$
\int_0^T\int_\Omega \mathbf{B}:\nabla\varphi d\mathbf{x}dt = \int_0^T\int_\Omega \mathbf{A}\left(\nabla\mathbf{w}\right):\nabla\varphi d\mathbf{x}dt.
$$

Proof. Since (6.23) is valid for all $\varphi \in V$, we can choose $\varphi = \mathbf{w}-\tilde{\varphi}$. Subtracting this equation from (6.27) yields

$$
\int_0^T\int_\Omega \left(\mathbf{B} - \mathbf{A}\left(\nabla\tilde{\varphi}\right)\right):\left(\nabla\mathbf{w} - \nabla\tilde{\varphi}\right)d\mathbf{x}dt \geq 0.
$$

Setting $\tilde{\varphi} = \beta - \varepsilon\varphi$ with $\varepsilon > 0$ and $\varphi \in V$ arbitrary gives

$$\varepsilon \int_0^T \int_\Omega (\mathbf{B} - \mathbf{A}(\nabla \mathbf{w} - \varepsilon \nabla \varphi)) : \nabla \varphi d\mathbf{x} dt \geq 0.$$

Division by ε and taking the limit $\varepsilon \to 0$ yield

$$\int_0^T \int_\Omega (\mathbf{B} - \mathbf{A}(\nabla \mathbf{w})) : \nabla \varphi d\mathbf{x} dt \geq 0. \tag{6.29}$$

If $\varphi \in V$, then also $-\varphi \in V$. Thus, if the integral is positive for φ, then it is negative for $-\varphi$ which is a contradiction to (6.29). Hence

$$\int_0^T \int_\Omega (\mathbf{B} - \mathbf{A}(\nabla \mathbf{w})) : \nabla \varphi d\mathbf{x} dt = 0 \quad \forall \varphi \in V.$$

□

Theorem 6.12. Existence of a weak solution. *Problem (6.4) possesses at least one solution* $\mathbf{w} \in V$ *for arbitrary* $\mathbf{f} \in L^2(0, T; L^2(\Omega))$ *and* $\mathbf{w}_0 \in W^{1,3}_{0,div}(\Omega)$.

Proof. It was shown in the previous lemmata that there is a $\mathbf{w} \in V$ which fulfils the weak formulation of (6.4). Since $\mathbf{w}$ is given as a limit of a sequence $\{\mathbf{w}^n\}$ with $\mathbf{w}^n(0, \mathbf{x}) = \mathbf{w}_0(\mathbf{x})$ for all n, it follows $\mathbf{w}(0, \mathbf{x}) = \mathbf{w}_0(\mathbf{x})$. □

Remark 6.13. Generalizations. In Ladyzhenskaya [Lad67], the existence of a solution has been proved for the weak formulation of the form

$$\int_0^T (\mathbf{w}_t + (\mathbf{w} \cdot \nabla)\mathbf{w}, \mathbf{v}) + ((\nu + \nu_S \|\nabla \mathbf{w}\|_F^\mu) \nabla \mathbf{w}, \nabla \mathbf{v})\, dt = \int_0^T (\mathbf{f}, \mathbf{v})\, dt$$

with $\mu \geq 2/5$. The restriction on μ comes from the application of the Sobolev embedding to obtain inequality (6.11). Du and Gunzburger [DG91] could generalize the result of Ladyzhenskaya to $\mu \geq 1/5$ by deriving new a priori error estimates. □

Theorem 6.14. Uniqueness of the weak solution. *Under the same assumptions as in Theorem 6.12, the weak solution of (6.4) is unique in* V.

Proof. Let us assume that there are two weak solutions $\mathbf{w}', \mathbf{w}'' \in V$ of (6.4) and denote $\tilde{\mathbf{w}} = \mathbf{w}' - \mathbf{w}''$. Thus, $\tilde{\mathbf{w}} \in V$ and $\tilde{\mathbf{w}}(0, \mathbf{x}) = \mathbf{0}$. Subtracting (6.4) for $\mathbf{w} = \mathbf{w}', \mathbf{v} = \tilde{\mathbf{w}}$ and $\mathbf{w} = \mathbf{w}'', \mathbf{v} = \tilde{\mathbf{w}}$ gives

$$\begin{aligned} 0 = \int_0^T & (\tilde{\mathbf{w}}_t, \tilde{\mathbf{w}}) + (\mathbf{A}(\nabla \mathbf{w}') - \mathbf{A}(\nabla \mathbf{w}''), \nabla \mathbf{w}' - \nabla \mathbf{w}'') \\ & + b(\mathbf{w}', \mathbf{w}', \tilde{\mathbf{w}}) - b(\mathbf{w}'', \mathbf{w}'', \tilde{\mathbf{w}})\, dt. \end{aligned}$$

This can be rewritten in the following form

$$\begin{aligned} 0 = \int_0^T & \frac{d}{dt}(\tilde{\mathbf{w}}, \tilde{\mathbf{w}}) + 2(\mathbf{A}(\nabla \mathbf{w}') - \mathbf{A}(\nabla \mathbf{w}''), \nabla \mathbf{w}' - \nabla \mathbf{w}'') \\ & + 2b(\tilde{\mathbf{w}}, \mathbf{w}', \tilde{\mathbf{w}})\, dt. \end{aligned}$$

Here, $b(\mathbf{w}'', \tilde{\mathbf{w}}, \tilde{\mathbf{w}}) = 0$ has been used, which follows from (5.4). The monotonicity (6.26) of $\mathbf{A}(\cdot)$, Hölder's inequality (2.7) and the Sobolev embedding (2.13) lead to

$$\begin{aligned}&\int_0^T \frac{d}{dt} \|\tilde{\mathbf{w}}\|_{L^2(\Omega)}^2 + 2\nu \|\nabla\tilde{\mathbf{w}}\|_{L^2(\Omega)}^2 \, dt \\ &\le c \int_0^T \|\tilde{\mathbf{w}}\|_{L^2(\Omega)} \|\nabla\mathbf{w}'\|_{L^3(\Omega)} \|\nabla\tilde{\mathbf{w}}\|_{L^2(\Omega)} \, dt. \end{aligned} \tag{6.30}$$

This can be estimated further by Young's inequality (2.6)

$$\begin{aligned}&\int_0^T \frac{d}{dt} \|\tilde{\mathbf{w}}\|_{L^2(\Omega)}^2 + 2\nu \|\nabla\tilde{\mathbf{w}}\|_{L^2(\Omega)}^2 \, dt \\ &\le \int_0^T 2\nu \|\nabla\tilde{\mathbf{w}}\|_{L^2(\Omega)}^2 + \frac{c^2}{8\nu} \|\tilde{\mathbf{w}}\|_{L^2(\Omega)}^2 \|\nabla\mathbf{w}'\|_{L^3(\Omega)}^2 \, dt,\end{aligned}$$

which gives

$$\|\tilde{\mathbf{w}}(T,\mathbf{x})\|_{L^2(\Omega)}^2 \le c \int_0^T \|\tilde{\mathbf{w}}\|_{L^2(\Omega)}^2 \|\nabla\mathbf{w}'\|_{L^3(\Omega)}^2 \, dt.$$

Gronwall's lemma (Lemma 2.1) now gives $\|\tilde{\mathbf{w}}(T,\mathbf{x})\|_{L^2(\Omega)}^2 \le 0$ for all T which proves the uniqueness of the solution in $L^\infty(0,T;L^2(\Omega))$. Since $H^1(0,T;L^2(\Omega)) \subset L^\infty(0,T;L^2(\Omega))$ by a Sobolev embedding theorem in one dimension, the solution is unique in $H^1(0,T;L^2(\Omega))$.

An alternative application of Young's inequality to (6.30) yields

$$\begin{aligned}&\int_0^T \frac{d}{dt} \|\tilde{\mathbf{w}}\|_{L^2(\Omega)}^2 + 2\nu \|\nabla\tilde{\mathbf{w}}\|_{L^2(\Omega)}^2 \, dt \\ &\le \int_0^T \nu \|\nabla\tilde{\mathbf{w}}\|_{L^2(\Omega)}^2 + \frac{c^2}{4\nu} \|\tilde{\mathbf{w}}\|_{L^2(\Omega)}^2 \|\nabla\mathbf{w}'\|_{L^3(\Omega)}^2 \, dt.\end{aligned}$$

Applying the result $\|\tilde{\mathbf{w}}(t,\mathbf{x})\|_{L^2(\Omega)}^2 = 0$ for almost all t gives

$$\int_0^T \nu \|\nabla\tilde{\mathbf{w}}\|_{L^2(\Omega)}^2 \, dt \le 0$$

from which $\|\nabla\tilde{\mathbf{w}}\|_{L^2(\Omega)} = 0$ a.e. in $(0,T)$ follows. Hence, $\|\nabla\tilde{\mathbf{w}}\|_{L^3(\Omega)} = 0$ a.e. in $(0,T)$ and the uniqueness of the weak solution in $L^2\left(0,T;W^{1,3}_{0,div}(\Omega)\right)$ is proved. Since T is finite, $L^3\left(0,T;W^{1,3}_{0,div}(\Omega)\right) \subset L^2\left(0,T;W^{1,3}_{0,div}(\Omega)\right)$ and the solution is unique in $L^3\left(0,T;W^{1,3}_{0,div}(\Omega)\right)$. $\square$

Theorem 6.15. Stability of the weak solution. *Let the assumptions of Theorem 6.12 be fulfilled and let* $\mathbf{w}', \mathbf{w}'' \in V$ *solutions of (6.4) with different initial data and different right hand sides* $\mathbf{f}', \mathbf{f}''$. *Then*

$$
\begin{aligned}
&\|\mathbf{w}' - \mathbf{w}''\|_{L^\infty(0,T;L^2(\Omega))} \\
&\le \left(\|\mathbf{w}'(0,\mathbf{x}) - \mathbf{w}''(0,\mathbf{x})\|^2_{L^2(\Omega)} + \frac{1}{2c_1} \int_0^T \|\mathbf{f}' - \mathbf{f}''\|^2_{L^2(\Omega)} \, dt \right) \\
&\quad \times \exp\left(c_2 \|\nabla \mathbf{w}'\|^2_{L^2(0,T;L^3(\Omega))} + \frac{c_1}{2} T \right),
\end{aligned}
$$

with $c_1, c_2 > 0$ *and* c_1 *can be chosen arbitrarily.*

If $\mathbf{f}' = \mathbf{f}''$, *then*

$$
\begin{aligned}
&\|\mathbf{w}' - \mathbf{w}''\|_{L^\infty(0,T;L^2(\Omega))} \\
&\le \|\mathbf{w}'(0,\mathbf{x}) - \mathbf{w}''(0,\mathbf{x})\|^2_{L^2(\Omega)} \exp\left(c_2 \|\nabla \mathbf{w}'\|^2_{L^2(0,T;L^3(\Omega))} \right).
\end{aligned}
$$

Proof. The proof starts in the same way as the proof of Theorem 6.14. One obtains

$$
\begin{aligned}
\|\tilde{\mathbf{w}}(T,\mathbf{x})\|^2_{L^2(\Omega)} &\le \|\tilde{\mathbf{w}}(0,\mathbf{x})\|^2_{L^2(\Omega)} + \int_0^T \frac{1}{2c_1} \|\mathbf{f}' - \mathbf{f}''\|^2_{L^2(\Omega)} \, dt \\
&\quad + \int_0^T \frac{c^2}{4\nu} \|\tilde{\mathbf{w}}\|^2_{L^2(\Omega)} \|\nabla \mathbf{w}'\|^2_{L^3(\Omega)} + \frac{c_1}{2} \|\tilde{\mathbf{w}}\|^2_{L^2(\Omega)} \, dt.
\end{aligned}
$$

Setting $c^2/(4\nu) = c_2$ and applying Gronwall's lemma (Lemma 2.1) proves the statement of the theorem. □

Remark 6.16. Other analytical investigations of the Smagorinsky model. Parés [Par92] studied the existence and uniqueness of a weak solution of a Smagorinsky model which differs from (6.1) in some aspects. First, the deformation tensor formulation of the viscous term and the deformation tensor formulation of the Smagorinsky model (4.3) are considered. Second, homogeneous Dirichlet boundary conditions are prescribed only at a part of the boundary $\Gamma_{\text{diri,hom}}$ with $\text{meas}(\Gamma_{\text{diri,hom}}) > 0$. On the rest of the boundary, slip with friction and penetration with resistance boundary conditions are given, e.g., the linear conditions described in Section 5.2.4. These boundary conditions lead to an additional term in the weak formulation of the momentum equation of the Smagorinsky model, see Section 5.2. The existence proof uses the Galerkin method in the same way as described here. In addition, estimates for the additional term coming from the boundary conditions have to be proved.

The Smagorinsky model can be used to stabilize the dominating convection in the stationary Navier-Stokes equations. Parés [Par92] considered the stationary Smagorinsky model with the same features as the time dependent one. He could prove the existence of weak solutions and the uniqueness in the case of small data. Du and Gunzburger [DG91] studied the stationary

Smagorinsky model with gradient formulation of the viscous term, $\|\nabla \mathbf{w}\|_F$ and homogeneous Dirichlet boundary conditions. They proved the uniqueness of a weak solution for small data. □

Remark 6.17. Comparison to results for the Navier-Stokes equations. For the Navier-Stokes equations in 3d, the uniqueness of a less regular solution (usually called weak solution, although the use of this term is not uniquely in the literature) and the existence of a more regular solution (often called strong solution) is not proved up to now, e.g., see Temam [Tem77], Galdi [Gal00] or Sohr [Soh01]. The outstanding importance of this problem is reflected by its inclusion into the *Clay Mathematics Institute Millenium Prize Problems.* The possibility of proving existence und uniqueness for the weak solution of the Smagorinsky model comes from the fact that the Smagorinsky model introduces a regularising artificial viscosity into the Navier-Stokes equations. The complete theory for the Smagorinsky model shows also that the complexity of this model is smaller than the complexity of the Navier-Stokes equations. □

Remark 6.18. A model with a different kind of turbulent viscosity. Layton and Lewandowski [LL02] studied model (5.1) with

$$\nu_T = c\delta \left\|\mathbf{w} - \overline{\mathbf{w}}\right\|_2$$

and $A = 0$. The turbulent viscosity in this case looks like the Iliescu-Layton subgrid scale model (4.30), but a different filter than convolution with the Gaussian is considered in [LL02]. It could be proved that this model has a solution in a weak sense. Uniqueness and regularity of the solutions are still an open problem. □

6.2 The Taylor LES Model

This section summarises results on the existence and uniqueness of a weak solution of the Taylor LES model with Smagorinsky subgrid scale term. The presented results were obtained by Coletti [Col98].

The strong formulation of the Taylor LES model with Smagorinsky subgrid scale term (4.3) and homogeneous Dirichlet boundary conditions investigated in [Col98] is

$$\begin{aligned} \mathbf{w}_t - \nabla\cdot\left(\left(\nu + \nu_S \|\nabla \mathbf{w}\|_F\right)\nabla \mathbf{w}\right) + (\mathbf{w}\cdot\nabla)\,\mathbf{w} & \\ +\nabla r + \nabla\cdot\frac{\delta^2}{2\gamma}\left(\nabla\mathbf{w}\nabla\mathbf{w}^T\right) &= \mathbf{f} \quad \text{in } (0,T]\times\Omega, \\ \nabla\cdot\mathbf{w} &= 0 \quad \text{in } [0,T]\times\Omega, \\ \mathbf{w} &= \mathbf{0} \quad \text{on } [0,T]\times\Gamma, \\ \mathbf{w}(0,\cdot) &= \mathbf{w}_0 \text{ in } \Omega, \\ \int_\Omega r d\mathbf{x} &= 0 \quad \text{in } (0,T], \end{aligned} \tag{6.31}$$

with $\nu_S > 0$.

Let $W^{1,3}_{0,div}(\Omega)$ and V be defined as in (6.2) and (6.3). The weak formulation of (6.31) reads as follows:

Find $\mathbf{w} \in V$ such that $\mathbf{w}(0,\mathbf{x}) = \mathbf{w}_0 \in W^{1,3}_{0,div}(\Omega)$ and for all $\mathbf{v} \in V$

$$\int_0^T \Bigg[(\mathbf{w}_t + (\mathbf{w}\cdot\nabla)\mathbf{w}, \mathbf{v}) + ((\nu + \nu_S \|\nabla\mathbf{w}\|_F)\nabla\mathbf{w}, \nabla\mathbf{v}) \tag{6.32}$$

$$- \frac{\delta^2}{2\gamma}\left(\left(\nabla\mathbf{w}\nabla\mathbf{w}^T\right), \nabla\mathbf{v}\right)\Bigg] dt = \int_0^T (\mathbf{f}, \mathbf{v})\, dt.$$

The existence of a weak solution of (6.32) is proved by the Galerkin method described in Section 6.1. The most parts of the proof by Ladyzhenskaya can be adopted literally. The new aspects to prove are a priori error estimates like Lemmata 6.1 - 6.3 and the monotonicity of the operator $\mathbf{A} : L^3(\Omega) \to L^{3/2}(\Omega)$

$$\mathbf{A}(\nabla\mathbf{w}) = (\nu + \nu_S \|\nabla\mathbf{w}\|_F)\nabla\mathbf{w} - \frac{\delta^2}{2\gamma}\left(\nabla\mathbf{w}\nabla\mathbf{w}^T\right) \tag{6.33}$$

similar to Lemma 6.9. Since the analysis for the Taylor LES model will not be presented in detail and the a priori error estimates will not be used at other places, we like to skip them here and refer to Coletti [Col98, Theorems 14 and 15]. In contrast, the monotonicity of $\mathbf{A}$ is also important for analysing finite element discretisations of the Taylor LES model in Section 8.2.

Lemma 6.19. *Let*

$$\nu_S \geq \delta^2/\gamma. \tag{6.34}$$

Then, for arbitrary functions $\mathbf{w}', \mathbf{w}'' \in W^{1,3}(\Omega)$ *holds the inequality*

$$\int_\Omega (\mathbf{A}(\nabla\mathbf{w}') - \mathbf{A}(\nabla\mathbf{w}'')) : (\nabla\mathbf{w}' - \nabla\mathbf{w}'')\, d\mathbf{x} \geq \nu \|\nabla\mathbf{w}' - \nabla\mathbf{w}''\|^2_{L^2(\Omega)} \tag{6.35}$$

with the operator $\mathbf{A}$ *defined in (6.33).*

Proof. The way to prove (6.35) is the same as to prove (6.26). The term coming from the Taylor LES leads to

$$\frac{\partial}{\partial\tau}\left(\nabla\mathbf{w}\nabla\mathbf{w}^T\right)_{ij} = \frac{\partial}{\partial\tau}\left(\sum_{l=1}^d \frac{\partial\mathbf{w}_i^\tau}{\partial x_l}\frac{\partial\mathbf{w}_j^\tau}{\partial x_l}\right)$$

$$= \sum_{l=1}^d \left(\frac{\partial\mathbf{w}_i'}{\partial x_l} - \frac{\partial\mathbf{w}_i''}{\partial x_l}\right)\frac{\partial\mathbf{w}_j^\tau}{\partial x_l} + \frac{\partial\mathbf{w}_i^\tau}{\partial x_l}\left(\frac{\partial\mathbf{w}_j'}{\partial x_l} - \frac{\partial\mathbf{w}_j''}{\partial x_l}\right).$$

In comparison to Lemma 6.9, the additional term

$$-\frac{\delta^2}{2\gamma}\int_0^1 \sum_{i,j,l=1}^{d}\left[\frac{\partial \mathbf{w}_j^\tau}{\partial x_l}\left(\frac{\partial \mathbf{w}_i'}{\partial x_l}-\frac{\partial \mathbf{w}_i''}{\partial x_l}\right)+\frac{\partial \mathbf{w}_i^\tau}{\partial x_l}\left(\frac{\partial \mathbf{w}_j'}{\partial x_l}-\frac{\partial \mathbf{w}_j''}{\partial x_l}\right)\right]$$
$$\times\left(\frac{\partial \mathbf{w}_i'}{\partial x_j}-\frac{\partial \mathbf{w}_i''}{\partial x_j}\right)d\tau \qquad (6.36)$$

has to be estimated. Two applications of the Cauchy-Schwarz inequality prove

$$\sum_{i,j,k=1}^{n} a_{ik}b_{kj}c_{ji} \le \|A\|_F\,\|B\|_F\,\|C\|_F\,,$$

where A, B, C are $n\times n$ matrices. Using this inequality, (6.36) can be estimated from below by

$$-\frac{\delta^2}{\gamma}\int_0^1 \|\nabla \mathbf{w}^\tau\|_F\,\|\nabla \mathbf{w}'-\nabla \mathbf{w}''\|_F^2\,d\tau.$$

Altogether, one obtains

$$\begin{aligned}&(\mathbf{A}(\nabla \mathbf{w}')-\mathbf{A}(\nabla \mathbf{w}'')):(\nabla \mathbf{w}'-\nabla \mathbf{w}'')\\ &\ge \nu(\nabla \mathbf{w}'-\nabla \mathbf{w}''):(\nabla \mathbf{w}'-\nabla \mathbf{w}'')\\ &\quad+\int_0^1\left(\nu_S-\frac{\delta^2}{\gamma}\right)\|\nabla \mathbf{w}^\tau\|_F\,\|\nabla \mathbf{w}'-\nabla \mathbf{w}''\|_F^2\,d\tau,\end{aligned}$$

see the proof of Lemma 6.9. The second term is non-negative if $\nu_S \ge \delta^2/\gamma$ which proves the lemma. □

Remark 6.20. The dominance of the Smagorinsky subgrid scale term. Using the standard choices $\nu_S = c_S\delta^2$ and $\gamma = 6$ one obtains the condition $c_S \ge 1/6$. Thus, the Smagorinsky term dominates the Taylor LES model term. This contradicts the derivation of these terms where the latter is a model of a term which is formally of order δ^2 and the former is substituted for a term which is formally of order δ^4. The dominance of the Smagorinsky subgrid scale term is used in the analysis of the Taylor LES model. The additional Taylor LES term is considered as a small perturbation of the Smagorinsky model which can be treated within the framework of the analysis of the Smagorinsky model. Numerical tests, Section 10.2, suggest that a similar analytical result like the following Theorem 6.21 cannot be expected for the Taylor LES model with a non-dominating Smagorinsky subgrid scale term. □

The main result of Coletti [Col98] on the existence and uniqueness of a weak solution of the Taylor LES model is the following theorem.

Theorem 6.21. Existence, uniqueness and stability of a weak solution for all $T<\infty$. *If*

- $\mathbf{w}_0 \in W^{1,3}_{0,div}(\Omega)$,
- $\mathbf{f} \in L^2(0,T;L^2(\Omega))$, $\mathbf{f}_t \in L^2(0,T;L^2(\Omega))$,

- *(6.34) is fulfilled,*

then a weak solution $\mathbf{w} \in V$ *of (6.32) exists. The weak solution of (6.32) is unique in* V*. Let* $\mathbf{w}', \mathbf{w}'' \in V$ *solutions of (6.32) with different initial data and different right hand sides* $\mathbf{f}', \mathbf{f}''$*. Then*

$$\begin{aligned}
&\|\mathbf{w}' - \mathbf{w}''\|_{L^\infty(0,T;L^2(\Omega))} \\
&\le \left(\|\mathbf{w}'(0,\mathbf{x}) - \mathbf{w}''(0,\mathbf{x})\|_{L^2(\Omega)}^2 + \frac{1}{2c_1} \int_0^T \|\mathbf{f}' - \mathbf{f}''\|_{L^2(\Omega)}^2 \, dt \right) \\
&\quad \times \exp\left(c_2 \|\nabla \mathbf{w}'\|_{L^2(0,T;L^3(\Omega))}^2 + \frac{c_1}{2} T \right),
\end{aligned}$$

with $c_1, c_2 > 0$ *and* c_1 *can be chosen arbitrarily. If* $\mathbf{f}' = \mathbf{f}''$*, then*

$$\begin{aligned}
&\|\mathbf{w}' - \mathbf{w}''\|_{L^\infty(0,T;L^2(\Omega))} \\
&\le \|\mathbf{w}'(0,\mathbf{x}) - \mathbf{w}''(0,\mathbf{x})\|_{L^2(\Omega)}^2 \exp\left(c_2 \|\nabla \mathbf{w}'\|_{L^2(0,T;L^3(\Omega))}^2 \right).
\end{aligned}$$

Remark 6.22. The Taylor LES model without turbulent viscosity. The existence, uniqueness and stability of a strong solution of (6.31) for $\nu_S = 0$ and small data was proved by Coletti [Col97] or [Col98, Chapter 3]. In the existence theorem, it is assumed that $\partial\Omega$ is sufficiently smooth, that $\mathbf{w}_0$ satisfies certain compatibility conditions and that the smallness conditions

$$\begin{aligned}
\|\mathbf{w}_0\|_{H^3(\Omega)} &\le \varepsilon^2, & \|\mathbf{f}\|_{L^2(0,T;H^2(\Omega))} &\le \varepsilon^2, \\
\|\mathbf{f}\|_{L^\infty([0,T];H^1(\Omega))} &\le \varepsilon^2, & \|\mathbf{f}_t\|_{L^2(0,T;L^2(\Omega))} &\le \varepsilon^2
\end{aligned} \tag{6.37}$$

with $\varepsilon \in (0, \varepsilon_0]$ are fulfilled. Then, there exists a solution of (6.31) for $\nu_S = 0$ with

$$\begin{aligned}
\mathbf{w} &\in C^0\left([0,T]; H^3(\Omega)\right) \cap L^2\left(0,T; H^4(\Omega)\right), \\
r &\in C^0\left([0,T]; H^2(\Omega)\right) \cap L^2\left(0,T; H^3(\Omega) \cap L_0^2(\Omega)\right).
\end{aligned}$$

The proof is based on Schauder's fixed point theorem. Defining an appropriate set of functions, it is proved that a fixed point iteration converges, provided ε is sufficiently small. The restriction on ε arises essentially from the stability estimates for auxiliary problems. As example, the stationary Stokes problem

$$\begin{aligned}
-\nu\Delta\mathbf{v} + \nabla s &= \mathbf{f} \text{ in } \Omega, \\
\nabla\cdot\mathbf{v} &= 0 \text{ in } \Omega, \\
\mathbf{v} &= \mathbf{0} \text{ on } \Gamma, \\
\int_\Omega s d\mathbf{x} &= 0.
\end{aligned} \tag{6.38}$$

with $\mathbf{f} \in H^1(\Omega)$ and $\|\mathbf{f}\|_{H^1(\Omega)} \le \varepsilon^2$ is considered. The a priori error estimate

$$\|\mathbf{v}\|_{H^3(\Omega)} + \|\nabla s\|_{H^1(\Omega)} \le c \|\mathbf{f}\|_{H^1(\Omega)} \tag{6.39}$$

is used, e.g., see Girault and Raviart [GR86, Theorem I.5.4, Remark I.5.6]. The constant c in (6.39) is known to be of moderate size for $\nu = 1$. Denoting the constant for $\nu = 1$ by c_1, scaling arguments show $c = c_1/\nu$ for (6.38). The estimate (6.39) now implies

$$\|\mathbf{v}\|_{L^\infty(0,T;H^3(\Omega))} \le \frac{c_1}{\nu}\|\mathbf{f}\|_{L^\infty(0,T;H^1(\Omega))} \le \frac{c_1}{\nu}\varepsilon^2 = \left(\frac{c_1\varepsilon}{\nu}\right)\varepsilon.$$

In the proof, it is required $\|\mathbf{v}\|_{L^\infty(0,T;H^3(\Omega))} \le \varepsilon$ which is fulfilled if

$$\frac{c_1\varepsilon}{\nu} \le 1 \quad \Leftrightarrow \quad \varepsilon \le \frac{\nu}{c_1}.$$

That means, ε is of the size of ν. Since ν is small for turbulent flows, the smallness assumptions (6.37) are in general not given in applications. Uniqueness and stability are proved under essentially the same assumptions as existence. The proofs are based on the application of Gronwall's lemma.

Under more restrictive assumptions on the size of the data, the periodicity of the solution and the time independence of the solution for a time independent right hand side have been proved, too □

6.3 The Rational LES Model

The key in the analysis for proving the existence of a weak solution of the Smagorinsky model and the Taylor LES model with Smagorinsky subgrid scale term is the monotonicity of the operator **A** defined in (6.25) and (6.33), respectively. The monotonicity of those operators was proved in Lemma 6.9 and 6.19. We will show in this section that the ideas of these proofs do not carry over to the rational LES model.

Let the operator $\mathbf{A} : L^3(\Omega) \to L^{3/2}(\Omega)$ given by

$$\mathbf{A}(\nabla\mathbf{w}) = (\nu + \nu_S\|\nabla\mathbf{w}\|_F)\nabla\mathbf{w} - \frac{\delta^2}{2\gamma}\mathbf{B}\left(\nabla\mathbf{w}\nabla\mathbf{w}^T\right) \tag{6.40}$$

where the operator $\mathbf{B} : L^{3/2}(\Omega) \to L^{3/2}(\Omega)$ is linear and bounded. The adjoint operator is denoted by $\mathbf{B}^* : L^3(\Omega) \to L^3(\Omega)$. The Taylor LES model is obtained by choosing **B** as the identity operator

$$\begin{array}{cc} \mathbf{B} : L^{3/2}(\Omega) \to L^{3/2}(\Omega) & \mathbf{B}^* : L^3(\Omega) \to L^3(\Omega) \\ \mathbf{v} \to \mathbf{v} & \mathbf{v} \to \mathbf{v}. \end{array} \tag{6.41}$$

The rational LES model with convolution is given by choosing B as the convolution operator with g_δ

$$\begin{array}{cc} \mathbf{B} : L^{3/2}(\Omega) \to L^{3/2}(\Omega) & \mathbf{B}^* : L^3(\Omega) \to L^3(\Omega) \\ \mathbf{v} \to g_\delta * \mathbf{v} & \mathbf{v} \to g_\delta * \mathbf{v}. \end{array} \tag{6.42}$$

This convolution operator has to be understood in the sense described in Remark 4.9. In the rational LES model with auxiliary problem, $\mathbf{B}$ is defined by the solution of the auxiliary problem with homogeneous Neumann boundary conditions, see Remark 4.7,

$$\begin{aligned} \mathbf{B} : L^{3/2}(\Omega) &\to L^{3/2}(\Omega) \qquad & \mathbf{B}^* : L^3(\Omega) &\to L^3(\Omega) \\ \mathbf{v} &\to \left(I - \frac{\delta^2}{4\gamma}\right)^{-1}\mathbf{v} & \mathbf{v} &\to \left(I - \frac{\delta^2}{4\gamma}\right)^{-1}\mathbf{v}. \end{aligned} \tag{6.43}$$

Now, the techniques of proving Lemma 6.9 and 6.19 are applied to the operator $\mathbf{A}$ defined in (6.40). We use the same notations as in those lemmata. As in the proof of Lemma 6.19, one obtains

$$\frac{\partial}{\partial\tau}\mathbf{B}\left(\nabla\mathbf{w}\nabla\mathbf{w}^T\right)_{ij} = \mathbf{B}\left(\sum_{l=1}^{d}\left(\frac{\partial\mathbf{w}_i'}{\partial x_l} - \frac{\partial\mathbf{w}_i''}{\partial x_l}\right)\frac{\partial\mathbf{w}_j^\tau}{\partial x_l} + \frac{\partial\mathbf{w}_i^\tau}{\partial x_l}\left(\frac{\partial\mathbf{w}_j'}{\partial x_l} - \frac{\partial\mathbf{w}_j''}{\partial x_l}\right)\right).$$

Using this formula gives

$$\begin{aligned}
&\int_\Omega \left(\mathbf{A}(\nabla\mathbf{w}') - \mathbf{A}(\nabla\mathbf{w}'')\right) : (\nabla\mathbf{w}' - \nabla\mathbf{w}'')\,d\mathbf{x} \\
&\geq \int_\Omega \Bigg\{\nu(\nabla\mathbf{w}' - \nabla\mathbf{w}'') : (\nabla\mathbf{w}' - \nabla\mathbf{w}'') + \int_0^1 \Bigg(\nu_S \|\nabla\mathbf{w}^\tau\|_F \|\nabla\mathbf{w}' - \nabla\mathbf{w}''\|_F^2 \\
&\quad -\frac{\delta^2}{2\gamma}\sum_{i,j,l=1}^{d}\mathbf{B}\left(\left(\frac{\partial\mathbf{w}_i'}{\partial x_l} - \frac{\partial\mathbf{w}_i''}{\partial x_l}\right)\frac{\partial\mathbf{w}_j^\tau}{\partial x_l} + \frac{\partial\mathbf{w}_i^\tau}{\partial x_l}\left(\frac{\partial\mathbf{w}_j'}{\partial x_l} - \frac{\partial\mathbf{w}_j''}{\partial x_l}\right)\right) \\
&\quad \times\left(\frac{\partial\mathbf{w}_i'}{\partial x_j} - \frac{\partial\mathbf{w}_i''}{\partial x_j}\right)\Bigg)d\tau\Bigg\}d\mathbf{x} \\
&= \int_\Omega \Bigg\{\nu(\nabla\mathbf{w}' - \nabla\mathbf{w}'') : (\nabla\mathbf{w}' - \nabla\mathbf{w}'') + \int_0^1 \Bigg(\nu_S \|\nabla\mathbf{w}^\tau\|_F \|\nabla\mathbf{w}' - \nabla\mathbf{w}''\|_F^2 \\
&\quad -\frac{\delta^2}{2\gamma}\sum_{i,j,l=1}^{d}\left[\left(\frac{\partial\mathbf{w}_i'}{\partial x_l} - \frac{\partial\mathbf{w}_i''}{\partial x_l}\right)\frac{\partial\mathbf{w}_j^\tau}{\partial x_l} + \frac{\partial\mathbf{w}_i^\tau}{\partial x_l}\left(\frac{\partial\mathbf{w}_j'}{\partial x_l} - \frac{\partial\mathbf{w}_j''}{\partial x_l}\right)\right] \\
&\quad \times\mathbf{B}^*\left(\frac{\partial\mathbf{w}_i'}{\partial x_j} - \frac{\partial\mathbf{w}_i''}{\partial x_j}\right)\Bigg)d\tau\Bigg\}d\mathbf{x} \\
&\geq \int_\Omega \Bigg\{\nu(\nabla\mathbf{w}' - \nabla\mathbf{w}'') : (\nabla\mathbf{w}' - \nabla\mathbf{w}'') + \int_0^1 \Bigg(\nu_S \|\nabla\mathbf{w}^\tau\|_F \|\nabla\mathbf{w}' - \nabla\mathbf{w}''\|_F^2 \\
&\quad -\frac{\delta^2}{\gamma}\|\nabla\mathbf{w}^\tau\|_F \|\nabla\mathbf{w}' - \nabla\mathbf{w}''\|_F \|\mathbf{B}^*(\nabla\mathbf{w}' - \nabla\mathbf{w}'')\|_F\Bigg)d\tau\Bigg\}d\mathbf{x}
\end{aligned}$$

$$= \int_\Omega \left\{ \nu \left(\nabla \mathbf{w}' - \nabla \mathbf{w}''\right) : \left(\nabla \mathbf{w}' - \nabla \mathbf{w}''\right) + \left(\int_0^1 \|\nabla \mathbf{w}^\tau\|_F \, d\tau \right) \right.$$

$$\left. \times \|\nabla \mathbf{w}' - \nabla \mathbf{w}''\|_F \left(\nu_S \|\nabla \mathbf{w}' - \nabla \mathbf{w}''\|_F - \frac{\delta^2}{\gamma} \|\mathbf{B}^* \left(\nabla \mathbf{w}' - \nabla \mathbf{w}''\right)\|_F \right) \right\} d\mathbf{x}.$$

For the Taylor LES model, where $\mathbf{B}^*$ is the identity, it is obvious that the second term is non-negative if $\nu_S \geq \delta^2/\gamma$. This is a point property holding for every $\mathbf{x} \in \Omega$. But the operators $\mathbf{B}$ given in (6.42) and (6.43) do not allow such a local argument. Using these operators, there is a global dependence of all points $\mathbf{x} \in \Omega$ among each other. It is easy to see that $\|\left(\nabla \mathbf{w}' - \nabla \mathbf{w}''\right)(\mathbf{x})\|_F = 0$ and $\|\mathbf{B}^* \left(\nabla \mathbf{w}' - \nabla \mathbf{w}''\right)(\mathbf{x})\|_F \neq 0$ can happen for $\mathbf{x} \in \Omega$. Thus, there is no choice of ν_S which guarantees the point-wise non-negativity of the third factor in the second term.

The considerations above do not state that the operator $\mathbf{A}$ defined in (6.40) with the operator $\mathbf{B}$ given in (6.42) or (6.43) is not monotone. It is only shown that in these cases the arguments used for proving the monotonicity of (6.25) and (6.33) fail. Up to now, a counter example for the monotonicity property is not known.

Remark 6.23. The rational LES model without subgrid scale term in a space periodic setting and small time intervals. The existence and uniqueness of a weak solution of the rational LES model with auxiliary problem without subgrid scale term, $\nu_T = 0$ in (5.1), have been studied by Berselli et al. [BGLI02]. They consider the case of a space periodic setting, i.e. $\Omega = (0, L)^3$ and periodic boundary conditions on $\partial\Omega$. Periodic boundary conditions are applied in the auxiliary problem as well. Berselli et al. could prove the existence and uniqueness of a solution in a certain space for small time intervals following the ideas developed in the book by Temam [Tem95]. The proofs uses the Galerkin method in a similar way as described in Section 6.1.2. Instead of testing with the basis function $\mathbf{v}^l$ like in (6.14), the finite dimensional system is tested with $\mathbf{w}^n - \delta^2/(4\gamma)\Delta \mathbf{w}^n$. Choosing these test functions, one can take advantage of special properties of the Laplacian in the space periodic setting, see Temam [Tem95, Section 2.2]. In contrast to Lemma 6.4, the existence of a solution $\mathbf{w}^n$ can only be established for small $T = \mathcal{O}\left(\delta^4\right)$.The proofs continues by deriving an a priori error estimate which controls even the $H^2(\Omega)$ norm of $\mathbf{w}^n$. The next step consists in proving the (weak) convergence of a subsequence of $\{\mathbf{w}^n\}$ in several norms to a function $\mathbf{w}$, similar to Lemma 6.5. With these convergence considerations it is shown that $\mathbf{w}$ solves the rational LES model with auxiliary problem without subgrid scale term. Since there is no non-linear viscous term involved in this model, monotonicity considerations are not necessary.

The numerical tests in Sections 10.3.3 and 11.1 indicate that long time solutions of the rational LES model without subgrid scale term in general either cannot be expected for small ν or they are instable with respects to small perturbations as they always occur in numerical methods. □

7

Discretisation of the LES Models

This chapter deals with the discretisation of the LES models of the Navier-Stokes equations

$$
\begin{aligned}
\mathbf{w}_t - \nabla \cdot ((2\nu + \nu_T)\,\mathbb{D}(\mathbf{w})) + (\mathbf{w} \cdot \nabla)\,\mathbf{w} & \\
+\nabla r + \nabla \cdot \frac{\delta^2}{2\gamma}\left(A\left(\nabla \mathbf{w} \nabla \mathbf{w}^T\right)\right) &= \mathbf{f} \quad \text{in } (0,T] \times \Omega, \\
\nabla \cdot \mathbf{w} &= 0 \quad \text{in } [0,T] \times \Omega, \\
\mathbf{w}(0,\cdot) &= \mathbf{w}_0 \text{ in } \Omega,
\end{aligned}
\tag{7.1}
$$

where A is given by the approximation of the Fourier transform of the Gaussian filter, ν_T is the turbulent viscosity and to simplify the notations we use $\mathbf{f}$ instead of $\overline{\mathbf{f}}$. System (7.1) has to be completed with boundary conditions. Depending on the boundary conditions, the additive constant of the pressure has to be fixed, see Section 5.3.

For discretising system (7.1), we apply the following strategy:

1. *Discretisation of (7.1) in time.* We will use a second order implicit time stepping scheme. The time discretisation leads in each discrete time step to a non-linear system of equations.
2. *Variational formulation and linearisation.* The non-linear system of equations is reformulated as variational problem and the non-linear variational problem is linearised by a fixed point iteration.
3. *Discretisation of the linear systems in space.* The linear system of equations arising in each step of the fixed point iteration is discretised by a finite element discretisation using an inf-sup stable pair of finite element spaces.

The individual steps in this strategy are described in detail in the following sections.

The discretisations described in this chapter have been implemented in the code *MooNMD* (*M*athematics and *o*bject-*o*riented *N*umerics in *M*ag*D*eburg[1]), for details on the philosophy behind this code see John and Matthies [JM03].

[1] *MD* is the abbreviation of Magdeburg on the license plate of German cars.

7.1 Discretisation in Time by the Crank-Nicolson or the Fractional-Step θ-Scheme

A number of numerical studies show that implicit time stepping schemes of second order perform very good in incompressible flow computations, e.g., see the evaluation of benchmark computations for laminar flow problems, Schäfer and Turek [ST96a], or tests presented in the book by Turek [Tur99]. For this reason, we are going to present here two of the most commonly used schemes, the Crank-Nicolson scheme and (two variants of) the fractional-step θ-scheme. A detailed discussion of these discretisations applied in the numerical solution of incompressible Navier-Stokes equations can be found in the book by Turek [Tur99, Chapter 3.2].

Let Δt_n be the current time step from t_{n-1} to t_n, i.e. $\Delta t_n = t_n - t_{n-1}$. We denote quantities at time level t_k by a subscript k. To describe the time stepping scheme for the LES models of the incompressible Navier-Stokes equations, we introduce a general time step of the form

$$
\begin{aligned}
&\mathbf{w}_k + \theta_1 \Delta t_n \Bigg[-\nabla \cdot ((2\nu + \nu_T)\,\mathbb{D}(\mathbf{w}_k)) + (\mathbf{w}_k \cdot \nabla)\,\mathbf{w}_k \\
&\quad + \nabla \cdot \frac{\delta^2}{2\gamma}\left(A\left(\nabla \mathbf{w}_k \nabla \mathbf{w}_k^T\right)\right)\Bigg] + \Delta t_k \nabla r_k \\
&= \mathbf{w}_{k-1} - \theta_2 \Delta t_n \Bigg[-\nabla \cdot (2\nu + \nu_T)\,\mathbb{D}(\mathbf{w}_{k-1}) + (\mathbf{w}_{k-1} \cdot \nabla)\,\mathbf{w}_{k-1} \\
&\quad + \nabla \cdot \frac{\delta^2}{2\gamma}\left(A\left(\nabla \mathbf{w}_{k-1} \nabla \mathbf{w}_{k-1}^T\right)\right)\Bigg] + \theta_3 \Delta t_n \mathbf{f}_{k-1} + \theta_4 \Delta t_n \mathbf{f}_k, \\
&\nabla \cdot \mathbf{w}_k = 0,
\end{aligned}
\tag{7.2}
$$

with the parameters $\theta_1, \ldots, \theta_4$. The time step (7.2) allows the implementation of a number of time stepping schemes by one single formula and the choice between the schemes by setting four parameters.

Three well known one-step θ-schemes are obtained by appropriate choices of these parameters, see Table 7.1.

Table 7.1. One-step θ-schemes

	θ_1	θ_2	θ_3	θ_4	t_{k-1}	t_k	Δt_k
forward Euler scheme	0	1	1	0	t_{n-1}	t_n	Δt_n
backward Euler scheme	1	0	0	1	t_{n-1}	t_n	Δt_n
Crank-Nicolson scheme	0.5	0.5	0.5	0.5	t_{n-1}	t_n	Δt_n

The fractional-step θ-scheme is obtained by three steps of form (7.2). We want to present two variants of this scheme. Let

$$\theta = 1 - \frac{\sqrt{2}}{2}, \quad \tilde{\theta} = 1 - 2\theta, \quad \tau = \frac{\tilde{\theta}}{1-\theta}, \quad \eta = 1 - \tau.$$

The two variants are presented in Table 7.2. Variant 2 requires the evalua-

Table 7.2. The two variants of the fractional-step θ-schemes

	θ_1	θ_2	θ_3	θ_4	t_{k-1}	t_k	Δt_k
variant 1	$\tau\theta$	$\eta\theta$	$\eta\theta$	$\tau\theta$	t_{n-1}	$t_{n-1} + \theta \Delta t_n$	$\theta \Delta t_n$
	$\eta\tilde{\theta}$	$\tau\tilde{\theta}$	$\tau\tilde{\theta}$	$\eta\tilde{\theta}$	$t_{n-1} + \theta \Delta t_n$	$t_n - \theta \Delta t_n$	$\tilde{\theta} \Delta t_n$
	$\tau\theta$	$\eta\theta$	$\eta\theta$	$\tau\theta$	$t_n - \theta \Delta t_n$	t_n	$\theta \Delta t_n$
variant 2	$\tau\theta$	$\eta\theta$	θ	0	t_{n-1}	$t_{n-1} + \theta \Delta t_n$	$\theta \Delta t_n$
	$\eta\tilde{\theta}$	$\tau\tilde{\theta}$	0	$\tilde{\theta}$	$t_{n-1} + \theta \Delta t_n$	$t_n - \theta \Delta t_n$	$\tilde{\theta} \Delta t_n$
	$\tau\theta$	$\eta\theta$	θ	0	$t_n - \theta \Delta t_n$	t_n	$\theta \Delta t_n$

tion of $\mathbf{f}$ only at the times t_{n-1} and $t_n - \theta \Delta t_n$ whereas variant 1 needs the evaluation of $\mathbf{f}$ in addition at $t_{n-1} + \theta \Delta t_n$ and at t_n. Both variants are second order schemes but variant 2 does not integrate second order polynomials (with respect to t) exactly. However, most of other fundamental properties, like stability, are the same for both variants. If not mentioned otherwise, we will use variant 1 in the numerical tests.

There are a number of investigations of the time discretisations introduced above applied to the Navier-Stokes equations, see Gresho and Sani [GS00, Section 3.16] or Emmrich [Emm01, Section 4.1] for a survey of the present state of art. The Crank-Nicolson scheme was studied by Temam [Tem77], Heywood and Rannacher [HR90] and Bause [Bau97] at the already spatially discretised Navier-Stokes equations (with a finite element method). One can prove, under a number of assumptions on the smoothness of the data, that the error between the time discrete and the time-continuous finite element solution in $L^2(\Omega)$ behaves like $(\Delta t)^2$ for the equidistant time step Δt. The fractional-step θ-scheme was investigated analytically by Temam [Tem77], Klouček and Rys [KR94] and Müller-Urbaniak [MU94]. A second order error estimate similar to the Crank-Nicolson scheme was proved in [MU94]. Investigations of time discretisations to systems of form (7.1) are not yet available.

The Crank-Nicolson and the fractional-step θ-scheme are already well tested and compared for the Navier-Stokes equations, see Emmrich [Emm01] for an overview. The Crank-Nicolson scheme is A-stable whereas the fractional-step θ-scheme is even strongly A-stable. That means, the Crank-Nicolson scheme may lead to numerical oscillations in problems with rough initial data or boundary conditions, as observed in the direct numerical simulations in Section 11.1. These oscillations are damped out only if sufficiently small time steps are used. Compared to the fractional-step θ-scheme, often a smaller time step has to be chosen for the Crank-Nicolson scheme to ensure robustness. If the time steps are sufficiently small, both schemes lead in general to solutions with no significant differences and with comparable numerical costs, Turek [Tur99]. Altogether, the fractional-step θ-scheme is considered at the moment as one of the best time stepping schemes for incompressible flow problems on the basis of accuracy and reliability, [Tur99].

7.2 The Variational Formulation and the Linearisation of the Time-Discrete Problem

The solution of (7.2) will be approximated by a finite element method. Finite element methods are popular and successful spatial discretisations used in computational fluid dynamics. The basis of the finite element method is a variational formulation of (7.2). Let (V_a, Q_a) be a pair of ansatz spaces with functions defined on Ω and (V_t, Q_t) a pair of test spaces. The choice of appropriate spaces is discussed in Section 5.3. The derivation of the variational problem is done in the usual way by multiplying the equations in (7.2) with test functions, integrating on Ω and applying integration by parts. The variational problem is to find $(\mathbf{w}_k, r_k) \in (V_a, Q_a)$ such that for all $(\mathbf{v}, q) \in (V_t, Q_t)$

$$
\begin{aligned}
&(\mathbf{w}_k, \mathbf{v}) + \theta_1 \Delta t_n \Bigg[\left((2\nu + \nu_T) \mathbb{D}(\mathbf{w}_k), \mathbb{D}(\mathbf{v}) \right) + \left((\mathbf{w}_k \cdot \nabla) \mathbf{w}_k, \mathbf{v} \right) \\
&\quad - \left(\frac{\delta^2}{2\gamma} \left(A \left(\nabla \mathbf{w}_k \nabla \mathbf{w}_k^T \right) \right), \nabla \mathbf{v} \right) \\
&\quad + \int_{\partial\Omega} \left(-(2\nu + \nu_T) \mathbb{D}(\mathbf{w}_k) + \frac{\delta^2}{2\gamma} \left(A \left(\nabla \mathbf{w}_k \nabla \mathbf{w}_k^T \right) \right) \right) \mathbf{n}_{\partial\Omega} \cdot \mathbf{v} ds \Bigg] \\
&\quad - \Delta t_k \left(r_k, \nabla \cdot \mathbf{v} \right) + \Delta t_k \int_{\partial\Omega} r_k \mathbf{v} \cdot \mathbf{n}_{\partial\Omega} ds \\
&= (\mathbf{w}_{k-1}, \mathbf{v}) + \theta_3 \Delta t_n (\mathbf{f}_{k-1}, \mathbf{v}) + \theta_4 \Delta t_n (\mathbf{f}_k, \mathbf{v}) \\
&\quad - \theta_2 \Delta t_n \Bigg[\left((2\nu + \nu_T) \mathbb{D}(\mathbf{w}_{k-1}), \mathbb{D}(\mathbf{v}) \right) + \left((\mathbf{w}_{k-1} \cdot \nabla) \mathbf{w}_{k-1}, \mathbf{v} \right) \\
&\quad - \left(\frac{\delta^2}{2\gamma} \left(A \left(\nabla \mathbf{w}_{k-1} \nabla \mathbf{w}_{k-1}^T \right) \right), \nabla \mathbf{v} \right) \\
&\quad + \int_{\partial\Omega} \left(-(2\nu + \nu_T) \mathbb{D}(\mathbf{w}_{k-1}) + \frac{\delta^2}{2\gamma} \left(A \left(\nabla \mathbf{w}_{k-1} \nabla \mathbf{w}_{k-1}^T \right) \right) \right) \mathbf{n}_{\partial\Omega} \cdot \mathbf{v} ds \Bigg], \\
&0 = (\nabla \cdot \mathbf{w}_k, q).
\end{aligned} \tag{7.3}
$$

Noting that $(\theta_1 + \theta_2) \Delta t_n = \Delta t_k$, we use the approximation

$$
\begin{aligned}
\Delta t_k \int_{\partial\Omega} r_k \mathbf{v} \cdot \mathbf{n}_{\partial\Omega} ds &= \theta_1 \Delta t_n \int_{\partial\Omega} r_k \mathbf{v} \cdot \mathbf{n}_{\partial\Omega} ds + \theta_2 \Delta t_n \int_{\partial\Omega} r_k \mathbf{v} \cdot \mathbf{n}_{\partial\Omega} ds \\
&\approx \theta_1 \Delta t_n \int_{\partial\Omega} r_k \mathbf{v} \cdot \mathbf{n}_{\partial\Omega} ds + \theta_2 \Delta t_n \int_{\partial\Omega} r_{k-1} \mathbf{v} \cdot \mathbf{n}_{\partial\Omega} ds
\end{aligned}
$$

and considering instead of (7.3) the problem

$$
\begin{aligned}
&(\mathbf{w}_k, \mathbf{v}) + \theta_1 \Delta t_n \Bigg[((2\nu + \nu_T)\,\mathbb{D}(\mathbf{w}_k), \mathbb{D}(\mathbf{v})) + ((\mathbf{w}_k \cdot \nabla)\,\mathbf{w}_k, \mathbf{v}) \\
&\quad - \left(\frac{\delta^2}{2\gamma} \left(A \left(\nabla \mathbf{w}_k \nabla \mathbf{w}_k^T \right)\right), \nabla \mathbf{v} \right) \\
&\quad + \int_{\partial\Omega} \left(-(2\nu + \nu_T)\,\mathbb{D}(\mathbf{w}_k) + \frac{\delta^2}{2\gamma} \left(A \left(\nabla \mathbf{w}_k \nabla \mathbf{w}_k^T \right)\right) + r_k \mathbb{I} \right) \mathbf{n}_{\partial\Omega} \cdot \mathbf{v} ds \Bigg] \\
&\quad - \Delta t_k \left(r_k, \nabla \cdot \mathbf{v} \right) \\
&= (\mathbf{w}_{k-1}, \mathbf{v}) + \theta_3 \Delta t_n (\mathbf{f}_{k-1}, \mathbf{v}) + \theta_4 \Delta t_n (\mathbf{f}_k, \mathbf{v}) \qquad (7.4) \\
&\quad - \theta_2 \Delta t_n \Bigg[((2\nu + \nu_T)\,\mathbb{D}(\mathbf{w}_{k-1}), \mathbb{D}(\mathbf{v})) + ((\mathbf{w}_{k-1} \cdot \nabla)\,\mathbf{w}_{k-1}, \mathbf{v}) \\
&\quad - \left(\frac{\delta^2}{2\gamma} \left(A \left(\nabla \mathbf{w}_{k-1} \nabla \mathbf{w}_{k-1}^T \right)\right), \nabla \mathbf{v} \right) \\
&\quad + \int_{\partial\Omega} \Bigg(-(2\nu + \nu_T)\,\mathbb{D}(\mathbf{w}_{k-1}) + \frac{\delta^2}{2\gamma} \left(A \left(\nabla \mathbf{w}_{k-1} \nabla \mathbf{w}_{k-1}^T \right)\right) \\
&\quad + r_{k-1} \mathbb{I} \Bigg) \mathbf{n}_{\partial\Omega} \cdot \mathbf{v} ds \Bigg], \\
0 &= (\nabla \cdot \mathbf{w}_k, q) .
\end{aligned}
$$

The appearance of the boundary integrals in the variational formulation depends on the prescribed boundary conditions. The integrals vanish on all parts of the boundary with Dirichlet or outflow boundary conditions. For slip with friction and penetration with resistance boundary conditions, the boundary integral on Γ_{sfpr} can be transformed in the form given in (5.10). Thus, system (7.4) can be written as

$$
\begin{aligned}
&(\mathbf{w}_k, \mathbf{v}) + \theta_1 \Delta t_n \Bigg[((2\nu + \nu_T)\,\mathbb{D}(\mathbf{w}_k), \mathbb{D}(\mathbf{v})) + ((\mathbf{w}_k \cdot \nabla)\,\mathbf{w}_k, \mathbf{v}) \\
&\quad - \left(\frac{\delta^2}{2\gamma} \left(A \left(\nabla \mathbf{w}_k \nabla \mathbf{w}_k^T \right)\right), \nabla \mathbf{v} \right) \\
&\quad + \int_{\Gamma_{\text{sfpr}}} \left(\alpha^{-1} (\mathbf{w}_k \cdot \mathbf{n}_{\partial\Omega}) (\mathbf{v} \cdot \mathbf{n}_{\partial\Omega}) + \beta \sum_{i=1}^{d-1} (\mathbf{w}_k \cdot \boldsymbol{\tau}_i) (\mathbf{v} \cdot \boldsymbol{\tau}_i) \right) ds \Bigg] \\
&\quad - \Delta t_k \left(r_k, \nabla \cdot \mathbf{v} \right)
\end{aligned}
$$

$$
\begin{aligned}
&= (\mathbf{w}_{k-1}, \mathbf{v}) + \theta_3 \Delta t_n (\mathbf{f}_{k-1}, \mathbf{v}) + \theta_4 \Delta t_n (\mathbf{f}_k, \mathbf{v}) \\
&\quad - \theta_2 \Delta t_n \Big[((2\nu + \nu_T) \mathbb{D}(\mathbf{w}_{k-1}), \mathbb{D}(\mathbf{v})) + ((\mathbf{w}_{k-1} \cdot \nabla) \mathbf{w}_{k-1}, \mathbf{v}) \\
&\quad - \left(\frac{\delta^2}{2\gamma} (A(\nabla \mathbf{w}_{k-1} \nabla \mathbf{w}_{k-1}^T)), \nabla \mathbf{v} \right) \\
&\quad + \int_{\Gamma_{\text{sfpr}}} \left(\alpha^{-1} (\mathbf{w}_{k-1} \cdot \mathbf{n}_{\partial\Omega})(\mathbf{v} \cdot \mathbf{n}_{\partial\Omega}) + \beta \sum_{i=1}^{d-1} (\mathbf{w}_{k-1} \cdot \boldsymbol{\tau}_i)(\mathbf{v} \cdot \boldsymbol{\tau}_i) \right) ds \Big], \\
0 &= (\nabla \cdot \mathbf{w}_k, q).
\end{aligned}
$$

Problem (7.5) possesses the following non-linearities:

- the non-linear viscous term $(\nu_T \mathbb{D}(\mathbf{w}_k), \mathbb{D}(\mathbf{v}))$ since in general ν_T depends on $\mathbf{w}_k$,
- the non-linear convective term $((\mathbf{w}_k \cdot \nabla) \mathbf{w}_k, \mathbf{v})$,
- the non-linear LES model $\left(A(\nabla \mathbf{w}_k \nabla \mathbf{w}_k^T), \nabla \mathbf{v}\right)$.

In the computations presented in this monograph, we circumvent the third non-linearity by using the approximation

$$
A(\nabla \mathbf{w}_k \nabla \mathbf{w}_k^T) \approx A(\nabla \mathbf{w}_{k-1} \nabla \mathbf{w}_{k-1}^T).
$$

That means, the operator A is applied to the solution of the previous sub time step and the result is kept unchanged during the non-linear iteration in the current sub time step. The simplified treatment of the LES model is mainly motivated by the large numerical costs which might be connected with the application of the operator A, especially in the rational LES model with convolution. In addition, this approach avoids the evaluation of the LES term on coarser levels of the multigrid method which will be used as solver for the discrete saddle point problems, see Section 9.3. For the non-linear viscous term, we consider the Smagorinsky model (4.3) and the Iliescu-Layton model (4.31). Whereas the Smagorinsky model will be treated completely non-linear

$$
\nu_T(\mathbf{w}_k) = c_S \delta^2 \|\mathbb{D}(\mathbf{w}_k)\|_F,
$$

the semi-implicit form of the Iliescu-Layton model

$$
\nu_T(\mathbf{w}_k, \mathbf{w}_{k-1}) = c_S \delta \left\| \mathbf{w}_k - \left(I - \frac{\delta^2}{4\gamma} \Delta \right)^{-1} \mathbf{w}_{k-1} \right\|_2 \tag{7.5}
$$

is used. The advantage of the semi-implicit form (7.5) is that the auxiliary problem has to be solved only once in each discrete time t_k for each component of the velocity vector. These two simplifications, together with $(\theta_1 + \theta_2) \Delta t_n = \Delta t_k$, lead to the final form of the variational problem considered in this monograph: find $(\mathbf{w}_k, r_k) \in (V_a, Q_a)$ such that for all $(\mathbf{v}, q) \in (V_t, Q_t)$

$$
\begin{aligned}
&(\mathbf{w}_k, \mathbf{v}) + \theta_1 \Delta t_n \Bigg[\left((2\nu + \nu_T(\mathbf{w}_k, \mathbf{w}_{k-1})) \mathbb{D}(\mathbf{w}_k), \mathbb{D}(\mathbf{v}) \right) + \left((\mathbf{w}_k \cdot \nabla) \mathbf{w}_k, \mathbf{v} \right) \\
&\quad + \int_{\Gamma_{\text{sfpr}}} \left(\alpha^{-1} (\mathbf{w}_k \cdot \mathbf{n}_{\partial\Omega})(\mathbf{v} \cdot \mathbf{n}_{\partial\Omega}) + \beta \sum_{i=1}^{d-1} (\mathbf{w}_k \cdot \boldsymbol{\tau}_i)(\mathbf{v} \cdot \boldsymbol{\tau}_i) \right) ds \Bigg] \\
&\quad - \Delta t_k (r_k, \nabla \cdot \mathbf{v}) \\
&= (\mathbf{w}_{k-1}, \mathbf{v}) + \theta_3 \Delta t_n (\mathbf{f}_{k-1}, \mathbf{v}) + \theta_4 \Delta t_n (\mathbf{f}_k, \mathbf{v}) \qquad (7.6) \\
&\quad - \theta_2 \Delta t_n \Bigg[\left((2\nu + \nu_T(\mathbf{w}_{k-1})) \mathbb{D}(\mathbf{w}_{k-1}), \mathbb{D}(\mathbf{v}) \right) + \left((\mathbf{w}_{k-1} \cdot \nabla) \mathbf{w}_{k-1}, \mathbf{v} \right) \\
&\quad + \int_{\Gamma_{\text{sfpr}}} \left(\alpha^{-1} (\mathbf{w}_{k-1} \cdot \mathbf{n}_{\partial\Omega})(\mathbf{v} \cdot \mathbf{n}_{\partial\Omega}) + \beta \sum_{i=1}^{d-1} (\mathbf{w}_{k-1} \cdot \boldsymbol{\tau}_i)(\mathbf{v} \cdot \boldsymbol{\tau}_i) \right) ds \Bigg] \\
&\quad + \Delta t_k \left(\frac{\delta^2}{2\gamma} \left(A \left(\nabla \mathbf{w}_{k-1} \nabla \mathbf{w}_{k-1}^T \right) \right), \nabla \mathbf{v} \right), \\
0 &= (\nabla \cdot \mathbf{w}_k, q).
\end{aligned}
$$

System (7.6) is solved iteratively starting with an initial guess $\left(\mathbf{w}_k^0, r_k^0\right)$. Given $(\mathbf{w}_k^m, r_k^m)$, the iterate $\left(\mathbf{w}_k^{m+1}, r_k^{m+1}\right)$ is computed by solving

$$
\begin{aligned}
&\left(\mathbf{w}_k^{m+1}, \mathbf{v}\right) + \theta_1 \Delta t_n \Bigg[\left((2\nu + \nu_T(\mathbf{w}_k^m, \mathbf{w}_{k-1})) \mathbb{D}\left(\mathbf{w}_k^{m+1}\right), \mathbb{D}(\mathbf{v}) \right) \\
&\quad + \left((\mathbf{w}_k^m \cdot \nabla) \mathbf{w}_k^{m+1}, \mathbf{v} \right) \\
&\quad + \int_{\Gamma_{\text{sfpr}}} \left(\alpha^{-1} \left(\mathbf{w}_k^{m+1} \cdot \mathbf{n}_{\partial\Omega}\right)(\mathbf{v} \cdot \mathbf{n}_{\partial\Omega}) + \beta \sum_{i=1}^{d-1} \left(\mathbf{w}_k^{m+1} \cdot \boldsymbol{\tau}_i\right)(\mathbf{v} \cdot \boldsymbol{\tau}_i) \right) ds \Bigg] \\
&\quad - \Delta t_k \left(r_k^{m+1}, \nabla \cdot \mathbf{v}\right) \\
&= (\mathbf{w}_{k-1}, \mathbf{v}) + \theta_3 \Delta t_n (\mathbf{f}_{k-1}, \mathbf{v}) + \theta_4 \Delta t_n (\mathbf{f}_k, \mathbf{v}) \qquad (7.7) \\
&\quad - \theta_2 \Delta t_n \Bigg[\left((2\nu + \nu_T(\mathbf{w}_{k-1})) \mathbb{D}(\mathbf{w}_{k-1}), \mathbb{D}(\mathbf{v}) \right) + \left((\mathbf{w}_{k-1} \cdot \nabla) \mathbf{w}_{k-1}, \mathbf{v} \right) \\
&\quad + \int_{\Gamma_{\text{sfpr}}} \left(\alpha^{-1} (\mathbf{w}_{k-1} \cdot \mathbf{n}_{\partial\Omega})(\mathbf{v} \cdot \mathbf{n}_{\partial\Omega}) + \beta \sum_{i=1}^{d-1} (\mathbf{w}_{k-1} \cdot \boldsymbol{\tau}_i)(\mathbf{v} \cdot \boldsymbol{\tau}_i) \right) ds \Bigg] \\
&\quad + \Delta t_k \left(\frac{\delta^2}{2\gamma} \left(A \left(\nabla \mathbf{w}_{k-1} \nabla \mathbf{w}_{k-1}^T \right) \right), \nabla \mathbf{v} \right), \\
0 &= \left(\nabla \cdot \mathbf{w}_k^{m+1}, q\right),
\end{aligned}
$$

$m = 0, 1, 2, \ldots$. That means, the linearisation is done by a fixed point iteration. If not mentioned otherwise, the initial guess is chosen to be the solution of the previous time step $\left(\mathbf{w}_k^0, r_k^0\right) = (\mathbf{w}_{k-1}, r_{k-1})$.

7.3 The Discretisation in Space

The linear system (7.7) is discretised by the Galerkin finite element method. Choosing finite element ansatz spaces (V_a^h, Q_a^h) and test spaces (V_t^h, Q_t^h), the goal is to find an approximation $(\mathbf{w}^h, r^h) \in (V_a^h, Q_a^h)$ to $(\mathbf{w}_k^{m+1}, r_k^{m+1})$ such that for all $(\mathbf{v}^h, q^h) \in (V_t^h, Q_t^h)$

$$
\begin{aligned}
&(\mathbf{w}^h, \mathbf{v}^h) + \theta_1 \Delta t_n \Bigg[\left(\left(2\nu + \nu_T\left(\mathbf{w}_{\text{old}}^h, \mathbf{w}_{k-1}^h\right)\right) \mathbb{D}\left(\mathbf{w}^h\right), \mathbb{D}\left(\mathbf{v}^h\right)\right) \\
&\quad + \left(\left(\mathbf{w}_{\text{old}}^h \cdot \nabla\right) \mathbf{w}^h, \mathbf{v}^h\right) \\
&\quad + \int_{\Gamma_{\text{sfpr}}} \left(\alpha^{-1} \left(\mathbf{w}^h \cdot \mathbf{n}_{\partial\Omega}\right)\left(\mathbf{v}^h \cdot \mathbf{n}_{\partial\Omega}\right) + \beta \sum_{i=1}^{d-1} \left(\mathbf{w}^h \cdot \boldsymbol{\tau}_i\right)\left(\mathbf{v}^h \cdot \boldsymbol{\tau}_i\right) \right) ds \Bigg] \\
&\quad - \Delta t_k \left(r^h, \nabla \cdot \mathbf{v}^h\right) \\
&= \left(\mathbf{w}_{k-1}^h, \mathbf{v}^h\right) + \theta_3 \Delta t_n \left(\mathbf{f}_{k-1}^h, \mathbf{v}^h\right) + \theta_4 \Delta t_n \left(\mathbf{f}_k^h, \mathbf{v}^h\right) \qquad (7.8) \\
&\quad - \theta_2 \Delta t_n \Bigg[\left(\left(2\nu + \nu_T\left(\mathbf{w}_{k-1}^h\right)\right) \mathbb{D}\left(\mathbf{w}_{k-1}^h\right), \mathbb{D}\left(\mathbf{v}^h\right)\right) + \left(\left(\mathbf{w}_{k-1}^h \cdot \nabla\right) \mathbf{w}_{k-1}^h, \mathbf{v}^h\right) \\
&\quad + \int_{\Gamma_{\text{sfpr}}} \left(\alpha^{-1} \left(\mathbf{w}_{k-1}^h \cdot \mathbf{n}_{\partial\Omega}\right)\left(\mathbf{v}^h \cdot \mathbf{n}_{\partial\Omega}\right) + \beta \sum_{i=1}^{d-1} \left(\mathbf{w}_{k-1}^h \cdot \boldsymbol{\tau}_i\right)\left(\mathbf{v}^h \cdot \boldsymbol{\tau}_i\right) \right) ds \Bigg] \\
&\quad + \Delta t_k \left(\frac{\delta^2}{2\gamma} \left(A^h \left(\nabla \mathbf{w}_{k-1}^h \left(\nabla \mathbf{w}_{k-1}^h\right)^T \right)\right), \nabla \mathbf{v}^h \right) \\
&0 = \left(\nabla \cdot \mathbf{w}^h, q^h\right).
\end{aligned}
$$

In (7.8), $\mathbf{w}_{\text{old}}^h$ is the current finite element approximation of the velocity, $\mathbf{w}_{k-1}^h$ the finite element solution for the velocity from the previous sub time step, $\mathbf{f}_{k-1}^h, \mathbf{f}_k^h$ finite element approximations of $\mathbf{f}$ at the previous and current sub time steps, respectively, and A^h is a discrete approximation of A as described in Sections 7.7 and 7.8.

Remark 7.1. Approximation of curvilinear domains. If Ω is a domain with a boundary which does not consists of straight lines or plain faces only, it has to be approximated first by a domain Ω^h which allows the application of the finite element method. Strictly speaking, all inner products in (7.8) have to be formulated on Ω^h and the possible new boundary part Γ_{sfpr}^h. However, in order not to overload the notations, we will keep the notations of (7.8) also for approximations of Ω. □

In the application of the finite element method, the spaces $V_a^h, Q_a^h, V_t^h, Q_t^h$ are equipped with bases $\{\mathbf{v}_i^a\}_{i=1}^{dN_v}, \{q_i^a\}_{i=1}^{N_p}, \ldots$, and the unknown functions are represented by linear combinations

$$
\mathbf{w}^h = \sum_{i=1}^{dN_v} w_i \mathbf{v}_i^a, \quad r^h = \sum_{i=1}^{N_p} r_i q_i^a.
$$

Here, N_v is the dimension of one component of the velocity spaces V_a^h, V_t^h, see also Remark 7.2, and N_p is the dimension of Q_a^h, Q_t^h. It is sufficient to choose as test functions in (7.8) the functions of the bases of V_t^h and Q_t^h. In this way, a linear system of equations

$$\begin{pmatrix} \mathcal{A}\left(\mathbf{w}_{\text{old}}^h\right) & B \\ C & 0 \end{pmatrix} \begin{pmatrix} w \\ r \end{pmatrix} = \begin{pmatrix} f \\ 0 \end{pmatrix} \tag{7.9}$$

is obtained for the unknown coefficients $w = (w_1, \ldots, w_{dN_v})^T$ and $r = (r_1, \ldots, r_{N_p})^T$. The dimension of the system is $(dN_v + N_p) \times (dN_v + N_p)$. The solution of (7.9) is described in Chapter 9.

Remark 7.2. Discrete gradient and deformation tensor formulation of the viscous term. The matrix block $\mathcal{A}$ is determined by the bases $\{\mathbf{v}_i^a\}_{i=1}^{dN_v}, \{\mathbf{v}_i^t\}_{i=1}^{dN_v}$. The basis functions are chosen in the standard way that each basis function does not vanish in one component of the velocity only, i.e.

$$\{\mathbf{v}_i^a\}_{i=1}^{dN_v} = \left\{\begin{pmatrix} v_i^a \\ 0 \\ 0 \end{pmatrix}\right\}_{i=1}^{N_v} \cup \left\{\begin{pmatrix} 0 \\ v_i^a \\ 0 \end{pmatrix}\right\}_{i=N_v+1}^{2N_v} \cup \left\{\begin{pmatrix} 0 \\ 0 \\ v_i^a \end{pmatrix}\right\}_{i=2N_v+1}^{3N_v}.$$

Using the gradient formulation of the viscous term, the matrix entry a_{ij} of $\mathcal{A}$ does not vanish if $\left(\nabla \mathbf{v}_j^a, \nabla \mathbf{v}_i^t\right) \neq 0$. In this case, $\mathcal{A}$ has the block form

$$\mathcal{A} = \begin{pmatrix} A_{11} & 0 & 0 \\ 0 & A_{11} & 0 \\ 0 & 0 & A_{11} \end{pmatrix}, \quad A_{11} \in \mathbb{R}^{N_v \times N_v}.$$

It is sufficient to store the matrix A_{11}. The deformation tensor formulation of the viscous term, which is used in this monograph, leads to a non-vanishing entry a_{ij} of $\mathcal{A}$ if $\left(\mathbb{D}\left(\mathbf{v}_j^a\right), \mathbb{D}\left(\mathbf{v}_i^t\right)\right)$ does not vanish. Then, $\mathcal{A}$ becomes

$$\mathcal{A} = \begin{pmatrix} A_{11} & A_{12} & A_{13} \\ A_{21} & A_{22} & A_{23} \\ A_{31} & A_{32} & A_{33} \end{pmatrix}, \quad A_{ij} \in \mathbb{R}^{N_v \times N_v}, \tag{7.10}$$

where the blocks A_{ij} are in general mutually different. Thus, from the numerical point of view, the deformation tensor formulation has the following disadvantages in comparison to the gradient formulation:

- the required memory to store $\mathcal{A}$ is much larger (about six times in our implementation),
- matrix-vector products with $\mathcal{A}$ are approximately three times as expensive (in terms of floating point operations),
- the time for assembling the matrix $\mathcal{A}$ is much larger and it becomes a considerable part of the total computing time. Of course, in the non-linear iteration only those parts of (7.10) are assembled new which changes

from one iteration step to the next. These are the three diagonal blocks. One assembling of these blocks in 3d takes about the same time as one multigrid cycle.

Nevertheless, we will use the deformation tensor formulation because it is correct from the physical point of view. □

7.4 Inf-Sup Stable Pairs of Finite Element Spaces

In this section, the finite element spaces are introduced which will be used in the computations. To avoid technical details which are not necessary for this introduction, we assume that the pair of ansatz and test finite element spaces coincide and denote them by $\left(V^h, Q^h\right)$. This is given if the problem possesses homogeneous Dirichlet boundary conditions. Otherwise, ansatz and test finite element spaces are different only in the values at the boundary of their functions. For a detailed description of finite element spaces used in computational fluid dynamics, we refer, e.g., to Ciarlet [Cia78], Girault and Raviart [GR86], Gunzburger [Gun89], Brezzi and Fortin [BF91], Brenner and Scott [BS94], Gresho and Sani [GS00] or the paper of Fortin [For93].

We denote by $\mathcal{T}_h$ a decomposition of Ω into triangles or quadrilaterals if $d = 2$ and tetrahedra or hexahedra if $d = 3$. This decomposition might be possible only if Ω is approximated by a simpler domain, cf. Remark 7.1. We only consider admissible triangulations in the usual sense, e.g., see Ciarlet [Cia91, Hypotheses ($\mathcal{T}_h$1) - ($\mathcal{T}_h$5)]. The mesh cells of $\mathcal{T}_h$ are denoted by K. The diameter of a mesh cell K is defined as the longest distance of points of $\overline{K}$, were $\overline{K}$ is the closure of K. This diameter is denoted by h_K and we set $h := \max_{K\in\mathcal{T}_h}\{h_K\}$.

A finite element is given by a quadruple $\left(\hat{K}, P\left(\hat{K}\right), \Sigma\left(P\right), F_K\right)$, where $\hat{K}$ is a reference mesh cell, $P\left(\hat{K}\right)$ a finite dimensional space of functions defined on $\hat{K}$ with $\dim\left(P\left(\hat{K}\right)\right) = n$, $\Sigma\left(P\right)$ a set of n linear functionals φ_i, $1 \le i \le n$, defined over $P\left(\hat{K}\right)$ which are $P\left(\hat{K}\right)$-unisolvent and

$$F_K \,:\, \overline{\hat{K}} \to \overline{K}, \quad \hat{\mathbf{x}} \to \mathbf{x}$$

the reference map from $\overline{\hat{K}}$ onto $\overline{K} = F_K\left(\overline{\hat{K}}\right)$. This definition of the finite element differs somewhat from the definition given in Ciarlet [Cia91, p. 93f.]. There, a finite element is defined directly on the mesh cell K. We like to emphasize with our notation that we use so-called mapped finite elements, i.e. the functions and functionals are given on the reference mesh cell and the finite element on an arbitrary mesh cell K is defined by mapping them to K with the reference map F_K. The functionals in the finite element spaces which are considered in this monograph are values of functions or mean values on faces or mesh cells, see Section 9.3.1 for some examples.

We will consider first finite elements on simplicial mesh cells. The reference mesh cell for triangles is the triangle with the vertices $(0,0),(1,0),(0,1)$ and for tetrahedra the tetrahedron with the vertices $(0,0,0),(1,0,0),(0,1,0),(0,0,1)$. For simplicial mesh cells, F_K is an affine map. The space $P\left(\hat{K}\right)$ consists of polynomials up to a given degree m. We define

$$\begin{aligned} P_m\left(\hat{K}\right) &:= \left\{ \sum_{i,j=0}^{i+j\le m} a_{ij} x_1^i x_2^j \right\} && \text{if } d=2, \\ P_m\left(\hat{K}\right) &:= \left\{ \sum_{i,j,l=0}^{i+j+l\le m} a_{ijl} x_1^i x_2^j x_3^l \right\} && \text{if } d=3 \end{aligned} \tag{7.11}$$

and

$$P_m(K) := \left\{ p = \hat{p} \circ F_K^{-1} \ : \ \hat{p} \in P_m\left(\hat{K}\right) \right\}.$$

The global finite element spaces on simplicial grids which will be used are

$$\begin{aligned} P_0 &:= \left\{ v \in L^2(\Omega) \ : \ v|_K \in P_0(K) \right\}, \\ P_m &:= \left\{ v \in H^1(\Omega) \ : \ v|_K \in P_m(K) \right\}, m \ge 1, \\ P_1^{\mathrm{nc}} &:= \{ v \in L^2(\Omega) \ : \ v|_K \in P_1(K), v \text{ is continuous at the barycentres} \\ &\qquad \text{of the faces of } K \}. \end{aligned}$$

The space P_1^{nc} is the well known non-conforming linear or Crouzeix/Raviart finite element space, [CR73].

The reference mesh cell for quadrilaterals is the square $(-1,1)^2$ and for hexahedra the cube $(-1,1)^3$. The map F_K is in general bilinear if $d=2$ and trilinear if $d=3$. It is affine only if the quadrilateral is a parallelogram or the hexahedron a parallelepiped. Note that in three dimensions the faces of a hexahedron K are in general not planar. A face of K contains the images of four vertices of $(-1,1)^3$ which determine a face of the reference mesh cell. The images of these four vertices are four different points in $\mathbb{R}^3$ such that not necessary a plane must exist containing all four of them. Besides the spaces $P_m\left(\hat{K}\right)$ defined in (7.11), we define for the description of the spaces on quadrilateral and hexahedral mesh cells

$$\begin{aligned} Q_m\left(\hat{K}\right) &:= \left\{ \sum_{i,j=0}^{m} a_{ij} x_1^i x_2^j \right\} && \text{if } d=2, \\ Q_m\left(\hat{K}\right) &:= \left\{ \sum_{i,j,l=0}^{m} a_{ijl} x_1^i x_2^j x_3^l \right\} && \text{if } d=3, \\ Q_m(K) &:= \left\{ p = \hat{p} \circ F_K^{-1} \ : \ \hat{p} \in Q_m\left(\hat{K}\right) \right\}. \end{aligned}$$

The global finite element spaces used on quadrilateral and hexahedral grids are

$$\begin{aligned}
Q_0 &:= \left\{v \in L^2(\Omega) \; : \; v|_K \in Q_0(K)\right\}, \\
Q_m &:= \left\{v \in H^1(\Omega) \; : \; v|_K \in Q_m(K)\right\}, m \geq 1, \\
P_m^{\text{disc}} &:= \left\{v \in L^2(\Omega) \; : \; v|_K \in P_m(K)\right\}, m \geq 1.
\end{aligned}$$

In addition, we use the class of spaces Q_1^{rot}, which are the spaces of non-conforming piecewise bilinears introduced by Rannacher and Turek [RT92] and studied comprehensively by Schieweck [Sch97]. The local spaces are defined as follows

$$\begin{aligned}
Q_1^{\text{rot}}\left(\hat{K}\right) &:= \{\hat{p} \; : \; \hat{p} \in \text{span}\{1, \hat{x}_1, \hat{x}_2, \hat{x}_1^2 - \hat{x}_2^2\}\} \qquad \text{if } d = 2, \\
Q_1^{\text{rot}}\left(\hat{K}\right) &:= \{\hat{p} \; : \; \hat{p} \in \text{span}\{1, \hat{x}_1, \hat{x}_2, \hat{x}_3, \hat{x}_1^2 - \hat{x}_2^2, \hat{x}_2^2 - \hat{x}_3^2\}\} \text{ if } d = 3, \\
Q_1^{\text{rot}}(K) &:= \left\{p = \hat{p} \circ F_K^{-1} \; : \; \hat{p} \in Q_1^{\text{rot}}\left(\hat{K}\right)\right\}.
\end{aligned} \tag{7.12}$$

To explain the construction of the global space Q_1^{rot}, we denote by $\mathcal{E}(K)$ the set of all $(d-1)$ dimensional faces of K. Then, the mean value oriented finite element space Q_1^{rot} in two dimensions is given by

$$\begin{aligned}
Q_1^{\text{rot}} := \Big\{ & v \in L^2(\Omega) \; : \; v|_K \in Q_1^{\text{rot}}(K), \\
& \int_E v|_K ds = \int_E v|_{K'} ds \;\; \forall E \in \mathcal{E}(K) \cap \mathcal{E}(K') \Big\},
\end{aligned}$$

$d = 2$, and the point value oriented finite element space Q_1^{rot} by

$$\begin{aligned}
Q_1^{\text{rot}} := \{ & v \in L^2(\Omega) \; : \; v|_K \in Q_1^{\text{rot}}(K), v \text{ is continuous at the barycentres} \\
& \text{of the faces of } K\},
\end{aligned}$$

$d = 2, 3$. We use the mean value oriented Q_1^{rot} finite element space in two dimensions and the point value oriented Q_1^{rot} finite element space in three dimensions. For $d = 3$, the integrals on the faces of mesh cells, whose equality is required in the mean value oriented Q_1^{rot} finite element space, involve a weighting function which depends on the particular mesh cell K. The computation of these weighting functions for all mesh cells is an additional computational overhead. That's why, Schieweck [Sch97, p. 21] suggests to use for $d = 3$ the simpler point value oriented form of the Q_1^{rot} finite element.

In the computations presented in this monograph, the triangulation $\mathcal{T}_h$ originates from an initial triangulation $\mathcal{T}_0$ by a number of uniform refinement steps. We use the standard red refinement. In 2d, triangles and quadrilaterals are divided into four mesh cells. In 3d, hexahedra are divided into eight hexahedra and tetrahedra into six tetrahedra. There are three possibilities of the red refinement of a tetrahedron. One has to take care that these possibilities are applied in an appropriate order, e.g., see Bey [Bey95].

Remark 7.3. The interpolation error of the finite element spaces. Let $\{\mathcal{T}_h\}$ be a family of uniform triangulations with the corresponding mesh size parameters $\{h\}$ and let Ω be a domain which can be triangulated without boundary approximations. Then, the interpolation error for sufficiently smooth functions of the spaces P_m and Q_m, $m \geq 1$, is $\mathcal{O}\left(h^{m+1}\right)$ in the $L^p\left(\Omega\right)$ norm and $\mathcal{O}\left(h^m\right)$ in the $W^{1,p}\left(\Omega\right)$ norm, $1 \leq p \leq \infty$, see Brenner and Scott [BS94, Section 4.4]. The same holds for P_m^{disc}, $m \geq 1$, with the $H^1\left(\Omega\right)$ norm replaced by the cell-wise computed $H^1\left(\Omega\right)$ norm. However, the interpolation error for the P_m^{disc} finite element spaces deteriorates on meshes which does not tend asymptotically to parallelogram or parallelepiped meshes, see Matthies [Mat01]. The interpolation errors of the lowest order non-conforming finite element spaces P_1^{nc} and Q_1^{rot} are $\mathcal{O}\left(h^2\right)$ in the $L^2\left(\Omega\right)$ norm and $\mathcal{O}\left(h\right)$ in the cell-wise computed $H^1\left(\Omega\right)$ norm. □

Only such pairs of finite element spaces $\left(V^h, Q^h\right)$ are used in the computations which fulfil the inf-sup or Babuška-Brezzi stability condition, i.e. there exists a constant $\kappa > 0$ independent of the triangulation such that

$$\inf_{q^h \in Q^h} \sup_{\mathbf{v}^h \in V^h} \frac{\left(\nabla \cdot \mathbf{v}^h, q^h\right)}{\|\nabla \mathbf{v}^h\|_{L^2(\Omega)} \|q^h\|_{L^2(\Omega)}} \geq \kappa. \tag{7.13}$$

For $V^h \in \{P_1^{\text{nc}}, Q_1^{\text{rot}}\}$, the $L^2\left(\Omega\right)$ norm of the gradient of $\mathbf{v}^h$ has to be replaced by the cell-wise computed $L^2\left(\Omega\right)$ norm of $\nabla \mathbf{v}^h$. Condition (7.13) guarantees the unique solvability of the discrete systems. Loosely speaking, (7.13) states that the number of degrees of freedom of the discrete velocity space has to be sufficiently large in comparison to the number of degrees of freedom of the discrete pressure space. Note that (7.13) can be proved in general only with some restrictions on the family of triangulations, e.g., that in a triangular triangulation all triangles posses at least one vertex not on $\partial\Omega$. However, it is common use to speak also of inf-sup stable pairs of finite element spaces if such a property of the triangulation is not given. This will be done also in this monograph.

In this monograph, the following pairs of inf-sup stable finite element spaces on quadrilateral or hexahedral meshes are considered (*discrete velocity space/discrete pressure space*):

- Q_1^{rot}/Q_0, Q_2/P_1^{disc}, Q_2/Q_1, Q_3/P_2^{disc}, Q_3/Q_2.

On simplicial meshes, we use the spaces

- P_1^{nc}/P_0, P_2/P_1, P_3/P_2.

As usual, the facts that the velocity is a vector-valued function and the finite element space of the pressure might be intersected with $L_0^2\left(\Omega\right)$ are not indicated in these notations.

Remark 7.4. Motivation for using inf-sup stable pairs of finite element spaces. The use of pairs of finite element spaces which do not fulfil (7.13), e.g., like Q_1/Q_0 or Q_1/Q_1, can lead to instabilities in the discrete pressure (checkerboard instabilities), see Girault and Raviart [GR86, Chapter II, Section 3.3] or Fortin [For93] for an example. The use of such discretisations requires additional correction terms in the variational formulation of the equations. The parameters involved in the correction terms have a considerable influence on the computed solution and they have to be chosen appropriately. In addition, an artificial boundary condition for the pressure has to be chosen with the consequence that the discrete pressure is often polluted near the boundary (boundary layers), e.g., see Fortin [For93]. The use of inf-sup stable pairs of finite element spaces avoids all these difficulties. □

Remark 7.5. Comments on the pairs of finite element spaces considered in this monograph. A comprehensive overview of finite element spaces used in incompressible flow computations can be found in Gresho and Sani [GS00, Section 3.13.2]. This remark describes briefly some advantages and drawbacks of the pairs of finite element spaces considered in this monograph and contains some comments on personal experiences with them.

Q_1^{rot}/Q_0, P_1^{nc}/P_0 – *the lowest order non-conforming pairs of finite element spaces*

The use of the Q_1^{rot}/Q_0 pair of finite element spaces is favoured by Turek [Tur99]. There is also a lot of experience of using the P_1^{nc}/P_0 pair of finite element spaces in the numerical solution of the Navier-Stokes equations. The advantages of these discretisations are as follows:

- The discrete systems can be solved very efficiently. This property is used in the construction of the multiple discretisation multilevel method presented in Section 9.3.3.
- Analytical results, e.g., Becker and Rannacher [BR94], as well as numerical studies, e.g., Turek [Tur99], show that the accuracy of these discretisations and the rate of convergence of multigrid solvers do not detoriate on highly anisotropic meshes, i.e. meshes with very thin and long mesh cells. There are applications where the use of such meshes is necessary, e.g., if layers have to be resolved.
- The non-conforming finite element spaces are well suited for parallel computing since degrees of freedom are connected to faces of the mesh cells. Thus, on a parallel computer, the degrees of freedom on the interfaces (boundary of a subdomain of Ω which is stored on a processor) have to be stored on two processors only. That means, only one communication suffices to interchange information between the same degree of freedom on different processors. This is not true for degrees of freedom which are connected to vertices of mesh cells. In this case, there exist some degrees of freedom which are stored on more than two processors, the so-called crosspoints.

The drawbacks of these pairs of finite element spaces are the followings:

- In our experience, the numerical results are often considerable more inaccurate than with higher order pairs of finite element spaces, see John and Matthies [JM01] and John [Joh02a]. This is the main reason why we use the

lowest order non-conforming pairs of finite element spaces only as auxiliary tool in the solution of the discrete systems, see Section 9.3.3.
- The analysis of discretisations with non-conforming finite element spaces is sometimes much more complicated than of discretisations with conforming spaces. There are even widely used techniques for conforming finite element spaces whose application to non-conforming spaces is so complicated that it is not practicable, e.g., the streamline diffusion stabilisation for convection dominated problems, see John et al. [JMT97, JMST98].

Q_2/P_1^{disc}, Q_2/Q_1, P_2/P_1 – *the second order pairs of finite element spaces*

These are the most popular pairs of finite element spaces in applications. The pairs Q_2/Q_1 and P_2/P_1 are well known as Taylor-Hood elements. The numerical results of incompressible flow computations obtained with the second order pairs are often considerably more accurate than with the pairs of lowest order non-conforming finite element spaces. However, a global gain of accuracy arising from a higher order convergence of the discrete solution to the solution of the continuous problem can only be expected if the solution of the continuous problem is smooth, e.g., if $\mathbf{u} \in H^2(\Omega)$ and $p \in H^1(\Omega)$ in the steady state Navier-Stokes equations. A global higher order of convergence will not be given in general if the solution of the continuous problem possesses singularities, e.g., at re-entrant corners. In this case, at best a higher order of convergence can occur locally, in subregions away from the singularities. At any rate, the results obtained with the second order pairs of finite element spaces will be as accurate as the results computed with the pairs of lowest order non-conforming finite element spaces.

The most important difference between the Taylor-Hood elements and Q_2/P_1^{disc} is that the former pairs use a continuous pressure approximation and the latter pair a discontinuous one. The discontinuous pressure approximation has been proved to be advantageous. The main reason is that much more efficient solvers can be constructed if a discontinuous pressure approximation is used. This is explained in detail for a coupled multigrid method in Section 9.3.3. In addition, Gresho and Sani [GS00] (Q_2/P_1^{disc} is called Q_2/P_{-1}) and Fortin [For93] (Q_2/P_1^{disc} is called Q_2/P_1) state that discontinuous pressure approximations ensure also a better conservation of mass.

Second order pairs of finite element spaces, in particular with discontinuous pressure approximation, e.g. Q_2/P_1^{disc}, are in our opinion the best choice based as well on the accuracy of the computed solution as on the efficiency in solving the discrete systems of equations.

Q_3/P_2^{disc}, Q_3/Q_2, P_3/P_2 – *the third order pairs of finite element spaces*

The pairs Q_3/Q_2 and P_3/P_2 are also called often Taylor-Hood elements. The third order finite elements can be found mainly in research codes and they are used to support results from numerical analysis. If the solution of the continuous problem is sufficiently smooth, one obtains more accurate results than with the second order pairs of finite element spaces. But solving the discrete problems becomes often considerable more complicated and takes much longer, e.g., see John [Joh02a]. The differences between the discretisations with discontinuous and continuous pressure approximation are the same as for second order elements. □

Remark 7.6. Isoparametric finite elements. If Ω has a part of its boundary which is curved, the use of standard finite elements leads to an additional boundary approximation error. This error might dominate the approximation error of the finite element space. If this happens, the extra costs of using higher order finite elements do not pay. One remedy is the use of isoparametric finite elements which provide a better boundary approximation than standard finite elements. The idea of isoparametric finite elements is to extend the space of admissible reference transformations F_K. This gives a greater flexibility, e.g., the image of an edge of the reference triangle can be a curved line. In general, one does not obtain an exact representation of the boundary but an approximation such that its error is of the same order as the interpolation error of the finite element space. The idea and the analysis of isoparametric finite elements are well described in the survey paper by Ciarlet [Cia91, Chapter VI]. The considerable gain of accuracy using isoparametric finite elements in the computation of parameters of incompressible flows which are relevant in applications is demonstrated in John and Matthies [JM01] and John [Joh02a]. □

Remark 7.7. Stabilisation of convection dominated flow problems. The Galerkin discretisation of convection dominated equations proves to be in general unstable. The computed solutions show spurious oscillations. There are various ways for stabilising a dominating convection like upwind schemes, the streamline diffusion method or the introduction of artificial viscosity, see Roos et al. [RST96] for an overview. In the LES models, a stabilisation is obtained by the turbulent viscosity term which acts as a non-linear artificial viscosity, see Remark 4.2. Only for the non-conforming discretisations of lowest order, an additional upwind stabilisation is used in our computations, see Section 7.5. □

7.5 The Upwind Stabilisation for Lowest Order Non-Conforming Finite Elements

The upwind discretisation of the non-linear convective term $b\left(\mathbf{w}_{\text{old}}^h, \mathbf{w}^h, \mathbf{v}^h\right)$ for non-conforming finite element spaces of lowest order will be an auxiliary tool which is used in the efficient numerical solution of the discrete systems obtained with higher order finite element discretisations, see Section 9.3.3. We found the multiple discretisation multilevel method described in this section to be more efficient if an additional upwind stabilisation is applied.

The upwind discretisation is a standard way to stabilise the dominating convective term in the P_1^{nc}/P_0 and the Q_1^{rot}/Q_0 finite element discretisations of the Navier-Stokes equations, e.g., see Roos et al. [RST96]. The upwind technique for non-conforming finite element discretisations of scalar convection-diffusion equations was originally proposed by Ohmori and Ushijima [OU84]. It was generalised and analysed for the P_1^{nc}/P_0 finite element discretisation of the Navier-Stokes equations by Schieweck and Tobiska [ST96b]. Rannacher and Turek [RT92] studied upwind discretisations of the Navier-Stokes equations on quadrilateral meshes. Their work was generalised by Schieweck [Sch97] and extended to three dimensions. All papers which are mentioned contain optimal error estimates of first order in a discrete $H^1(\Omega)$ norm for the velocity and the $L^2(\Omega)$ norm for the pressure.

Upwind discretisation applied to the Navier-Stokes equations are known to lead to discrete systems which can be solved very efficiently by multigrid methods, e.g., see Turek [Tur99], John [Joh99] and John and Tobiska [JT00]. The great disadvantage of these discretisations is their low accuracy, not only measured in norms of Sobolev spaces but also with respect to quantities which are important in applications, see John and Matthies [JM01] or John [Joh02a]. That's why, we use them merely for enhancing the efficiency of a multilevel method and we want to give only a short description.

Let $\mathbf{w}_{\mathrm{old}}^h, \mathbf{w}^h, \mathbf{v}^h$ be non-conforming finite element functions of lowest order. The degrees of freedom of such functions can be assigned to nodes B_i which are situated on the faces of the mesh cells, see Figure 7.1. Let K_1, K_2 be the mesh cells which share the face F_i with node B_i. By connecting the barycentres of K_1, K_2 with the end points of the edge F_i if $d = 2$ (edges of the face F_i if $d = 3$), a control volume ω_i around B_i is defined. Let Γ_{ik} be the face of the control volume between B_i and a neighbour node B_k with the unit normal $\mathbf{n}_{ik}$ directed outward with respect to ω_i. We denote by Λ_i the set of all indices $l \neq i$ for which the nodes B_l and B_i posses a common face of ω_i. To obtain a stable discretisation, it is necessary to take the local direction of the convection into account. The local direction between B_i and B_k is given by the flux across Γ_{ik}

$$\int_{\Gamma_{ik}} \mathbf{w}_{\mathrm{old}}^h \cdot \mathbf{n}_{ik} ds.$$

The information which is contained in the flow is transported into the direction of the convection (downwind direction). A stable discretisation of the convective term in B_i is obtained by weighting the upwind neighbours of B_i more than the downwind neighbours in order to guarantee an adequate transport of information.

A detailed derivation of the upwind discretisation can be found in Roos et al. [RST96]. The upwind discretisation of the convective term is given by

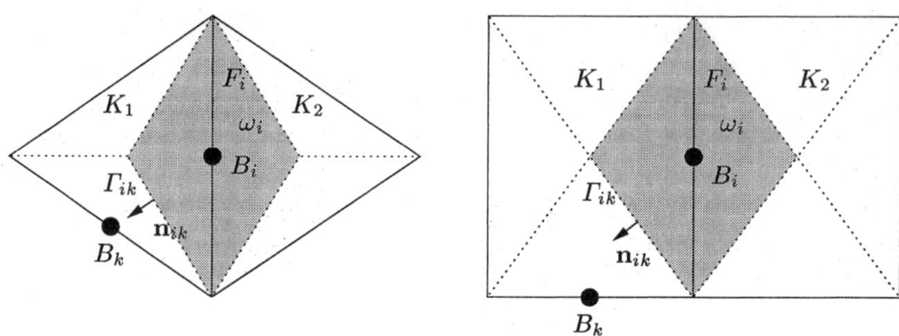

Fig. 7.1. Control volumes for the upwind discretisation

$$\begin{aligned}
b\left(\mathbf{w}_{\text{old}}^h, \mathbf{w}^h, \mathbf{v}^h\right) &\approx b^{\text{upw}}\left(\mathbf{w}_{\text{old}}^h, \mathbf{w}^h, \mathbf{v}^h\right) \\
&:= \sum_{i=1}^{N_v} \sum_{k\in\Lambda_i} \int_{\Gamma_{ik}} \left[\left(\mathbf{w}_{\text{old}}^h \cdot \mathbf{n}_{ik}\right)\left(1-\lambda_{ik}\left(\mathbf{w}_{\text{old}}^h\right)\right)\left(\mathbf{w}^h\left(B_k\right)-\mathbf{w}^h\left(B_i\right)\right)\right] \\
&\qquad\qquad \cdot \mathbf{v}^h\left(B_i\right) d\mathbf{s},
\end{aligned}$$

where N_v is the number of nodes with velocity degrees of freedom. Let S_{ik} be the barycentre of Γ_{ik} and $|\Gamma_{ik}|$ be the $(d-1)$ dimensional measure. The integral on the face Γ_{ik} is approximated by

$$\begin{aligned}
&\int_{\Gamma_{ik}} \left[\left(\mathbf{w}_{\text{old}}^h \cdot \mathbf{n}_{ik}\right)\left(1-\lambda_{ik}\left(\mathbf{w}_{\text{old}}^h\right)\right)\left(\mathbf{w}^h\left(B_k\right)-\mathbf{w}^h\left(B_i\right)\right)\right] \cdot \mathbf{v}^h\left(B_i\right) d\mathbf{s} \\
&\approx \left|\Gamma_{ik}\right| \Big\{ \left[\left(\mathbf{w}_{\text{old}}^h\left(S_{ik}\right) \cdot \mathbf{n}_{ik}\right)\left(1-\lambda_{ik}\left(\mathbf{w}_{\text{old}}^h\left(S_{ik}\right)\right)\right)\left(\mathbf{w}^h\left(B_k\right)-\mathbf{w}^h\left(B_i\right)\right)\right] \\
&\qquad \cdot \mathbf{v}^h\left(B_i\right) \Big\}.
\end{aligned}$$

The value $\lambda_{ik}\left(\mathbf{w}_{\text{old}}^h\left(S_{ik}\right)\right)$ is given by

$$\lambda_{ik}\left(\mathbf{w}_{\text{old}}^h\right) = \Phi(t), \quad \text{with} \quad t = \frac{1}{2\nu}\int_{\Gamma_{ik}} \mathbf{w}_{\text{old}}^h \cdot \mathbf{n}_{ik} d\mathbf{s},$$

where the function $\Phi(t)$ must fulfil certain requirements which can be found in [RST96]. The most common choices of $\Phi(t)$ lead to the so-called simple or sharp upwind discretisation and the Samarskij upwind discretisation. Numerical tests for the Navier-Stokes equations by John [Joh97] show that the results with Samarskij upwinding are somewhat more accurate and the behaviour of a multigrid method is basically the same. The rate of convergence using Samarskij upwinding might be slightly worse. In the multiple discretisation multilevel method described in Section 9.3.3, the result of the non-conforming upwind finite element discretisation of lowest order should be an approximation of the result with a higher order discretisation. For the reason of a probably better approximation, we prefer to use the Samarskij upwinding given by

$$\Phi(t) = \begin{cases} (0.5+t)/(1+t) & \text{if } t \geq 0, \\ 1/(2-2t) & \text{if } t < 0. \end{cases}$$

7.6 The Implementation of the Slip With Friction and Penetration With Resistance Boundary Condition

In this section, the implementation of the boundary integral term in (7.8)

$$\int_{\Gamma_{\mathrm{sfpr}}} \alpha^{-1}\left(\mathbf{w}^h \cdot \mathbf{n}_{\partial\Omega}\right)\left(\mathbf{v}^h \cdot \mathbf{n}_{\partial\Omega}\right) + \beta(\delta,\nu)\left[\left(\mathbf{w}^h \cdot \boldsymbol{\tau}_1\right)\left(\mathbf{v}^h \cdot \boldsymbol{\tau}_1\right) + \left(\mathbf{w}^h \cdot \boldsymbol{\tau}_2\right)\left(\mathbf{v}^h \cdot \boldsymbol{\tau}_2\right)\right] ds \tag{7.14}$$

coming from the slip with friction and penetration with resistance boundary condition is described.

The explicit way to treat (7.14) is to replace $\mathbf{w}^h$ by the current approximation of $\mathbf{w}^h$ and to put the boundary integral on the right hand side of the discrete momentum equation. We have implemented in MooNMD an implicit treatment of (7.14) and describe the implementation for $d=3$. The modifications in the two dimensional case are obvious. Let $\{\mathbf{v}_j^a\}_{j=1}^{3N_v}$ be a basis of the velocity ansatz space and $\mathbf{v}_i^t$ be a test function which does not vanish on Γ_{sfpr}. The basis functions of the ansatz spaces are ordered such that $\mathbf{v}_j^a$ with $(i-1)N_v < j \leq iN_v$ vanishes in all components but the i-th, see Remark 7.2. With

$$\mathbf{w}^h = \sum_{j=1}^{3N_v} w_j \mathbf{v}_j^a, \quad \mathbf{v}_j^a = \begin{pmatrix} v_{j1}^a \\ v_{j2}^a \\ v_{j3}^a \end{pmatrix}, \quad \mathbf{v}_i^t = \begin{pmatrix} v_{i1}^t \\ v_{i2}^t \\ v_{i3}^t \end{pmatrix}, \quad \mathbf{n}_{\partial\Omega} = \begin{pmatrix} n_1 \\ n_2 \\ n_3 \end{pmatrix},$$

we obtain

$$\begin{aligned} &\int_{\Gamma_{\mathrm{sfpr}}} \alpha^{-1}\left(\mathbf{w}^h \cdot \mathbf{n}_{\partial\Omega}\right)\left(\mathbf{v}_i^t \cdot \mathbf{n}_{\partial\Omega}\right) ds \\ &= \sum_{j=1}^{3N_v} \alpha^{-1} w_j \int_{\Gamma_{\mathrm{sfpr}}} \left(v_{j1}^a n_1 + v_{j2}^a n_2 + v_{j3}^a n_3\right)\left(v_{i1}^t n_1 + v_{i2}^t n_2 + v_{i3}^t n_3\right) ds. \end{aligned}$$

This gives for $i \leq N_v$, i.e. for $v_{i2}^t = v_{i3}^t = 0$

$$\begin{aligned} &\int_{\Gamma_{\mathrm{sfpr}}} \alpha^{-1}\left(\mathbf{w}^h \cdot \mathbf{n}_{\partial\Omega}\right)\left(\mathbf{v}_i^t \cdot \mathbf{n}_{\partial\Omega}\right) ds \\ &= \sum_{j=1}^{3N_v} \alpha^{-1} w_j \left(\int_{\Gamma_{\mathrm{sfpr}}} v_{j1}^a v_{i1}^t n_1^2 ds + \int_{\Gamma_{\mathrm{sfpr}}} v_{j2}^a v_{i1}^t n_1 n_2 ds + \int_{\Gamma_{\mathrm{sfpr}}} v_{j3}^a v_{i1}^t n_1 n_3 ds\right). \end{aligned}$$

Similar formulas are obtained for $N_v < i \le 2N_v$ and $2N_v < i \le 3N_v$. The system matrix of the linear discrete saddle point problem has the form

$$\begin{pmatrix} A_{11} & A_{12} & A_{13} & B_1 \\ A_{21} & A_{22} & A_{23} & B_2 \\ A_{31} & A_{32} & A_{33} & B_3 \\ C_1 & C_2 & C_3 & 0 \end{pmatrix}.$$

If the integral

$$\alpha^{-1} \int_{\Gamma_{\mathrm{sfpr}}} v_{jl}^a v_{ik}^t n_l n_k ds, \quad k, l = 1, 2, 3,$$

does not vanish, it gives a contribution to the matrix entry $(A_{kl})_{ij}$.

The other two terms of (7.14) are treated in a similar way. Let

$$\boldsymbol{\tau}_1 = \begin{pmatrix} \tau_{11} \\ \tau_{12} \\ \tau_{13} \end{pmatrix}, \quad \boldsymbol{\tau}_2 = \begin{pmatrix} \tau_{21} \\ \tau_{22} \\ \tau_{23} \end{pmatrix},$$

then the contribution to the matrix entry $(A_{kl})_{ij}$ from (7.14) has the form

$$\alpha^{-1} \int_{\Gamma_{\mathrm{sfpr}}} v_{jl}^a v_{ik}^t n_l n_k ds + \beta \int_{\Gamma_{\mathrm{sfpr}}} v_{jl}^a v_{ik}^t \left(\tau_{1l}\tau_{1k} + \tau_{2l}\tau_{2k}\right) ds. \qquad (7.15)$$

Remark 7.8. Independence of (7.15) from the choice of the tangential vectors. For $d = 2$, a tangential vector is immediately given by $\boldsymbol{\tau}_1 = (n_2, -n_1)^T$. The only alternative tangential vector is $-\boldsymbol{\tau}_1$ and it is obvious that the value of (7.15) does not depend on the choice of the tangential vector.

For $d = 3$, let $\boldsymbol{\tau}_1, \boldsymbol{\tau}_2$ be two arbitrary vectors which span the tangential plane such that $\{\mathbf{n}_{\partial\Omega}, \boldsymbol{\tau}_1, \boldsymbol{\tau}_2\}$ is a system of orthonormal vectors. One alternative choice of the system of tangential vectors is to reflect one of them, giving, e.g., the system $\{\mathbf{n}_{\partial\Omega}, \boldsymbol{\tau}_1, -\boldsymbol{\tau}_2\}$. Clearly, this alternative choice leaves (7.15) unchanged. Another way of changing the original system of tangential vectors is to rotate them in the tangential plane around the axis $\mathbf{n}_{\partial\Omega}$ by the angle θ. This operation transforms the original system of orthonormal vectors into $\{\mathbf{n}_{\partial\Omega}, \boldsymbol{\tau}_1 \cos\theta + \boldsymbol{\tau}_2 \sin\theta, \boldsymbol{\tau}_2 \cos\theta - \boldsymbol{\tau}_1 \sin\theta\}$. It is a straightforward calculation to check that this operation also does not change (7.15). Altogether, also for $d = 3$ one can choose arbitrary tangential vectors $\boldsymbol{\tau}_1, \boldsymbol{\tau}_2$ such that $\{\mathbf{n}_{\partial\Omega}, \boldsymbol{\tau}_1, \boldsymbol{\tau}_2\}$ is a system of orthonormal vectors. □

The tangential vectors are chosen in the computations presented in this monograph as follows.

Algorithm 7.9. Computation of $\boldsymbol{\tau}_1 = (\tau_{11}, \tau_{12}, \tau_{13})$ and $\boldsymbol{\tau}_2 = (\tau_{21}, \tau_{22}, \tau_{23})$. Given the normal vector $\mathbf{n}_{\partial\Omega} = (\mathtt{n}_1, \mathtt{n}_2, \mathtt{n}_3)$ with $\|\mathbf{n}_{\partial\Omega}\|_2 = 1$. Then, there is at least one component $\mathtt{n}_i$ with $|\mathtt{n}_i| \ge 0.5$.

1. if $(|\mathtt{n}_1| \ge 0.5$ OR $|\mathtt{n}_2| \ge 0.5)$

$\quad\mathrm{n} := \sqrt{\mathrm{n}_1^2 + \mathrm{n}_2^2}$
$\quad\tau_{11} := \mathrm{n}_2/\mathrm{n}$
$\quad\tau_{12} := -\mathrm{n}_1/\mathrm{n}$
$\quad\tau_{13} := 0$
$\quad\tau_{21} := -\tau_{12}\mathrm{n}_3$
$\quad\tau_{22} := \tau_{11}\mathrm{n}_3$
$\quad\tau_{23} := \tau_{12}\mathrm{n}_1 - \tau_{11}\mathrm{n}_2$
`else`
$\quad\mathrm{n} := \sqrt{\mathrm{n}_2^2 + \mathrm{n}_3^2}$
$\quad\tau_{11} := 0$
$\quad\tau_{12} := -\mathrm{n}_3/\mathrm{n}$
$\quad\tau_{13} := \mathrm{n}_2/\mathrm{n}$
$\quad\tau_{21} := \tau_{13}\mathrm{n}_2 - \tau_{12}\mathrm{n}_3$
$\quad\tau_{22} := -\tau_{13}\mathrm{n}_1$
$\quad\tau_{23} := \tau_{12}\mathrm{n}_1$
`endif`

The distinction of the two cases in Algorithm 7.9 ensures that a division by zero cannot happen.

Remark 7.10. Weak imposition of the no penetration constraint. The first term in (7.14) can be considered as a weak imposition of the no penetration constraint as penalty formulation with the penalty parameter α^{-1}. In examples with slip (with or without friction) and no penetration boundary conditions, we always applied the no penetration boundary condition weakly with the penalty parameter $\alpha = 10^{-12}$. □

7.7 The Discretisation of the Auxiliary Problem in the Rational LES Model

The rational LES model with auxiliary problem requires the solution of

$$\begin{aligned} -\frac{\delta^2}{4\gamma}\Delta\mathbb{X} + \mathbb{X} &= \nabla\mathbf{w}\nabla\mathbf{w}^T \text{ in } \Omega, \\ \frac{\partial\mathbb{X}}{\partial\mathbf{n}_{\partial\Omega}} &= \mathbf{0} \qquad\qquad \text{on } \partial\Omega, \end{aligned} \tag{7.16}$$

see Remark 4.7. The weak formulation of (7.16) is to find $\mathbb{X} \in H^1(\Omega)$ such that for all $\mathbb{Y} \in H^1(\Omega)$

$$\frac{\delta^2}{4\gamma}(\nabla\mathbb{X}, \nabla\mathbb{Y}) + (\mathbb{X}, \mathbb{Y}) = \left(\nabla\mathbf{w}\nabla\mathbf{w}^T, \mathbb{Y}\right). \tag{7.17}$$

This system is solved only once in each sub time step as described in Section 7.2. The right hand side of (7.17) is approximated by $\left(\nabla\mathbf{w}_{k-1}^h\left(\nabla\mathbf{w}_{k-1}^h\right)^T, \mathbb{Y}\right)$, where $\mathbf{w}_{k-1}^h$ is the finite element solution of the velocity from the previous sub time step. Then, (7.17) is discretised by the Galerkin finite element method: Find $\mathbb{X}^h \in V_{\mathrm{aux}}^h$ such that for all $\mathbb{Y}^h \in V_{\mathrm{aux}}^h$

$$\frac{\delta^2}{4\gamma}\left(\nabla\mathbb{X}^h, \nabla\mathbb{Y}^h\right) + \left(\mathbb{X}^h, \mathbb{Y}^h\right) = \left(\nabla\mathbf{w}_{k-1}^h \left(\nabla\mathbf{w}_{k-1}^h\right)^T, \mathbb{Y}^h\right). \tag{7.18}$$

The space V_{aux}^h can be chosen among the finite element spaces described in Section 7.4. In the computations presented in this monograph, the local polynomial degrees of the functions from V_{aux}^h and the functions form the velocity finite element space V^h are always the same. Both spaces differ in general in the boundary condition of their functions.

In the discretisation of (7.18) and the numerical solution of the discrete problem, the symmetry of the tensor on the right hand side is of course exploited. The choice of the quadrature rules as described in Section 7.9 ensures also that the right hand side of (7.18) is integrated exactly if all reference transformations are affine.

A system of form (7.16) with the right hand side replaced by $\mathbf{w}_{k-1}^h$ has to be solved if the Iliescu-Layton subgrid scale model (7.5) is used. This system is discretised in the same way as the system for the auxiliary problem. Test and ansatz space coincide with the velocity test and ansatz space (up to boundary conditions).

7.8 The Computation of the Convolution in the Rational LES Model

This section describes the operator

$$\left(A^h\left(\nabla\mathbf{w}_{k-1}^h \left(\nabla\mathbf{w}_{k-1}^h\right)^T\right)\right),$$

equation (7.8), in the case of the operator A being the convolution with g_δ.

If not mentioned otherwise, all finite element functions are extended trivially outside Ω. Exceptions occur only if problems with periodic boundary conditions are considered. Thus, we have

$$\begin{aligned} & g_\delta * \left(\nabla\mathbf{w}_{k-1}^h \left(\nabla\mathbf{w}_{k-1}^h\right)^T\right)(\mathbf{y}) \\ &= \int_\Omega g_\delta(\mathbf{y}-\mathbf{x})\left(\nabla\mathbf{w}_{k-1}^h \left(\nabla\mathbf{w}_{k-1}^h\right)^T\right)(\mathbf{x})\,d\mathbf{x}. \end{aligned} \tag{7.19}$$

This function will be approximated by a discrete function from a finite element space V_{conv}^h. If not mentioned otherwise, V_{conv}^h will be the same finite element space which is used for the discrete velocity (up to boundary conditions).

It is not needed to compute (7.19) for all $\mathbf{y} \in \overline{\Omega}$ but only for such $\mathbf{y}$ which are necessary to determine a function in V_{conv}^h uniquely. The geometric positions which are assigned to nodes of finite element functions on the reference cell $\hat{K}$ for the finite element spaces whose nodal functionals are values of functions can be found in every book on finite elements. For the lowest order non-conforming finite element spaces, we assign the barycentres of the $(d-1)$ dimensional faces of $\hat{K}$ to the nodes of the finite element functions. The image of each of the geometric positions under F_K, $K \in \mathcal{T}_h$, determines a position $\mathbf{y}$ for which (7.19) is computed. The fact that $g_\delta(\mathbf{x})$ is negligible for $\|\mathbf{x}\|_2 > \overline{\delta} = C\delta$ with C sufficiently large, see Figure 3.1, is used for another simplification of (7.19) which saves a considerable amount of computing time. In this simplification, the integral on Ω is replaced by the integral in the $\overline{\delta}$-neighbourhood $U(\mathbf{y}, \overline{\delta})$ of $\mathbf{y}$. In the computations, we have used $\overline{\delta} = \sqrt{2}\delta$. Altogether, we compute for all $\mathbf{y}$ which are necessary to determine a finite element function in V^h_{conv} uniquely

$$
\begin{aligned}
& g^h_\delta * \left(\nabla \mathbf{w}^h_{k-1} \left(\nabla \mathbf{w}^h_{k-1}\right)^T\right)(\mathbf{y}) \\
& := \int_{\Omega \cap U(\mathbf{y},\overline{\delta})} g_\delta(\mathbf{y}-\mathbf{x}) \left(\nabla \mathbf{w}^h_{k-1} \left(\nabla \mathbf{w}^h_{k-1}\right)^T\right)(\mathbf{x})\, d\mathbf{x}.
\end{aligned} \tag{7.20}
$$

The actual evaluation of (7.20) by quadrature rules is described in Section 7.9.

Remark 7.11. Choice of $U(\mathbf{y}, \overline{\delta})$. One aspect which is important for the efficient computation of (7.20) is the way of defining $U(\mathbf{y}, \overline{\delta})$. We study two variants:

1. Let $\mathbf{y} \in \overline{K} \in \mathcal{T}_h$. If $\mathbf{y}$ is in the interior of K, $\mathbf{y} \in K$, (7.20) is computed on K and all mesh cells which have a common $(d-1)$ dimensional face with K. In the case that $\mathbf{y}$ is on the boundary of K, $\mathbf{y} \in \partial K$, (7.20) is computed on all mesh cells K' with $\mathbf{y} \in \overline{K'}$ and all adjacent mesh cells which has a common $(d-1)$ dimensional face with one of the mesh cells K', see Figure 7.2 for typical domains of integration.

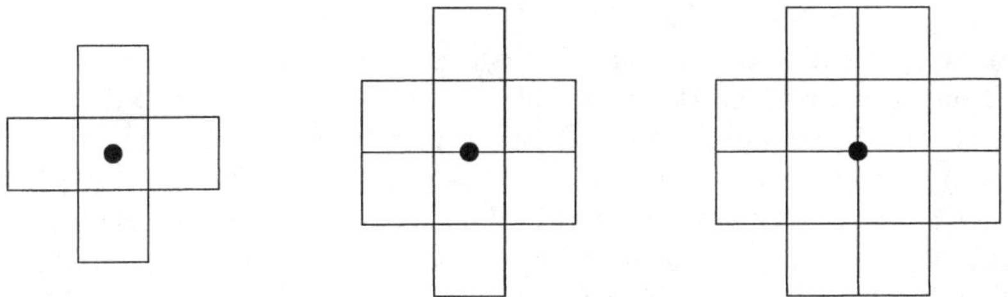

Fig. 7.2. Typical situation for $U(\mathbf{y}, \overline{\delta})$ in the first approach of computing (7.20), Remark 7.11

The advantage of this approach is that the computation of (7.20) can be performed by one single loop over the mesh cells and computations

in nearest neighbours for each mesh cell. The disadvantage is that in general the domain of integration does not cover the closed ball $\overline{B}(\mathbf{y},\overline{\delta})$ Thus, parts of the neighbourhood of $\mathbf{y}$ in which $g_\delta(\mathbf{y}-\mathbf{x})$ is small but not negligible are neglected in the evaluation of (7.20). The effect of this approximation becomes the larger the larger δ is in comparison to the mesh width h.
2. This variant computes (7.20) by a double loop over the mesh cells. The outer loop finds all $\mathbf{y}$ for which (7.20) has to be evaluated. The inner loop checks if the distance of the barycentre of a mesh cell to $\mathbf{y}$ is smaller than or equal to $\overline{\delta}$. Only if this is the case, the contribution of this mesh cell to (7.20) is computed. This variant is more accurate than the first variant but considerably more expensive because of the double loop over the mesh cells. On fine meshes, the costs of computing a convolution using this variant dominate the total computing time.

We present a comparison of both variants on a coarse mesh in Section 10.3. If not mentioned otherwise, the first variant was used in the numerical tests. □

The approximation of the convolution by a second order partial differential operator, Remark 4.8, is in general much more efficient with respect to the computing time than both variants presented in Remark 7.11. We incorporated different variants of the numerical approximation of the convolution operator into the numerical tests in order to evaluate the differences in the computed solutions.

7.9 The Evaluation of Integrals, Numerical Quadrature

The benefit of the accuracy of discretisations with higher order finite element spaces is only obtained if the arising integrals in the discrete formulation of the LES models (7.8) are evaluated sufficiently accurate. We will motivate in this section the choice of the quadrature rules which are used in the computations.

Let ϕ^h be a finite element function whose integral on Ω has to be computed, e.g., $\phi^h = \nabla \mathbf{w}^h : \nabla \mathbf{v}^h$. The general approach to compute this integral starts by splitting it into a sum of integrals on the mesh cells. Next, the integral on each mesh cell is transformed to the reference mesh cell. Last, the integrals on the reference mesh cell are approximated by a quadrature rule. Let $\mathcal{T}_h$ be the current triangulation, $\hat{K}$ the reference mesh cell, $\hat{\phi}^h(\hat{\mathbf{x}})$ be the transformation of $\phi^h(\mathbf{x})$ to $\hat{K}$, $J_K(\hat{\mathbf{x}})$ the Jacobian of the reference map F_K, and $(\hat{\mathbf{x}}_l,\theta_l), l = 1,\ldots,n$, a quadrature rule with quadrature points $\hat{\mathbf{x}}_l$ and weights θ_l. Then, the computation of the integral of ϕ^h on Ω reads as follows

$$\begin{aligned}\int_\Omega \phi^h(\mathbf{x})\,d\mathbf{x} &= \sum_{K\in\mathcal{T}_h}\int_K \phi^h(\mathbf{x})\,d\mathbf{x} = \sum_{K\in\mathcal{T}_h}\int_{\hat{K}} \hat{\phi}^h(\hat{\mathbf{x}})\,|\det J_K(\hat{\mathbf{x}})\,|d\hat{\mathbf{x}}\\ &\approx \sum_{K\in\mathcal{T}_h}\left(\sum_{l=1}^n \theta_l\hat{\phi}^h(\hat{\mathbf{x}}_l)\,|\det J_K(\hat{\mathbf{x}}_l)\,|\right). \end{aligned} \tag{7.21}$$

This approach requires the implementation of quadrature rules for reference mesh cells only.

By the definition of the finite element spaces in Section 7.4, the finite element functions are polynomials on $\hat{K}$. The choice of the quadrature rule in the computations presented in this monograph is determined by the polynomial degree of the non-linear convective term transformed to $\hat{K}$

$$\left(\hat{\mathbf{w}}_{\text{old}}^h \cdot \hat{\nabla}\right) \hat{\mathbf{w}}^h \cdot \hat{\mathbf{v}}^h. \tag{7.22}$$

This is the polynomial of the largest degree among the polynomials in (7.8).

Let $Q_m\left(\hat{K}\right)$ be the ansatz and test finite element space on the reference square or cube. Then, $\hat{\mathbf{w}}_{\text{old}}^h$ and $\hat{\mathbf{v}}^h$ are polynomials of degree dm and $\hat{\nabla}\hat{\mathbf{w}}^h$ of degree $dm-1$. Altogether, (7.22) possesses the degree $3dm-1$. The polynomial degree of (7.22) for $Q_1^{\text{rot}}\left(\hat{K}\right)$ is 5 in 2d and 3d. The construction of quadrature rules for the reference square or cube is simply done by tensor products of one dimensional quadrature rules which are exact for polynomials of degree $3m$ in 1d. We use Gaussian quadrature rules, called Gauss(l), which integrate polynomials of degree $2l-1$ exactly. From $3dm = d(2l-1)$, one can determine the order of the Gaussian quadrature rule which is sufficient for an exact integration of (7.22), see Table 7.3. For the $Q_1^{\text{rot}}\left(\hat{K}\right)$ finite element, the application of Gauss(2) suffices.

Using the ansatz spaces $P_m\left(\hat{K}\right)$ on the reference simplex, $\hat{\mathbf{w}}_{\text{old}}^h$ and $\hat{\mathbf{v}}^h$ have the polynomial degree m and $\hat{\nabla}\hat{\mathbf{w}}^h$ the degree $m-1$. Thus, the polynomial (7.22) has the degree $3m-1$. The degree of (7.22) for $P_1^{\text{nc}}\left(\hat{K}\right)$ is 2. The design of quadrature rules for higher order polynomials on simplicial reference cells is more difficult than on the unit square or unit cube. We give in Table 7.3 also the references where the quadrature rules which are used can be found in the literature. For an overview on quadrature rules, we refer to the papers by Cools and Rabinowitz [CR93] and by Cools [Coo99]. Nearly all quadrature rules which are used posses positive weights. Only the quadrature rule for polynomials of degree 8 on the tetrahedron has some negative weights.

Table 7.3. Number of quadrature points on the reference mesh cells and references to the literature

m	l	square	cube	degree of (7.22)	triangle		tetrahedron	
1	2	4	8	2	3		4	
2	4	16	64	5	7	[Str71]	15	[Str71]
3	5	25	125	8	15	[CH87]	43	[Bec92]

In general, we used the quadrature rules given above also for computing the discrete convolution. Tests with other quadrature rules on a hexahedral mesh showed only small differences in the computed results.

Remark 7.12. Quadrature errors. There are two sources of quadrature errors despite the choice of the quadrature rules as explained above.

1. The term $|\det J_K(\hat{\mathbf{x}})|$ is not constant such that the integrand in (7.21) is a higher order polynomial or, in general, not a polynomial at all. This happens if F_K is not an affine map. One situation where this occurs is the use of quadrilaterals or hexahedra which are not parallelograms or parallelepipeds. In these cases, F_K is a bilinear or trilinear map. Another situation is the use of isoparametric finite elements as explained in Remark 7.6. The use of quadrature rules for isoparametric triangles of degree 2 applied to the Laplace equation is studied in detail in Ciarlet [Cia91, Section 39]. It is shown that a quadrature rule which is exact for standard triangular finite elements of degree 2 suffices to maintain the optimal order of convergence in the error estimates. Also a numerical study of the two dimensional Navier-Stokes equations by John and Matthies [JM01] does not show a deterioration of the rates of convergence in computations using non-parallelogram mesh cells or isoparametric finite elements.
2. The function $\hat{\phi}^h(\hat{\mathbf{x}})$ in (7.21) is not a polynomial. This happens in the non-linear viscous term, e.g., in the term $\left\|\hat{\mathbb{D}}\left(\hat{\mathbf{w}}^h_{\text{old}}\right)\right\|_F \hat{\mathbb{D}}\left(\hat{\mathbf{w}}^h\right) : \hat{\mathbb{D}}\left(\hat{\mathbf{v}}^h\right)$ coming from the Smagorinsky subgrid scale model due to the square root in the definition of $\left\|\hat{\mathbb{D}}\left(\hat{\mathbf{w}}^h_{\text{old}}\right)\right\|_F$. Let $\hat{\mathbf{w}}^h_{\text{old}}, \hat{\mathbf{w}}^h$ be polynomials of degree dm on the reference square or cube. Then, $\hat{\mathbb{D}}\left(\hat{\mathbf{w}}^h\right), \hat{\mathbb{D}}\left(\hat{\mathbf{v}}^h\right)$ are also polynomials of degree dm and $\left\|\hat{\mathbb{D}}\left(\hat{\mathbf{w}}^h_{\text{old}}\right)\right\|_F$ is the square root of a polynomial of degree $2dm$. Thus, using the quadrature rule which is exact for polynomials of degree $3dm$ causes a quadrature error which does not dominate the approximation error of the finite element space. Similar considerations can be made for the Iliescu-Layton subgrid scale term (7.5) and for the reference simplices. Other terms which might lead to a non-polynomial integrand $\hat{\phi}^h(\hat{\mathbf{x}})$ are $\hat{\mathbf{f}}^h_{k-1} \cdot \hat{\mathbf{v}}^h$ and $\hat{\mathbf{f}}^h_k \cdot \hat{\mathbf{v}}^h$ since the right hand side $\mathbf{f}$ does not need to be a polynomial.

□

8

Error Analysis of Finite Element Discretisations of the LES Models

This chapter presents an error analysis of time-continuous finite element discretisations of the Smagorinsky model and the Taylor LES model.

The main part of this chapter, Section 8.1, deals with the Smagorinsky model and the analysis follows a paper by John and Layton [JL02]. It is often claimed that a straightforward finite element discretisation of this model can simulate the motion of turbulent flows with complexity independent of the viscosity ν or the Reynolds number Re, respectively, and depending only on the filter width δ. This phenomenon is investigated in Section 8.1 and finite element error estimates with constants independent of ν are proved. Assuming as little regularity of the solution of the continuous Smagorinsky model as possible, this proof requires the introduction of an additional constant viscosity $a_0(\delta) > 0$ into the Smagorinsky model, see Sections 8.1.4 and 8.1.5. If a higher regularity of this solution is assumed, an error estimate independent of ν can be proved for the standard Smagorinsky model (4.3), Section 8.1.6. The failure of the present analysis in some interesting cases is discussed in Section 8.1.7 and a numerical example which confirms the error estimates independent of ν is presented in Section 8.1.8.

The Taylor LES model with Smagorinsky subgrid scale term will be considered in Section 8.2. It was already seen in Section 6.2 that the unique solvability of this model can be proved under the assumption that the Smagorinsky subgrid scale term dominates the term coming from the Taylor LES model, see Remark 6.20 for a discussion of this situation. Thus, it is not surprisingly that under the same assumption a finite element error analysis of the Taylor LES model with Smagorinsky subgrid scale term is possible. Since the way to prove the error estimate is similar to the Smagorinsky model, only the most important steps and the final result are presented. The complete analysis can be found in Iliescu et al. [IJL02]. In contrast to the Smagorinsky model, the constants in the error estimate depend on ν.

8.1 The Smagorinsky Model

This section contains a finite element error analysis of the time-continuous discretisation of the Smagorinsky model which is based on John and Layton [JL02].

Let Ω be a bounded, simply connected domain in $\mathbb{R}^d$, $d = 2, 3$, with polygonal or polyhedral boundary. Thus, it is possible to decompose Ω into mesh cells without errors coming from boundary approximations. Suppose, the boundary $\partial\Omega$ is composed of faces (straight lines or parts of faces) $\Gamma_0, \ldots, \Gamma_J$ with meas $(\Gamma_0) > 0$.

We consider the Smagorinsky model with slip with linear friction and no penetration boundary conditions, see Section 5.2.4,

$$\begin{aligned} \mathbf{w}_t - \nabla \cdot ((2\nu + \nu_T)\, \mathbb{D}(\mathbf{w})) & \\ + (\mathbf{w} \cdot \nabla)\, \mathbf{w} + \nabla r &= \mathbf{f} \quad \text{in } (0, T] \times \Omega, \\ \nabla \cdot \mathbf{w} &= 0 \quad \text{in } [0, T] \times \Omega, \\ \mathbf{w} &= \mathbf{0} \quad \text{on } [0, T] \times \Gamma_0, \\ \mathbf{w} \cdot \mathbf{n}_{\partial\Omega} &= 0 \quad \text{on } [0, T] \times \Gamma_j, j = 1, \ldots, J, \\ \mathbf{w} \cdot \boldsymbol{\tau}_{j,k} + \beta^{-1} \mathbf{n}_{\partial\Omega}^T \mathbb{S}\, (\mathbf{w}, r)\, \boldsymbol{\tau}_{j,k} &= 0 \quad \text{on } [0, T] \times \Gamma_j, j = 1, \ldots, J, \\ \mathbf{w}\, (0, \cdot) &= \mathbf{w}_0 \text{ in } \Omega, \\ \int_\Omega r d\mathbf{x} &= 0 \quad \text{in } (0, T]. \end{aligned} \tag{8.1}$$

Here, $\{\boldsymbol{\tau}_{j,k}\}_{k=1}^{d-1}$ is an orthonormal system of tangential vectors in each point of Γ_j, $1, \ldots, J$, and $\mathbb{S}\, (\mathbf{w}, r)$ is the stress tensor

$$\mathbb{S}\, (\mathbf{w}, r) = (2\nu + \nu_T)\, \mathbb{D}(\mathbf{w}) - r\mathbb{I}.$$

The turbulent viscosity parameter is given by

$$\nu_T = a_0\, (\delta) + c_S \delta^2 \left\| \mathbb{D}(\mathbf{w}) \right\|_F, \quad a_0\, (\delta) \geq 0.$$

This is a generalisation of the definition given in Section 4.1.1.

Remark 8.1. Analysis of finite element discretisations for the stationary Smagorinsky model. Finite element methods for the stationary Smagorinsky model have been studied by Du and Gunzburger [DG90]. They could prove that the discrete solution converges to the solution of the continuous problem under minimal regularity assumptions on this solution. In addition, an optimal order finite element error estimate for $\left\| \mathbf{w} - \mathbf{w}^h \right\|_{H^1(\Omega)}$ is given. □

Remark 8.2. Finite element convergence analysis for the large eddies in stationary problems. A finite element convergence analysis for the large eddies defined by convolution with the Gaussian filter is available for the stationary Stokes and Navier-Stokes equations, see John and Layton [JL01] and Dunca et al. [DJL03a]. Under certain assumptions on the solution $(\mathbf{w}, r)$ of these equations, the finite element spaces and the filter width, it is shown that the filtered discrete velocity converges of higher order to $\overline{\mathbf{w}}$ in $L^2(\Omega)$ than the discrete velocity to $\mathbf{w}$, i.e. the convergence of $\left\|\overline{\mathbf{w}} - \overline{\mathbf{w}^h}\right\|_{L^2(\Omega)}$ for the mesh width $h \to 0$ is of higher order than the convergence of $\left\|\mathbf{w} - \mathbf{w}^h\right\|_{L^2(\Omega)}$. □

The analysis of the discretisation error covers inf-sup stable finite element discretisations in the so-called time-continuous case, i.e. a discretisation in time is not considered. This is a standard approach, e.g., see Heywood and Rannacher [HR82].

Let

$$
\begin{aligned}
V &= \left\{\mathbf{v} \in W^{1,3}(\Omega), \mathbf{v} = \mathbf{0} \text{ on } \Gamma_0, \mathbf{v}\cdot\mathbf{n}_{\partial\Omega} = 0 \text{ on } \Gamma_j, j = 1,\dots,J\right\},\\
Q &= L_0^2(\Omega).
\end{aligned}
$$

Since $\mathbf{v}\cdot\mathbf{n} = 0$ on $\partial\Omega$ for all $\mathbf{v} \in V$, Poincare's inequality (2.8) holds in V. In addition, we obtain also Korn's inequality in V

$$
\|\nabla\mathbf{v}\|_{L^3(\Omega)} \le C\|\mathbb{D}(\mathbf{v})\|_{L^3(\Omega)} \tag{8.2}
$$

by taking $|\mathbf{v}| = \|\mathbf{v}\|_{L^p(\Gamma_0)}$ in (2.10). The dual space of V is denoted by V^* and it is equipped with the norm

$$
\|\boldsymbol{\eta}\|_{V^*} := \sup_{\mathbf{v}\in V} \frac{\int_\Omega \boldsymbol{\eta}\cdot\mathbf{v}d\mathbf{x}}{\|\mathbb{D}(\mathbf{v})\|_{L^3(\Omega)}}. \tag{8.3}
$$

Note that $\|\mathbb{D}(\mathbf{v})\|_{L^3(\Omega)}$ defines a norm in V as a consequence of Poincare's and Korn's inequality. In the case $\partial\Omega = \Gamma_0$, it is $V^* = W^{-1,3/2}(\Omega)$ equipped with the norm (8.3).

We consider herein conform discretisations, i.e. the finite element spaces V^h and Q^h are subspaces of V and Q, respectively. The spaces of weakly divergence free and discrete divergence free functions are defined by

$$
\begin{aligned}
V_{\text{div}} &= \{\mathbf{v} \in V \;:\; (\nabla\cdot\mathbf{v}, q) = 0 \;\; \forall\, q \in Q\},\\
V_{\text{div}}^h &= \left\{\mathbf{v}^h \in V^h \;:\; \left(\nabla\cdot\mathbf{v}^h, q^h\right) = 0 \;\; \forall\, q^h \in Q^h\right\}.
\end{aligned}
$$

Using instead of a test function from Q in the definition of V_{div} a test function $q \in L^2(\Omega)$, the bilinear form still vanishes. Since every $q \in L^2(\Omega)$ admits a decomposition $q = q_0 + C$ with $q_0 \in Q$ and C is a constant, it follows

$$
\begin{aligned}
(\nabla\cdot\mathbf{v}, q) &= (\nabla\cdot\mathbf{v}, q_0 + C) = (\nabla\cdot\mathbf{v}, C) \\
&= -\int_\Gamma C\mathbf{v}\cdot\mathbf{n}_{\partial\Omega} ds - (\mathbf{v}, \nabla C) = 0 \quad \forall \mathbf{v} \in V_{\text{div}}, q \in L^2(\Omega).
\end{aligned} \tag{8.4}
$$

To simplify the notation, whenever $\boldsymbol{\tau}_j$ occurs, it will be understood that the term is summed over the two tangential vectors if $d = 3$, e.g.,

$$\|\mathbf{v}\cdot\boldsymbol{\tau}_j\|^2_{L^2(\Gamma_j)} := \|\mathbf{v}\cdot\boldsymbol{\tau}_{j,1}\|^2_{L^2(\Gamma_j)} + \|\mathbf{v}\cdot\boldsymbol{\tau}_{j,2}\|^2_{L^2(\Gamma_j)}\,.$$

In the error estimates, we will use a variation of Gronwall's lemma in differential form, Lemma 2.1.

Lemma 8.3. Variation of Gronwall's lemma in differential form. *Let $T \in \mathbb{R}^+ \cup \infty$, $f \in W^{1,1}(0,T)$ and $h, g, \lambda \in L^1(0,T)$ and $h(t), \lambda(t) \geq 0$ a.e. in $(0,T)$. Then,*

$$f'(t) + h(t) \leq g(t) + \lambda(t) f(t) \quad \text{a.e. in } [0,T] \tag{8.5}$$

implies for almost all $t \in [0,T]$

$$\begin{aligned} f(t) &+ \int_0^t h(s)\,ds \\ &\leq \exp\left(\int_0^t \lambda(\tau)\,d\tau\right) f(0) + \int_0^t \exp\left(\int_s^t \lambda(\tau)\,d\tau\right) g(s)\,ds. \end{aligned} \tag{8.6}$$

Moreover, if $g(t) \geq 0$ a.e. in $(0,T)$, it holds

$$f(t) + \int_0^t h(s)\,ds \leq \exp\left(\int_0^t \lambda(\tau)\,d\tau\right)\left(f(0) + \int_0^t g(s)\,ds\right). \tag{8.7}$$

Proof. From (8.5) follows a.e. in $[0,T]$

$$f'(s) - \lambda(s) f(s) + h(s) \leq g(s)\,.$$

The positivity of the exponential implies

$$\begin{aligned} \exp\left(-\int_0^s \lambda(\tau)\,d\tau\right) & \left(f'(s) - \lambda(s) f(s) + h(s)\right) \\ & \leq \exp\left(-\int_0^s \lambda(\tau)\,d\tau\right) g(s)\,. \end{aligned}$$

Integration on $(0,t) \subset [0,T]$ gives

$$\begin{aligned} \exp\left(-\int_0^t \lambda(\tau)\,d\tau\right) f(t) - f(0) &+ \int_0^t \exp\left(-\int_0^s \lambda(\tau)\,d\tau\right) h(s)\,ds \\ &\leq \int_0^t \exp\left(-\int_0^s \lambda(\tau)\,d\tau\right) g(s)\,ds. \end{aligned} \tag{8.8}$$

The monotonicity of the exponential implies

$$\exp\left(-\int_0^t \lambda(\tau)\,d\tau\right) \int_0^t h(s)\,ds \leq \int_0^t \exp\left(-\int_0^s \lambda(\tau)\,d\tau\right) h(s)\,ds.$$

Using this inequality to estimate the left hand side of (8.8) from below and multiplication of the resulting inequality with $\exp\left(\int_0^t \lambda(\tau)\,d\tau\right)$ prove (8.6).

If g is non-negative, we obtain

$$\int_0^t \exp\left(\int_s^t \lambda(\tau)\, d\tau\right) g(s)\, ds \le \exp\left(\int_0^t \lambda(\tau)\, d\tau\right) \int_0^t g(s)\, ds$$

from which (8.7) follows. □

8.1.1 The Variational Formulation and Stability Estimates

Now we turn to the variational formulation of (8.1). We consider a variational formulation with the skew-symmetric form of the non-linear convection term and with the possibility of a least squares stabilisation of the momentum equation.

The variational problem is to find $(\mathbf{w}, r) \in V \times Q$ such that

i) for all $t \in (0, T]$ and all $(\mathbf{v}, q) \in V \times Q$

$$(\mathbf{w}_t, \mathbf{v}) + a(\mathbf{w}, \mathbf{w}, \mathbf{v}) + b_s(\mathbf{w}, \mathbf{w}, \mathbf{v}) + (q, \nabla \cdot \mathbf{w}) - (r, \nabla \cdot \mathbf{v}) = (\mathbf{f}, \mathbf{v}) \quad (8.9)$$

with

$$\begin{aligned} a(\mathbf{u}, \mathbf{w}, \mathbf{v}) &= \alpha_s (\nabla \cdot \mathbf{w}, \nabla \cdot \mathbf{v}) \\ &\quad + \left(\left(2\nu + a_0(\delta) + c_S \delta^2 \|\mathbb{D}(\mathbf{u})\|_F\right) \mathbb{D}(\mathbf{w}), \mathbb{D}(\mathbf{v})\right) \\ &\quad + \sum_{j=1}^{J} \beta (\mathbf{w} \cdot \boldsymbol{\tau}_j, \mathbf{v} \cdot \boldsymbol{\tau}_j)_{\Gamma_j} \end{aligned}$$

where $\alpha_s > 0$ is given and

$$b_s(\mathbf{u}, \mathbf{w}, \mathbf{v}) = \frac{1}{2}\left(b(\mathbf{u}, \mathbf{w}, \mathbf{v}) - b(\mathbf{u}, \mathbf{v}, \mathbf{w})\right).$$

ii) $\mathbf{w}(0, \mathbf{x}) = \mathbf{w}_0(\mathbf{x})$.

The main properties of the non-linear viscous or Smagorinsky term $\left(\|\mathbb{D}(\mathbf{w})\|_F \mathbb{D}(\mathbf{w}), \mathbb{D}(\mathbf{v})\right)$ which are needed in the analysis are its strong monotonicity and a local Lipschitz continuity. These properties will be proved in Lemma 8.5. The following lemma gives an auxiliary result.

Lemma 8.4. *Let $A \in L^3(\Omega)$ with $A(\mathbf{x}) \in \mathbb{R}^{d\times d}$ for every $\mathbf{x} \in \Omega$, then*

$$\|A\|_{L^3(\Omega)} \le \| \|A\|_F \|_{L^3(\Omega)} \le c(d) \|A\|_{L^3(\Omega)}. \quad (8.10)$$

Proof. It is

$$\|A\|_{L^3(\Omega)}^3 = \int_\Omega \sum_{i,j=1}^{d} |a_{ij}|^3, \qquad \| \|A\|_F \|_{L^3(\Omega)}^3 = \int_\Omega \left(\sum_{i,j=1}^{d} a_{ij}^2\right)^{3/2}.$$

Since any matrix norms are equivalent, there are constants $0 < c_1(d) < c_2(d)$ such that

$$c_1(d)\left(\sum_{i,j=1}^{d} |a_{ij}|^3\right)^{1/3} \le \left(\sum_{i,j=1}^{d} a_{ij}^2\right)^{1/2} \le c_2(d)\left(\sum_{i,j=1}^{d} |a_{ij}|^3\right)^{1/3}.$$

From (2.5) follows with $p = 3/2$ that one can choose $c_1(d) = 1$. Raising the inequality of the matrix norms to the power 3 and integrating on Ω proves (8.10). □

Lemma 8.5. Properties of the non-linear viscous term. *There are constants* $\underline{C}$ *and* $\overline{C}$ *such that for all* $\mathbf{u}, \mathbf{v}, \mathbf{w} \in W^{1,3}(\Omega)$

$$\left(\|\mathbb{D}(\mathbf{u})\|_F \mathbb{D}(\mathbf{u}) - \|\mathbb{D}(\mathbf{v})\|_F \mathbb{D}(\mathbf{v}), \mathbb{D}(\mathbf{u}-\mathbf{v})\right) \ge \underline{C}\, \|\mathbb{D}(\mathbf{u}-\mathbf{v})\|_{L^3(\Omega)}^3 \quad (8.11)$$

(strong monotonicity) and

$$\begin{aligned} &\left(\|\mathbb{D}(\mathbf{u})\|_F \mathbb{D}(\mathbf{u}) - \|\mathbb{D}(\mathbf{v})\|_F \mathbb{D}(\mathbf{v}), \mathbb{D}(\mathbf{w})\right) \\ &\le \overline{C} C_L \|\mathbb{D}(\mathbf{u}-\mathbf{v})\|_{L^3(\Omega)} \|\mathbb{D}(\mathbf{w})\|_{L^3(\Omega)} \end{aligned} \quad (8.12)$$

with $C_L = \max\left\{\|\mathbb{D}(\mathbf{u})\|_{L^3(\Omega)}, \|\mathbb{D}(\mathbf{v})\|_{L^3(\Omega)}\right\}$ *(local Lipschitz continuity).*

Proof. The proof of the strong monotonicity starts like that of Lemma 6.9. In this lemma, the monotonicity of the Smagorinsky term has been proved. To prove the strong monotonicity, one has to estimate one term more carefully than it was done in Lemma 6.9, namely

$$\int_0^1 \sum_{i,j=1}^{d} \nu_S \|\nabla \mathbf{w}^\tau\|_F \left(\frac{\partial \mathbf{w}_i'}{\partial x_j} - \frac{\partial \mathbf{w}_i''}{\partial x_j}\right)^2 d\tau,$$

which was estimated from below by zero in this lemma. We start the estimate of this term using the equivalence of norms in finite dimensional spaces

$$\begin{aligned} &\int_0^1 \sum_{i,j=1}^{d} \nu_S \|\nabla \mathbf{w}^\tau\|_F \left(\frac{\partial \mathbf{w}_i'}{\partial x_j} - \frac{\partial \mathbf{w}_i''}{\partial x_j}\right)^2 d\tau \\ &\ge C\nu_S \int_0^1 \sum_{i,j=1}^{d} \left(\sum_{k,l=1}^{d} \left|\frac{\partial \mathbf{w}_k^\tau}{\partial x_l}\right|\right) \left(\frac{\partial \mathbf{w}_i'}{\partial x_j} - \frac{\partial \mathbf{w}_i''}{\partial x_j}\right)^2 d\tau \\ &= C\nu_S \sum_{i,j=1}^{d} \sum_{k,l=1}^{d} \left(\int_0^1 \left|\tau \frac{\partial \mathbf{w}_k'}{\partial x_l} + (1-\tau)\frac{\partial \mathbf{w}_k''}{\partial x_l}\right| d\tau\right) \left(\frac{\partial \mathbf{w}_i'}{\partial x_j} - \frac{\partial \mathbf{w}_i''}{\partial x_j}\right)^2. \end{aligned}$$

We apply now the estimate

$$\int_0^1 |\tau a + (1-\tau) b| \, d\tau \geq \frac{1}{4} |a - b| , \quad a, b \in \mathbb{R}.$$

To prove this estimate, one has to distinguish the three cases $a, b, \geq 0$, $a, b \leq 0$ and $a > 0, b \leq 0$. In the first and second case, one gets easily

$$\int_0^1 |\tau a + (1-\tau) b| \, d\tau = \frac{1}{2} |a + b| \geq \frac{1}{4} |a - b| .$$

In the third case, one obtains

$$\int_0^1 |\tau a + (1-\tau) b| \, d\tau = \frac{1}{2} \frac{a^2 + b^2}{a - b} \geq \frac{1}{4} (a - b) = \frac{1}{4} |a - b| .$$

Thus,

$$\begin{aligned}
&\int_0^1 \sum_{i,j=1}^{d} \nu_S \left\| \nabla \mathbf{w}^\tau \right\|_F \left(\frac{\partial \mathbf{w}_i'}{\partial x_j} - \frac{\partial \mathbf{w}_i''}{\partial x_j} \right)^2 d\tau \\
&\geq C \nu_S \sum_{i,j=1}^{d} \sum_{k,l=1}^{d} \left(\frac{1}{4} \left| \frac{\partial \mathbf{w}_k'}{\partial x_l} - \frac{\partial \mathbf{w}_k''}{\partial x_l} \right| \right) \left(\frac{\partial \mathbf{w}_i'}{\partial x_j} - \frac{\partial \mathbf{w}_i''}{\partial x_j} \right)^2 \\
&\geq C \nu_S \sum_{i,j=1}^{d} \left(\left| \frac{\partial \mathbf{w}_i'}{\partial x_j} - \frac{\partial \mathbf{w}_i''}{\partial x_j} \right| \right) \left(\frac{\partial \mathbf{w}_i'}{\partial x_j} - \frac{\partial \mathbf{w}_i''}{\partial x_j} \right)^2 \\
&= C \nu_S \sum_{i,j=1}^{d} \left| \frac{\partial \mathbf{w}_i'}{\partial x_j} - \frac{\partial \mathbf{w}_i''}{\partial x_j} \right|^3 .
\end{aligned}$$

In the last estimate, we have dropped non-negative terms. Combining this estimate with Lemma 6.9 proves the strong monotonicity of the Smagorinsky term.

The local Lipschitz continuity is proved by first applying Hölder's inequality. We obtain

$$\begin{aligned}
&\left(\| \mathbb{D}(\mathbf{u}) \|_F \, \mathbb{D}(\mathbf{u}) - \| \mathbb{D}(\mathbf{v}) \|_F \, \mathbb{D}(\mathbf{v}) , \mathbb{D}(\mathbf{w}) \right) \\
&= \left(\| \mathbb{D}(\mathbf{u}) \|_F \, \mathbb{D}(\mathbf{u}) - \| \mathbb{D}(\mathbf{u}) \|_F \, \mathbb{D}(\mathbf{v}) , \mathbb{D}(\mathbf{w}) \right) \\
&\quad + \left(\| \mathbb{D}(\mathbf{u}) \|_F \, \mathbb{D}(\mathbf{v}) - \| \mathbb{D}(\mathbf{v}) \|_F \, \mathbb{D}(\mathbf{v}) , \mathbb{D}(\mathbf{w}) \right) \\
&\leq \left\| \| \mathbb{D}(\mathbf{u}) \|_F \right\|_{L^3(\Omega)} \| \mathbb{D}(\mathbf{u} - \mathbf{v}) \|_{L^3(\Omega)} \| \mathbb{D}(\mathbf{w}) \|_{L^3(\Omega)} \\
&\quad + \left\| \| \mathbb{D}(\mathbf{u}) \|_F - \| \mathbb{D}(\mathbf{v}) \|_F \right\|_{L^3(\Omega)} \| \mathbb{D}(\mathbf{v}) \|_{L^3(\Omega)} \| \mathbb{D}(\mathbf{w}) \|_{L^3(\Omega)} .
\end{aligned}$$

We get from Lemma 8.4

$$\left\| \| \mathbb{D}(\mathbf{u}) \|_F \right\|_{L^3(\Omega)}^3 \leq C \| \mathbb{D}(\mathbf{u}) \|_{L^3(\Omega)}^3$$

and using in addition the triangle inequality yields

$$
\begin{aligned}
\left\| \|\mathbb{D}(\mathbf{u})\|_F - \|\mathbb{D}(\mathbf{v})\|_F \right\|_{L^3(\Omega)} &\leq \left\| \|\mathbb{D}(\mathbf{u}) - \mathbb{D}(\mathbf{v})\|_F \right\|_{L^3(\Omega)} \\
&\leq C \|\mathbb{D}(\mathbf{u}-\mathbf{v})\|_{L^3(\Omega)} .
\end{aligned}
$$

Combining all estimates proves the local Lipschitz continuity. □

Lemma 8.6. Leray's inequality for the solution of (8.9). *A solution of (8.9) satisfies*

$$
\begin{aligned}
\frac{1}{2} \|\mathbf{w}(T,\mathbf{x})\|_{L^2(\Omega)}^2 + \int_0^T \Bigg(\sum_{j=1}^J \beta \|\mathbf{w}\cdot\boldsymbol{\tau}_j\|_{L^2(\Gamma_j)}^2 + (2\nu + a_0(\delta)) \|\mathbb{D}(\mathbf{w})\|_{L^2(\Omega)}^2 \\
+ \underline{C} c_S \delta^2 \|\mathbb{D}(\mathbf{w})\|_{L^3(\Omega)}^3 \Bigg) dt \\
\leq \frac{1}{2} \|\mathbf{w}_0(\mathbf{x})\|_{L^2(\Omega)}^2 + \int_0^T (\mathbf{f},\mathbf{w})\, dt. \qquad (8.13)
\end{aligned}
$$

Proof. Choosing $(\mathbf{v}, q) = (\mathbf{w}, r)$ in (8.9) gives

$$
(\mathbf{w}_t, \mathbf{w}) + a(\mathbf{w},\mathbf{w},\mathbf{w}) + b_s(\mathbf{w},\mathbf{w},\mathbf{w}) + (r, \nabla\cdot\mathbf{w}) - (r, \nabla\cdot\mathbf{w}) = (\mathbf{f},\mathbf{w}) .
$$

The skew symmetric non-linear convective term vanishes. From the other tri-linear term vanishes the part $a_s(\nabla\cdot\mathbf{w}, \nabla\cdot\mathbf{w})$ because $\mathbf{w} \in V_{\text{div}}$, $\nabla\cdot\mathbf{w} \in L^2(\Omega)$ and (8.4). In addition,

$$
\left(\|\mathbb{D}(\mathbf{w})\|_F \mathbb{D}(\mathbf{w}), \mathbb{D}(\mathbf{v}) \right) \geq \underline{C} \|\mathbb{D}(\mathbf{w})\|_{L^3(\Omega)}^3
$$

by (8.11). Thus

$$
\begin{aligned}
(\mathbf{w}_t, \mathbf{w}) + (2\nu + a_0(\delta)) \|\mathbb{D}(\mathbf{w})\|_{L^2(\Omega)}^2 \\
+ \underline{C} c_S \delta^2 \|\mathbb{D}(\mathbf{w})\|_{L^3(\Omega)}^3 + \sum_{j=1}^J \beta \|\mathbf{w}\cdot\boldsymbol{\tau}_j\|_{L^2(\Gamma_j)}^2 \leq (\mathbf{f},\mathbf{w}) .
\end{aligned}
$$

Integration on $(0,T)$ yields the statement of the lemma. □

The discrete problem is defined in finite element spaces V^h and Q^h with $V^h \subset V$ and $Q^h \subset Q$. We assume that the spaces V^h and Q^h fulfil the inf-sup condition

$$
\inf_{\lambda^h \in Q^h} \sup_{\mathbf{v}^h \in V^h} \frac{(\lambda^h, \nabla\cdot\mathbf{v}^h)}{\|\lambda^h\|_{L^2(\Omega)} \left[\left\|\mathbb{D}(\mathbf{v}^h)\right\|_{L^2(\Omega)}^2 + \sum_{j=1}^J \left\|\mathbf{v}^h\cdot\boldsymbol{\tau}_j\right\|_{H^{1/2}(\Gamma_j)}^2 \right]^{1/2}} \geq C, \qquad (8.14)
$$

where $C > 0$ is independent of h.

Lemma 8.7. *If* $\left(V^h, Q^h\right)$ *satisfies*

$$\inf_{\lambda_h \in Q^h} \sup_{\mathbf{v}_h \in V^h \cap H_0^1(\Omega)} \frac{\left(\lambda^h, \nabla \cdot \mathbf{v}^h\right)}{\|\lambda^h\|_{L^2(\Omega)} \ \|\nabla \mathbf{v}^h\|_{L^2(\Omega)}} \geq C_1 > 0,$$

then (8.14) holds.

Proof. By trace theorem, Grisvard [Gri92], and the Poincaré inequality (2.8), follows for any $\lambda_h \ (\neq 0) \in Q^h$, $\mathbf{v}^h \ (\neq \mathbf{0}) \in V^h$

$$\frac{\left(\lambda^h, \nabla \cdot \mathbf{v}^h\right)}{\|\lambda^h\|_{L^2(\Omega)} \left[\|\mathbb{D}(\mathbf{v}^h)\|_{L^2(\Omega)}^2 + \sum_{j=1}^{J} \left\|\mathbf{v}^h \cdot \boldsymbol{\tau}_j\right\|_{H^{1/2}(\Gamma_j)}^2\right]^{1/2}}$$
$$\geq C \frac{\left(\lambda^h, \nabla \cdot \mathbf{v}^h\right)}{\|\lambda^h\|_{L^2(\Omega)} \|\mathbf{v}^h\|_{H^1(\Omega)}}$$
$$\geq C \frac{\left(\lambda^h, \nabla \cdot \mathbf{v}^h\right)}{\|\lambda^h\|_{L^2(\Omega)} \|\nabla \mathbf{v}^h\|_{L^2(\Omega)}}.$$

□

The continuous-in-time finite element method seeks to find $\left(\mathbf{w}^h, r^h\right) \in V^h \times Q^h$ such that

$$\begin{aligned} \left(\mathbf{w}_t^h, \mathbf{v}^h\right) + a\left(\mathbf{w}^h, \mathbf{w}^h, \mathbf{v}^h\right) + b_s\left(\mathbf{w}^h, \mathbf{w}^h, \mathbf{v}^h\right) \\ + \left(q^h, \nabla \cdot \mathbf{w}^h\right) - \left(r^h, \nabla \cdot \mathbf{v}^h\right) = \left(\mathbf{f}, \mathbf{v}^h\right) \end{aligned} \tag{8.15}$$

for all $\left(\mathbf{v}^h, q^h\right) \in V^h \times Q^h$ where $\mathbf{w}^h(0, \mathbf{x})$ is an approximation to $\mathbf{w}_0(\mathbf{x})$.

Lemma 8.8. Leray's inequality for the solution of (8.15). *A solution of (8.15) satisfies*

$$\frac{1}{2}\left\|\mathbf{w}^h(T, \mathbf{x})\right\|_{L^2(\Omega)}^2 + \int_0^T \left(\alpha_s \left\|\nabla \cdot \mathbf{w}^h\right\|_{L^2(\Omega)}^2 + \sum_{j=1}^{J} \beta \left\|\mathbf{w}^h \cdot \boldsymbol{\tau}_j\right\|_{L^2(\Gamma_j)}^2\right.$$
$$\left. + \left(2\nu + a_0(\delta)\right)\left\|\mathbb{D}\left(\mathbf{w}^h\right)\right\|_{L^2(\Omega)}^2 + \underline{C} c_S \delta^2 \left\|\mathbb{D}\left(\mathbf{w}^h\right)\right\|_{L^3(\Omega)}^3\right) dt$$
$$\leq \frac{1}{2}\|\mathbf{w}_0(\mathbf{x})\|_{L^2(\Omega)}^2 + \int_0^T \left(\mathbf{f}, \mathbf{w}^h\right) dt.$$

Proof. The proof proceeds in the same way as the proof of Lemma 8.6. □

We give now several stability estimates which are uniform in ν for the velocity components $\mathbf{w}$ of the solution of (8.9) and $\mathbf{w}^h$ of the solution of (8.15).

Theorem 8.9. Stability of $\mathbf{w}$. *The velocity component* $\mathbf{w}$ *of the solution of (8.9) satisfies for* $T > 0$

$$\frac{1}{2}\|\mathbf{w}(T,\mathbf{x})\|_{L^2(\Omega)}^2 + \int_0^T \left(\sum_{j=1}^J \beta \|\mathbf{w}\cdot\boldsymbol{\tau}_j\|_{L^2(\Gamma_j)}^2 + (2\nu + a_0(\delta))\|\mathbb{D}(\mathbf{w})\|_{L^2(\Omega)}^2 \right.$$
$$\left. + \frac{2}{3}\underline{C}_{CS}\delta^2\|\mathbb{D}(\mathbf{w})\|_{L^3(\Omega)}^3\right) dt \tag{8.16}$$
$$\leq \frac{1}{2}\|\mathbf{w}_0(\mathbf{x})\|_{L^2(\Omega)}^2 + \frac{2}{3}\left(\underline{C}_{CS}\right)^{-1/2}\delta^{-1}\left(\sup_{\mathbf{v}\in L^3(0,T;V)}\frac{\int_0^T(\mathbf{f},\mathbf{v})\,dt}{\|\mathbb{D}(\mathbf{v})\|_{L^3(0,T;L^3(\Omega))}}\right)^{3/2},$$

$$\frac{1}{2}\|\mathbf{w}(T,\mathbf{x})\|_{L^2(\Omega)}^2 + \int_0^T e^{T-t}\left(\sum_{j=1}^J \beta\|\mathbf{w}\cdot\boldsymbol{\tau}_j\|_{L^2(\Gamma_j)}^2\right.$$
$$\left. + (2\nu + a_0(\delta))\|\mathbb{D}(\mathbf{w})\|_{L^2(\Omega)}^2 + \underline{C}_{CS}\delta^2\|\mathbb{D}(\mathbf{w})\|_{L^3(\Omega)}^3\right) dt \tag{8.17}$$
$$\leq \frac{e^T}{2}\|\mathbf{w}_0(\mathbf{x})\|_{L^2(\Omega)}^2 + \frac{1}{2}\int_0^T e^{T-t}\|\mathbf{f}\|_{L^2(\Omega)}^2\,dt.$$

Proof. (8.16): This estimate is proved by an application of Young's inequality and Leray's inequality (8.13). It is

$$\int_0^T(\mathbf{f},\mathbf{w})\,dt = \|\mathbb{D}(\mathbf{w})\|_{L^3(0,T;L^3(\Omega))}\frac{\int_0^T(\mathbf{f},\mathbf{w})\,dt}{\|\mathbb{D}(\mathbf{w})\|_{L^3(0,T;L^3(\Omega))}}$$
$$\leq \|\mathbb{D}(\mathbf{w})\|_{L^3(0,T;L^3(\Omega))}\sup_{\mathbf{v}\in L^3(0,T;L^3(\Omega))}\frac{\int_0^T(\mathbf{f},\mathbf{v})\,dt}{\|\mathbb{D}(\mathbf{v})\|_{L^3(0,T;L^3(\Omega))}}$$
$$\leq \frac{\underline{C}_{CS}\delta^2}{3}\|\mathbb{D}(\mathbf{w})\|_{L^3(0,T;L^3(\Omega))}^3$$
$$+\frac{2}{3}\left(\underline{C}_{CS}\right)^{-1/2}\delta^{-1}\left(\sup_{\mathbf{v}\in L^3(0,T;L^3(\Omega))}\frac{\int_0^T(\mathbf{f},\mathbf{v})\,dt}{\|\mathbb{D}(\mathbf{v})\|_{L^3(0,T;L^3(\Omega))}}\right)^{3/2}.$$

(8.17): In the same way as in the proof of Lemma 8.6, we obtain with the Cauchy-Schwarz inequality and Young's inequality

$$\frac{1}{2}\frac{d}{dt}\|\mathbf{w}\|_{L^2(\Omega)}^2 + (2\nu + a_0(\delta))\|\mathbb{D}(\mathbf{w})\|_{L^2(\Omega)}^2$$
$$+\underline{C}_{CS}\delta^2\|\mathbb{D}(\mathbf{w})\|_{L^3(\Omega)}^3 + \sum_{j=1}^J\beta\|\mathbf{w}\cdot\boldsymbol{\tau}_j\|_{L^2(\Gamma_j)}^2 \leq \frac{\|\mathbf{f}\|_{L^2(\Omega)}^2}{2} + \frac{\|\mathbf{w}\|_{L^2(\Omega)}^2}{2}.$$

Multiplying this equality with e^{T-t}, integrating on $(0,T)$ and using

$$\int_0^T e^{T-t}\left(\frac{d}{dt}\left\|\mathbf{w}\right\|_{L^2(\Omega)}^2 - \left\|\mathbf{w}\right\|_{L^2(\Omega)}^2\right) dt$$
$$= \left\|\mathbf{w}\left(T,\mathbf{x}\right)\right\|_{L^2(\Omega)}^2 - e^T\left\|\mathbf{w}\left(0,\mathbf{x}\right)\right\|_{L^2(\Omega)}^2$$

concludes the proof. □

Theorem 8.10. Stability of $\mathbf{w}_t$. *Let $(\mathbf{w},r)$ be the solution of (8.9). Then, there is a constant C independent of ν such that for almost all $t\in[0,T]$*

$$\left\|\mathbf{w}_t\right\|_{V^*} \le C\Big(\left\|\mathbf{w}\right\|_{L^3(\Omega)}^2 + \left\|r\right\|_{L^{3/2}(\Omega)} + \left(2\nu + a_0\left(\delta\right)\right)\left\|\mathbb{D}\left(\mathbf{w}\right)\right\|_{L^{3/2}(\Omega)}$$
$$+ c_S\delta^2\left\|\mathbb{D}\left(\mathbf{w}\right)\right\|_{L^3(\Omega)}^2 + \left\|\mathbf{f}\right\|_{V^*}\Big),$$

$$\left\|\mathbf{w}_t\right\|_{L^{3/2}(0,T;V^*)}^{3/2} \le C\Big(\left\|\mathbf{w}\right\|_{L^3(0,T;L^3(\Omega))}^3 + \left\|r\right\|_{L^{3/2}(0,T;L^{3/2}(\Omega))}^{3/2}$$
$$+ \left(2\nu + a_0\left(\delta\right)\right)\left\|\mathbb{D}\left(\mathbf{w}\right)\right\|_{L^{3/2}(0,T;L^{3/2}(\Omega))}^{3/2}$$
$$+ c_S\delta^2\left\|\mathbb{D}\left(\mathbf{w}\right)\right\|_{L^3(0,T;L^3(\Omega))}^3 + \left\|\mathbf{f}\right\|_{L^{3/2}(0,T;V^*)}^{3/2}\Big).$$

Proof. Dividing (8.9) by $\left\|\mathbb{D}\left(\mathbf{v}\right)\right\|_{L^3(\Omega)}$, taking the supremum over $\mathbf{v}\in V$ and applying the triangle inequality give

$$\left\|\mathbf{w}_t\right\|_{V^*} \le \left\|\nabla\cdot\left(\mathbf{w}\mathbf{w}^T\right)\right\|_{V^*} + \left\|\nabla r\right\|_{V^*} + c_S\delta^2\left\|\nabla\cdot\left(\left\|\mathbb{D}\left(\mathbf{w}\right)\right\|_F\mathbb{D}\left(\mathbf{w}\right)\right)\right\|_{V^*} +$$
$$\left(2\nu + a_0\left(\delta\right)\right)\left\|\nabla\cdot\mathbb{D}\left(\mathbf{w}\right)\right\|_{V^*} + \left\|\mathbf{f}\right\|_{V^*}.$$

The definition of the norm (8.3), integration by parts, using $\mathbf{v}\cdot\mathbf{n}_{\partial\Omega}=0$ on $\partial\Omega$ for $\mathbf{v}\in V$, Hölder's inequality and Korn's inequality give, e.g.,

$$\left\|\nabla r\right\|_{V^*} = \sup_{\mathbf{v}\in V}\frac{\int_\Omega -r\left(\nabla\cdot\mathbf{v}\right)d\mathbf{x}}{\left\|\mathbb{D}\left(\mathbf{v}\right)\right\|_{L^3(\Omega)}} \le \sup_{\mathbf{v}\in V}\frac{\left\|r\right\|_{L^{3/2}(\Omega)}\left\|\nabla\cdot\mathbf{v}\right\|_{L^3(\Omega)}}{\left\|\mathbb{D}\left(\mathbf{v}\right)\right\|_{L^3(\Omega)}}$$
$$\le C\left\|r\right\|_{L^{3/2}(\Omega)}.$$

The other terms are estimated in the same way using also

$$\left\|\nabla\cdot\left(\mathbf{w}\mathbf{w}^T\right)\right\|_{L^{3/2}(\Omega)} = \left\|\mathbf{w}\right\|_{L^3(\Omega)}^2$$

and

$$\left\|\left\|\mathbb{D}\left(\mathbf{w}\right)\right\|_F\mathbb{D}\left(\mathbf{w}\right)\right\|_{L^{3/2}(\Omega)} = \left\|\mathbb{D}\left(\mathbf{w}\right)\right\|_{L^3(\Omega)}^2.$$

The second inequality follows by raising both sides of the first estimate to the power $3/2$ and integrating in time. □

Stability estimates for $\mathbf{w}^h$ are obtained in the same way as the stability estimates for $\mathbf{w}$ proved in Theorem 8.9.

Theorem 8.11. Stability of $\mathbf{w}^h$. *The velocity component* $\mathbf{w}^h$ *of the solution of (8.15) satisfies for* $T > 0$

$$
\begin{aligned}
&\frac{1}{2}\left\|\mathbf{w}^h(T,\mathbf{x})\right\|_{L^2(\Omega)}^2+\int_0^T\left(\alpha_s\left\|\nabla\cdot\mathbf{w}^h\right\|_{L^2(\Omega)}^2+\sum_{j=1}^J\beta\left\|\mathbf{w}^h\cdot\boldsymbol{\tau}_j\right\|_{L^2(\Gamma_j)}^2\right.\\
&\quad\left.+\left(2\nu+a_0(\delta)\right)\left\|\mathbb{D}\left(\mathbf{w}^h\right)\right\|_{L^2(\Omega)}^2+\frac{2}{3}\underline{C}_{CS}\delta^2\left\|\mathbb{D}\left(\mathbf{w}^h\right)\right\|_{L^3(\Omega)}^3\right)dt \qquad (8.18)\\
&\le\frac{1}{2}\left\|\mathbf{w}_0^h(\mathbf{x})\right\|_{L^2(\Omega)}^2+\frac{2}{3}\left(\underline{C}_{CS}\right)^{-1/2}\delta^{-1}\left(\sup_{\mathbf{v}\in L^3(0,T;V)}\frac{\int_0^T(\mathbf{f},\mathbf{v})\,dt}{\|\mathbb{D}(\mathbf{v})\|_{L^3(0,T;L^3(\Omega))}}\right)^{3/2},
\end{aligned}
$$

$$
\begin{aligned}
&\frac{1}{2}\left\|\mathbf{w}^h(T,\mathbf{x})\right\|_{L^2(\Omega)}^2+\int_0^T e^{T-t}\left(\alpha_s\left\|\nabla\cdot\mathbf{w}^h\right\|_{L^2(\Omega)}^2+\sum_{j=1}^J\beta\left\|\mathbf{w}^h\cdot\boldsymbol{\tau}_j\right\|_{L^2(\Gamma_j)}^2\right.\\
&\quad\left.+\left(2\nu+a_0(\delta)\right)\left\|\mathbb{D}\left(\mathbf{w}^h\right)\right\|_{L^2(\Omega)}^2+\underline{C}_{CS}\delta^2\left\|\mathbb{D}\left(\mathbf{w}^h\right)\right\|_{L^3(\Omega)}^3\right)dt \qquad (8.19)\\
&\le\frac{e^T}{2}\left\|\mathbf{w}_0^h(\mathbf{x})\right\|_{L^2(\Omega)}^2+\frac{1}{2}\int_0^T e^{T-t}\|\mathbf{f}\|_{L^2(\Omega)}^2\,dt.
\end{aligned}
$$

8.1.2 Goal of the Error Analysis and Outline of the Proof

The goal of the analysis is to estimate the error $\left\|\mathbf{w}-\mathbf{w}^h\right\|$ in appropriate norms under consideration of the following aspects:

- The error estimate should be independent of ν.
- The assumption on the regularity of the solution of the variational problem (8.9) should be as weak as possible.

The natural regularity for the weak solution of the Smagorinsky model is

$$\nabla\mathbf{w}\in L^3\left(0,T;L^3(\Omega)\right). \qquad (8.20)$$

This regularity is required in order that the Smagorinsky term is well defined, see Section 5.3. In addition, Ladyzhenskaya could show for (8.20) the unique solvability of the weak problem, see Section 6.1.

It turns out that in the presented analysis a ν-independent error estimate assuming (8.20) is only possible if $a_0(\delta) > 0$, see Sections 8.1.4 and 8.1.5.

Since the error analysis involves a lot of technical details, we like to give in advance the plan of the proof.

1. Subtract the variational formulation (8.9) and the discrete problem (8.15) with arbitrary test functions $\left(\mathbf{v}^h, q^h\right)\in V^h\times Q^h$. This gives the error equation (8.23).

2. Choose an arbitrary $\tilde{\mathbf{w}} \in V_{\text{div}}^h$ and split the error $\mathbf{e} = (\mathbf{w} - \tilde{\mathbf{w}}) - (\mathbf{w}^h - \tilde{\mathbf{w}}) = \eta - \phi^h$. Take ϕ^h as test function in (8.23) which gives (8.24).
3. Estimate the left hand side of the error equation (8.24) from below by the strong monotonicity (8.11) and the right hand side of (8.24) from above by the local Lipschitz continuity (8.12) of the Smagorinsky term.
4. Derive a differential inequality of the form
$$\frac{d}{dt}\left\|\phi^h\right\|_{L^2(\Omega)}^2 + \tilde{g}\left(\phi^h\right) \le g(\eta) + \lambda\left(t, \mathbf{w}, \mathbf{w}^h\right)\left\|\phi^h\right\|_{L^2(\Omega)}^2 \tag{8.21}$$
with $0 \le \tilde{g}\left(\phi^h\right), g(\eta), \lambda\left(t, \mathbf{w}, \mathbf{w}^h\right)$.
5. Estimate $g(\eta)$ from above uniformly in ν by the stability estimates proved in Section 8.1.1.
6. Show $\lambda\left(t, \mathbf{w}, \mathbf{w}^h\right) \in L^1(0,T)$ uniformly in ν by the stability estimates.
7. Apply Gronwall's lemma, Lemma 8.3, to (8.21) to obtain an error estimate for ϕ^h.
8. Get the error estimate for $\mathbf{w} - \mathbf{w}^h$ by the triangle inequality. The terms containing η are bounded independently of ν by interpolation error estimates of the finite element spaces.

8.1.3 The Error Equation

Let us first note that for standard piecewise polynomial finite element spaces it is known that the L^2 projection of a function in $L^p(\Omega), p \ge 2$, is in $L^p(\Omega)$ itself and the L^2 projection operator is stable in $L^p(\Omega), p \ge 2$, Crouzeix and Thomée [CT87].

Let $\mathbf{e} = \mathbf{w} - \mathbf{w}^h$ denote the error and $\tilde{\mathbf{w}} \in V_{\text{div}}^h$ be an approximation of $\mathbf{w}$. The error is decomposed into
$$\mathbf{e} = (\mathbf{w} - \tilde{\mathbf{w}}) - \left(\mathbf{w}^h - \tilde{\mathbf{w}}\right) = \eta - \phi^h. \tag{8.22}$$

By the choice of $\tilde{\mathbf{w}}$ follows $\phi^h \in V_{\text{div}}^h$. Subtracting (8.15) from (8.9) gives for all $\mathbf{v}^h \in V_{\text{div}}^h$ and $q^h \in Q_h$
$$\begin{aligned}\left(\mathbf{e}_t, \mathbf{v}^h\right) + a\left(\mathbf{w}, \mathbf{w}, \mathbf{v}^h\right) - a\left(\mathbf{w}^h, \mathbf{w}^h, \mathbf{v}^h\right) + b_s\left(\mathbf{w}, \mathbf{w}, \mathbf{v}_h\right)& \\ -b_s\left(\mathbf{w}^h, \mathbf{w}^h, \mathbf{v}_h\right) - \left(r - q^h, \nabla \cdot \mathbf{v}^h\right) = 0.&\end{aligned} \tag{8.23}$$

It follows with $\mathbf{v}^h = \phi^h$
$$\begin{aligned}&\left(\phi_t^h, \phi^h\right) + a\left(\mathbf{w}^h, \mathbf{w}^h, \phi^h\right) - a\left(\tilde{\mathbf{w}}, \tilde{\mathbf{w}}, \phi^h\right) \\ &= \left(\eta_t, \phi^h\right) + a\left(\mathbf{w}, \mathbf{w}, \phi^h\right) - a\left(\tilde{\mathbf{w}}, \tilde{\mathbf{w}}, \phi^h\right) + b_s\left(\mathbf{w}, \mathbf{w}, \phi^h\right) \\ &\quad -b_s\left(\mathbf{w}^h, \mathbf{w}^h, \phi^h\right) - \left(r - q^h, \nabla \cdot \phi^h\right).\end{aligned} \tag{8.24}$$

The monotonicity of $a(\cdot,\cdot,\cdot)$, (8.11), implies

$$
\begin{aligned}
& a\left(\mathbf{w}^h, \mathbf{w}^h, \boldsymbol{\phi}^h\right) - a\left(\tilde{\mathbf{w}}, \tilde{\mathbf{w}}, \boldsymbol{\phi}^h\right) \\
& \geq \alpha_s \left\| \nabla \cdot \boldsymbol{\phi}^h \right\|_{L^2(\Omega)}^2 + (2\nu + a_0(\delta)) \left\| \mathbb{D}\left(\boldsymbol{\phi}^h\right) \right\|_{L^2(\Omega)}^2 + \underline{C} c_S \delta^2 \left\| \mathbb{D}\left(\boldsymbol{\phi}^h\right) \right\|_{L^3(\Omega)}^3 \\
& \quad + \sum_{j=1}^{J} \beta \left\| \boldsymbol{\phi}^h \cdot \boldsymbol{\tau}_j \right\|_{L^2(\Gamma_j)}^2 .
\end{aligned}
$$

The local Lipschitz continuity of the trilinear form, (8.12), gives the estimate

$$
\begin{aligned}
& a\left(\mathbf{w}, \mathbf{w}, \boldsymbol{\phi}^h\right) - a\left(\tilde{\mathbf{w}}, \tilde{\mathbf{w}}, \boldsymbol{\phi}^h\right) \\
& \leq \alpha_s \left\| \nabla \cdot \boldsymbol{\phi}^h \right\|_{L^2(\Omega)} \left\| \nabla \cdot \boldsymbol{\eta} \right\|_{L^2(\Omega)} \\
& \quad + (2\nu + a_0(\delta)) \left\| \mathbb{D}\left(\boldsymbol{\phi}^h\right) \right\|_{L^2(\Omega)} \left\| \mathbb{D}(\boldsymbol{\eta}) \right\|_{L^2(\Omega)} \\
& \quad + \overline{C} C_L c_S \delta^2 \left\| \mathbb{D}\left(\boldsymbol{\phi}^h\right) \right\|_{L^3(\Omega)} \left\| \mathbb{D}(\boldsymbol{\eta}) \right\|_{L^3(\Omega)} \\
& \quad + \sum_{j=1}^{J} \beta \left\| \boldsymbol{\phi}^h \cdot \boldsymbol{\tau}_j \right\|_{L^2(\Gamma_j)} \left\| \boldsymbol{\eta} \cdot \boldsymbol{\tau}_j \right\|_{L^2(\Gamma_j)}
\end{aligned}
$$

with

$$
C_L = \max\left\{ \left\| \mathbb{D}(\mathbf{w}) \right\|_{L^3(\Omega)}, \left\| \mathbb{D}(\tilde{\mathbf{w}}) \right\|_{L^3(\Omega)} \right\}. \tag{8.25}
$$

The terms on the right hand side are estimated further by Young's inequality (2.6) and the definition of the norm in V^*, (8.3),

$$
\begin{aligned}
& \overline{C} C_L c_S \delta^2 \left\| \mathbb{D}\left(\boldsymbol{\phi}^h\right) \right\|_{L^3(\Omega)} \left\| \mathbb{D}(\boldsymbol{\eta}) \right\|_{L^3(\Omega)} \\
& \leq \frac{\underline{C} c_S \delta^2}{3} \left\| \mathbb{D}\left(\boldsymbol{\phi}^h\right) \right\|_{L^3(\Omega)}^3 + \frac{2 \overline{C}^{3/2} C_L^{3/2} c_S \delta^2}{3 \underline{C}^{1/2}} \left\| \mathbb{D}(\boldsymbol{\eta}) \right\|_{L^3(\Omega)}^{3/2},
\end{aligned}
$$

$$
\alpha_s \left\| \nabla \cdot \boldsymbol{\phi}^h \right\|_{L^2(\Omega)} \left\| \nabla \cdot \boldsymbol{\eta} \right\|_{L^2(\Omega)} \leq \frac{\alpha_s}{4} \left\| \nabla \cdot \boldsymbol{\phi}^h \right\|_{L^2(\Omega)}^2 + \alpha_s \left\| \nabla \cdot \boldsymbol{\eta} \right\|_{L^2(\Omega)}^2 ,
$$

$$
\begin{aligned}
& \left(\boldsymbol{\eta}_t, \boldsymbol{\phi}^h\right) \leq \left\| \boldsymbol{\eta}_t \right\|_{V^*} \left\| \mathbb{D}\left(\boldsymbol{\phi}^h\right) \right\|_{L^3(\Omega)} \\
& \quad \leq \frac{2}{3} \left(\underline{C} c_S \delta^2\right)^{-1/2} \left\| \boldsymbol{\eta}_t \right\|_{V^*}^{3/2} + \frac{\underline{C} c_S \delta^2}{3} \left\| \mathbb{D}\left(\boldsymbol{\phi}^h\right) \right\|_{L^3(\Omega)}^3 .
\end{aligned}
$$

Young's inequality is applied to the other terms with $p = q = 2$ and $t = 1$.

Inserting these estimates into (8.24), using the Cauchy-Schwarz inequality and collecting terms give in the case $\alpha_s > 0$

$$\begin{aligned}
&\frac{1}{2}\frac{d}{dt}\left\|\phi^h\right\|^2_{L^2(\Omega)} + \frac{\alpha_s}{2}\left\|\nabla\cdot\phi^h\right\|^2_{L^2(\Omega)} + \frac{2\nu + a_0(\delta)}{2}\left\|\mathbb{D}\left(\phi^h\right)\right\|^2_{L^2(\Omega)} \\
&\quad + \frac{\underline{C}c_S\delta^2}{3}\left\|\mathbb{D}\left(\phi^h\right)\right\|^3_{L^3(\Omega)} + \sum_{j=1}^{J}\frac{\beta}{2}\left\|\phi^h\cdot\boldsymbol{\tau}_j\right\|^2_{L^2(\Gamma_j)} \\
&\leq \frac{2}{3\left(\underline{C}c_S\right)^{1/2}\delta}\left\|\boldsymbol{\eta}_t\right\|^{3/2}_{V^*} + \alpha_s\left\|\nabla\cdot\boldsymbol{\eta}\right\|^2_{L^2(\Omega)} + \frac{2\nu + a_0(\delta)}{2}\left\|\mathbb{D}(\boldsymbol{\eta})\right\|^2_{L^2(\Omega)} \\
&\quad + \frac{2\overline{C}^{3/2}C_L^{3/2}c_S\delta^2}{3\underline{C}^{1/2}}\left\|\mathbb{D}(\boldsymbol{\eta})\right\|^{3/2}_{L^3(\Omega)} + \sum_{j=1}^{J}\frac{\beta}{2}\left\|\boldsymbol{\eta}\cdot\boldsymbol{\tau}_j\right\|^2_{L^2(\Gamma_j)} \\
&\quad + \left|b_s\left(\mathbf{w},\mathbf{w},\phi^h\right) - b_s\left(\mathbf{w}^h,\mathbf{w}^h,\phi^h\right)\right| + \frac{1}{\alpha_s}\left\|r - q^h\right\|^2_{L^2(\Omega)}.
\end{aligned} \tag{8.26}$$

In the case $\alpha_s = 0$, the last term in (8.24) can be estimated by Hölder's, Korn's and Young's inequality,

$$\begin{aligned}
\left(r - q^h, \nabla\cdot\phi^h\right) &\leq \left\|r - q^h\right\|_{L^{3/2}(\Omega)}\left\|\nabla\cdot\phi^h\right\|_{L^3(\Omega)} \\
&\leq \left\|r - q^h\right\|_{L^{3/2}(\Omega)}\left\|\mathbb{D}\left(\phi^h\right)\right\|_{L^3(\Omega)} \\
&\leq \frac{\varepsilon}{3}\left\|\mathbb{D}\left(\phi^h\right)\right\|^3_{L^3(\Omega)} + \frac{2}{3\varepsilon^{1/2}}\left\|r - q^h\right\|^{3/2}_{L^{3/2}(\Omega)}.
\end{aligned}$$

Choosing $\varepsilon = \underline{C}c_S\delta^2/2$ gives

$$\begin{aligned}
&\frac{1}{2}\frac{d}{dt}\left\|\phi^h\right\|^2_{L^2(\Omega)} + \frac{2\nu + a_0(\delta)}{2}\left\|\mathbb{D}\left(\phi^h\right)\right\|^2_{L^2(\Omega)} + \frac{\underline{C}c_S\delta^2}{6}\left\|\mathbb{D}\left(\phi^h\right)\right\|^3_{L^3(\Omega)} \\
&\quad + \sum_{j=1}^{J}\frac{\beta}{2}\left\|\phi^h\cdot\boldsymbol{\tau}_j\right\|^2_{L^2(\Gamma_j)} \\
&\leq \frac{2}{3\left(\underline{C}c_S\right)^{1/2}\delta}\left\|\boldsymbol{\eta}_t\right\|^{3/2}_{V^*} + \frac{2\nu + a_0(\delta)}{2}\left\|\mathbb{D}(\boldsymbol{\eta})\right\|^2_{L^2(\Omega)} \\
&\quad + \frac{2\overline{C}^{3/2}C_L^{3/2}c_S\delta^2}{3\underline{C}^{1/2}}\left\|\mathbb{D}(\boldsymbol{\eta})\right\|^{3/2}_{L^3(\Omega)} + \sum_{j=1}^{J}\frac{\beta}{2}\left\|\boldsymbol{\eta}\cdot\boldsymbol{\tau}_j\right\|^2_{L^2(\Gamma_j)} \\
&\quad + \left|b_s\left(\mathbf{w},\mathbf{w},\phi^h\right) - b_s\left(\mathbf{w}^h,\mathbf{w}^h,\phi^h\right)\right| + \frac{2\sqrt{2}}{3\underline{C}^{1/2}c_S^{1/2}\delta}\left\|r - q^h\right\|^{3/2}_{L^{3/2}(\Omega)}.
\end{aligned} \tag{8.27}$$

Inequalities (8.26) and (8.27) are the basic differential inequalities for the error estimates.

8.1.4 The Case $\nabla\mathbf{w} \in L^3\left(0,T;L^3(\Omega)\right)$ and $a_0(\delta) = 0$

We consider first the case $a_0(\delta) = 0$ and $\nabla\mathbf{w} \in L^3\left(0,T;L^3(\Omega)\right)$.

A straightforward calculation gives the decomposition

$$\begin{aligned} & b_s\left(\mathbf{w}, \mathbf{w}, \phi^h\right) - b_s\left(\mathbf{w}^h, \mathbf{w}^h, \phi^h\right) \\ & = b_s\left(\mathbf{w}, \eta, \phi^h\right) + b_s\left(\eta, \mathbf{w}^h, \phi^h\right) - b_s\left(\phi^h, \mathbf{w}^h, \phi^h\right), \end{aligned} \tag{8.28}$$

where $b_s\left(\mathbf{w}, \phi^h, \phi^h\right) = 0$ has been used.

The first term of (8.28) is bounded by using Hölder's, Korn's and Young's inequality

$$\begin{aligned} & \left|b_s\left(\mathbf{w}, \eta, \phi^h\right)\right| \\ & = \frac{1}{2}\left|b\left(w, \eta, \phi^h\right) - b\left(w, \phi^h, \eta\right)\right| \\ & \leq \frac{1}{2}\left(\left\|\phi^h\right\|_{L^2(\Omega)} \|\nabla \eta\|_{L^3(\Omega)} \|\mathbf{w}\|_{L^6(\Omega)} + \left\|\nabla \phi^h\right\|_{L^3(\Omega)} \|\mathbf{w}\|_{L^2(\Omega)} \|\eta\|_{L^6(\Omega)}\right) \\ & \leq \frac{1}{2}\left(\left\|\phi^h\right\|_{L^2(\Omega)} \|\nabla \eta\|_{L^3(\Omega)} \|\mathbf{w}\|_{L^6(\Omega)}\right. \\ & \quad \left. + C\left\|\mathbb{D}\left(\phi^h\right)\right\|_{L^3(\Omega)} \|\mathbf{w}\|_{L^2(\Omega)} \|\eta\|_{L^6(\Omega)}\right) \\ & \leq \frac{1}{4}\|\mathbf{w}\|_{L^6(\Omega)}^2 \left\|\phi^h\right\|_{L^2(\Omega)}^2 + \frac{1}{4}\|\nabla \eta\|_{L^3(\Omega)}^2 + \frac{\varepsilon_1}{6}\left\|\mathbb{D}\left(\phi^h\right)\right\|_{L^3(\Omega)}^3 \\ & \quad + \frac{C}{3\varepsilon_1^{1/2}} \|\mathbf{w}\|_{L^2(\Omega)}^{3/2} \|\eta\|_{L^6(\Omega)}^{3/2}. \end{aligned} \tag{8.29}$$

In the same way, we obtain

$$\begin{aligned} b_s\left(\eta, \mathbf{w}^h, \phi^h\right) \leq & \frac{1}{4}\left\|\nabla \mathbf{w}^h\right\|_{L^3(\Omega)}^2 \left\|\phi^h\right\|_{L^2(\Omega)}^2 + \frac{1}{4}\|\eta\|_{L^6(\Omega)}^2 \\ & + \frac{\varepsilon_1}{6}\left\|\mathbb{D}\left(\phi^h\right)\right\|_{L^3(\Omega)}^3 + \frac{C}{3\varepsilon_1^{1/2}} \left\|\mathbf{w}^h\right\|_{L^2(\Omega)}^{3/2} \|\eta\|_{L^6(\Omega)}^{3/2}. \end{aligned} \tag{8.30}$$

The third term of (8.28) is estimated by Hölder's inequality, the Sobolev embedding $W^{1,3}(\Omega) \to L^6(\Omega)$, Korn's and Young's inequality in the following way

$$\begin{aligned} b_s\left(\phi^h, \mathbf{w}^h, \phi^h\right) \leq & \frac{1}{2}\left(\left\|\nabla \mathbf{w}^h\right\|_{L^3(\Omega)} \left\|\phi^h\right\|_{L^6(\Omega)} \left\|\phi^h\right\|_{L^2(\Omega)}\right. \\ & \left. + \left\|\nabla \phi^h\right\|_{L^3(\Omega)} \left\|\phi^h\right\|_{L^2(\Omega)} \left\|\mathbf{w}^h\right\|_{L^6(\Omega)}\right) \\ \leq & C\left\|\nabla \mathbf{w}^h\right\|_{L^3(\Omega)} \left\|\phi^h\right\|_{W^{1,3}(\Omega)} \left\|\phi^h\right\|_{L^2(\Omega)} \\ & + C\left\|\mathbb{D}\left(\phi^h\right)\right\|_{L^3(\Omega)} \left\|\phi^h\right\|_{L^2(\Omega)} \left\|\mathbf{w}^h\right\|_{W^{1,3}(\Omega)} \end{aligned}$$

$$\leq \frac{\varepsilon_2}{6}\left\|\mathbb{D}\left(\phi^h\right)\right\|_{L^3(\Omega)}^3 + \frac{C}{\varepsilon_2^{1/2}}\left\|\nabla \mathbf{w}^h\right\|_{L^3(\Omega)}^{3/2}\left\|\phi^h\right\|_{L^2(\Omega)}^{3/2} \quad (8.31)$$
$$+\frac{\varepsilon_2}{6}\left\|\mathbb{D}\left(\phi^h\right)\right\|_{L^3(\Omega)}^3 + \frac{C}{\varepsilon_2^{1/2}}\left\|\phi^h\right\|_{L^2(\Omega)}^{3/2}\left\|\mathbb{D}\left(\mathbf{w}^h\right)\right\|_{L^3(\Omega)}^{3/2}.$$

To obtain the last inequality, we have used Poincare's inequality and Korn's inequality

$$\left\|\phi^h\right\|_{W^{1,3}(\Omega)} \leq C\left\|\nabla\phi^h\right\|_{L^3(\Omega)} \leq C\left\|\mathbb{D}\left(\phi^h\right)\right\|_{L^3(\Omega)}.$$

Collecting the terms of (8.29), (8.30) and (8.31) with $\varepsilon_1 = \varepsilon_2$ yields the estimate

$$\left|b_s\left(\mathbf{w},\mathbf{w},\phi^h\right) - b_s\left(\mathbf{w}^h,\mathbf{w}^h,\phi^h\right)\right|$$
$$\leq \left[\frac{1}{4}\|\nabla\boldsymbol{\eta}\|_{L^3(\Omega)}^2 + \frac{C}{\varepsilon_1^{1/2}}\left(\|\mathbf{w}\|_{L^2(\Omega)}^{3/2} + \left\|\mathbf{w}^h\right\|_{L^2(\Omega)}^{3/2}\right)\|\boldsymbol{\eta}\|_{L^6(\Omega)}^{3/2} + \frac{1}{4}\|\boldsymbol{\eta}\|_{L^6(\Omega)}^2\right]$$
$$+\frac{2\varepsilon_1}{3}\left\|\mathbb{D}\left(\phi^h\right)\right\|_{L^3(\Omega)}^3 + \frac{C}{\varepsilon_1^{1/2}}\left\|\mathbb{D}\left(\mathbf{w}^h\right)\right\|_{L^3(\Omega)}^{3/2}\left\|\phi^h\right\|_{L^2(\Omega)}^{3/2} \quad (8.32)$$
$$+\left[\frac{1}{4}\|\mathbf{w}\|_{L^6}^2 + \frac{1}{4}\left\|\nabla\mathbf{w}^h\right\|_{L^3}^2\right]\left\|\phi^h\right\|_{L^2(\Omega)}^2.$$

We consider the case $\alpha_s > 0$, i.e. error equation (8.26). The parameter ε_1 is chosen such that

$$\frac{2\varepsilon_1}{3} = \frac{1}{6}\underline{C}c_S\delta^2,$$

i.e. $\varepsilon_1 = \mathcal{O}\left(\delta^2\right)$ and $\varepsilon_1^{-1/2} = \mathcal{O}\left(\delta^{-1}\right)$. We obtain

$$\frac{1}{2}\frac{d}{dt}\left\|\phi^h\right\|_{L^2(\Omega)}^2 + \frac{\alpha_s}{2}\left\|\nabla\cdot\phi^h\right\|_{L^2(\Omega)}^2 + \nu\left\|\mathbb{D}\left(\phi^h\right)\right\|_{L^2(\Omega)}^2$$
$$+\frac{\underline{C}c_S\delta^2}{6}\left\|\mathbb{D}\left(\phi^h\right)\right\|_{L^3(\Omega)}^3 + \sum_{j=1}^{J}\frac{\beta}{2}\left\|\phi^h\cdot\boldsymbol{\tau}_j\right\|_{L^2(\Gamma_j)}^2$$
$$\leq \left[\frac{2}{3\left(\underline{C}c_S\right)^{1/2}\delta}\|\boldsymbol{\eta}_t\|_{V^*}^{3/2} + \alpha_s\|\nabla\cdot\boldsymbol{\eta}\|_{L^2(\Omega)}^2 + \nu\|\mathbb{D}(\boldsymbol{\eta})\|_{L^2(\Omega)}^2\right.$$
$$+\frac{2\overline{C}^{3/2}C_L^{3/2}c_S\delta^2}{3\underline{C}^{1/2}}\|\mathbb{D}(\boldsymbol{\eta})\|_{L^3(\Omega)}^{3/2} + \sum_{j=1}^{J}\frac{\beta}{2}\|\boldsymbol{\eta}\cdot\boldsymbol{\tau}_j\|_{L^2(\Gamma_j)}^2 + \frac{1}{4}\|\nabla\boldsymbol{\eta}\|_{L^3(\Omega)}^2$$
$$+C\delta^{-1}\left(\|\mathbf{w}\|_{L^2(\Omega)}^{3/2} + \left\|\mathbf{w}^h\right\|_{L^2(\Omega)}^{3/2}\right)\|\boldsymbol{\eta}\|_{L^6(\Omega)}^{3/2} + \frac{1}{4}\|\boldsymbol{\eta}\|_{L^6(\Omega)}^2$$
$$\left.+\frac{1}{\alpha_s}\left\|r - q^h\right\|_{L^2(\Omega)}^2\right] + C\delta^{-1}\left\|\mathbb{D}\left(\mathbf{w}^h\right)\right\|_{L^3(\Omega)}^{3/2}\left\|\phi^h\right\|_{L^2(\Omega)}^{3/2}$$
$$+\left[\frac{1}{4}\|\mathbf{w}\|_{L^6}^2 + \frac{1}{4}\left\|\nabla\mathbf{w}^h\right\|_{L^3}^2\right]\left\|\phi^h\right\|_{L^2(\Omega)}^2. \quad (8.33)$$

Remark 8.12. The failure of this approach. The last step of the proof would be the application of Gronwall's lemma, Lemma 8.3, for $f(t) = \left\|\phi^h(t)\right\|^2_{L^2(\Omega)}$. However, the term with $\left\|\phi^h\right\|^{3/2}_{L^2(\Omega)} = \left(\left\|\phi^h\right\|^2_{L^2(\Omega)}\right)^{3/4}$ in the right hand side of (8.33) does not fit into the basic inequality (8.5) of Lemma 8.3. The power 3/4 is too small. Thus, Gronwall's lemma cannot be applied and this way of analysis fails for the case $\nabla \mathbf{w} \in L^3(0,T;L^3(\Omega))$ and $a_0(\delta) = 0$. □

Remark 8.13. An alternative unsuccessful approach. An alternative analysis using the Gagliardo-Nirenberg inequality (2.11) leads to a term on the right hand side with the factor $\left\|\phi^h\right\|^{12/7}_{L^2(\Omega)} = \left(\left\|\phi^h\right\|^2_{L^2(\Omega)}\right)^{6/7}$. Since $6/7 < 1$, the final conclusion that Gronwall's lemma cannot be applied holds also for this approach.

Remark 8.14. Successful approaches requiring stronger assumptions or giving weaker results. An error estimate for $a_0(\delta) > 0$ can be achieved if a higher regularity of the solution is assumed, see Section 8.1.6. For $\nabla \mathbf{w} \in L^3(0,T;L^3(\Omega))$ and $a_0(\delta) = 0$, error estimates which are non-uniform in ν are also possible, see Section 8.1.7. □

8.1.5 The Case $\nabla \mathbf{w} \in L^3(0,T;L^3(\Omega))$ and $a_0(\delta) > 0$

Lemma 8.15. *Assume $\alpha_s > 0$ and $a_0(\delta) > 0$ for $\delta > 0$. Let*

$$\begin{aligned}\lambda(t) := \frac{1}{4}\|\mathbf{w}\|^2_{L^6(\Omega)} + \frac{1}{4}\left\|\nabla \mathbf{w}^h\right\|^2_{L^3(\Omega)} + \frac{C}{a_0(\delta)}\left\|\nabla \mathbf{w}^h\right\|^2_{L^3(\Omega)} \\ + C a_0(\delta)^{-1/2}\alpha_s^{-3/2}\left\|\mathbb{D}\left(\mathbf{w}^h\right)\right\|^3_{L^3(\Omega)}.\end{aligned}$$

Then there is a constant $C_1(\delta)$ independent of ν and h such that for $0 < T < \infty$

$$\|\lambda(t)\|_{L^1(0,T)} \le C_1(\delta).$$

Proof. By the Sobolev embedding $W^{1,3}(\Omega) \to L^6(\Omega)$, we have $\|\mathbf{w}\|_{L^6(\Omega)} \le C\|\mathbb{D}(\mathbf{w})\|_{L^3(\Omega)}$ which is bounded uniformly in ν by (8.16) and (8.17). By the stability estimates (8.18) and (8.19) follows $\left\|\mathbb{D}\left(\mathbf{w}^h\right)\right\|_{L^3(\Omega)} \in L^3(0,T)$ uniformly in ν and h. Since $L^3(0,T) \subset L^2(0,T)$, we have also $\left\|\mathbb{D}\left(\mathbf{w}^h\right)\right\|_{L^3(\Omega)} \in L^2(0,T)$ Thus, $\left\|\mathbb{D}\left(\mathbf{w}^h\right)\right\|^3_{L^3(\Omega)} \in L^1(0,T)$ and $\left\|\mathbb{D}\left(\mathbf{w}^h\right)\right\|^2_{L^3(\Omega)} \in L^1(0,T)$ uniformly in ν and h. □

Lemma 8.16. *Under the assumptions of Lemma 8.15 there is a constant $C_2(\delta)$ independent of ν and h such that*

$$\delta^{-1}\left(\left\|\mathbf{w}^h\right\|^{3/2}_{L^2(\Omega)} + \|\mathbf{w}\|^{3/2}_{L^2(\Omega)}\right) \le C_2(\delta).$$

Proof. The statement of the lemma follows for $\|\mathbf{w}\|_{L^2(\Omega)}$ by the stability estimates (8.16) and (8.17) and for $\left\|\mathbf{w}^h\right\|_{L^2(\Omega)}$ by (8.18) and (8.19). □

Theorem 8.17. *Assume $\alpha_s > 0$ and $a_0(\delta) > 0$. Then, the error $\mathbf{w} - \mathbf{w}^h$ satisfies for $T > 0$*

$$\begin{aligned}
&\left\|\mathbf{w} - \mathbf{w}^h\right\|^2_{L^\infty(0,T;L^2(\Omega))} + \alpha_s \left\|\nabla \cdot \left(\mathbf{w} - \mathbf{w}^h\right)\right\|^2_{L^2(0,T;L^2(\Omega))} \\
&\quad + \left(\nu + C a_0(\delta)\right) \left\|\mathbb{D}\left(\mathbf{w} - \mathbf{w}^h\right)\right\|^2_{L^2(0,T;L^2(\Omega))} \\
&\quad + \delta^2 \left\|\mathbb{D}\left(\mathbf{w} - \mathbf{w}^h\right)\right\|^3_{L^3(0,T;L^3(\Omega))} + \sum_{j=1}^{J} \beta \left\|\left(\mathbf{w} - \mathbf{w}^h\right) \cdot \boldsymbol{\tau}_j\right\|^2_{L^2(0,T;L^2(\Gamma_j))} \\
&\le C \exp\left(C_1(\delta)\right) \left\|\left(\mathbf{w} - \mathbf{w}^h\right)(0, \mathbf{x})\right\|^2_{L^2(\Omega)} \\
&\quad + C \inf_{\tilde{\mathbf{w}} \in V^h_{\mathrm{div}}, q^h \in Q^h} \mathcal{F}\left(\mathbf{w} - \tilde{\mathbf{w}}, r - q^h, \delta\right)
\end{aligned}$$

with

$$\begin{aligned}
&\mathcal{F}\left(\mathbf{w} - \tilde{\mathbf{w}}, r - q^h, \delta\right) \\
&= \|\mathbf{w} - \tilde{\mathbf{w}}\|^2_{L^\infty(0,T;L^2(\Omega))} + \delta^2 \|\mathbb{D}(\mathbf{w} - \tilde{\mathbf{w}})\|^3_{L^3(0,T;L^3(\Omega))} \\
&\quad + \exp\left(C_1(\delta)\right) \Bigg[\|(\mathbf{w} - \tilde{\mathbf{w}})(0, \mathbf{x})\|^2_{L^2(\Omega)} + \delta^{-1} \|(\mathbf{w} - \tilde{\mathbf{w}})_t\|^{3/2}_{L^{3/2}(0,T;V^*)} \\
&\quad + \alpha_s \|\nabla \cdot (\mathbf{w} - \tilde{\mathbf{w}})\|^2_{L^2(0,T;L^2(\Omega))} + \left(2\nu + a_0(\delta)\right) \|\mathbb{D}(\mathbf{w} - \tilde{\mathbf{w}})\|^2_{L^2(0,T;L^2(\Omega))} \\
&\quad + C(\delta) \|\mathbb{D}(\mathbf{w} - \tilde{\mathbf{w}})\|^{3/2}_{L^3(0,T;L^3(\Omega))} + \sum_{j=1}^{J} \beta \|(\mathbf{w} - \tilde{\mathbf{w}}) \cdot \boldsymbol{\tau}_j\|^2_{L^2(0,T;L^2(\Gamma_j))} \\
&\quad + \|\nabla(\mathbf{w} - \tilde{\mathbf{w}})\|^2_{L^2(0,T;L^3(\Omega))} + C_2(\delta) \|\mathbf{w} - \tilde{\mathbf{w}}\|^{3/2}_{L^{3/2}(0,T;L^6(\Omega))} \\
&\quad + \|\mathbf{w} - \tilde{\mathbf{w}}\|^2_{L^2(0,T;L^6(\Omega))} + \frac{1}{\alpha_s} \left\|r - q^h\right\|^2_{L^2(0,T;L^2(\Omega))} \Bigg]
\end{aligned}$$

and $C_1(\delta)$ and $C_2(\delta)$ defined in Lemma 8.15 and 8.16.

Proof. The key of the analysis is the estimate of

$$\begin{aligned}
&b_s\left(\mathbf{w}, \mathbf{w}, \boldsymbol{\phi}^h\right) - b_s\left(\mathbf{w}^h, \mathbf{w}^h, \boldsymbol{\phi}^h\right) \\
&= b_s\left(\mathbf{w}, \boldsymbol{\eta}, \boldsymbol{\phi}^h\right) + b_s\left(\boldsymbol{\eta}, \mathbf{w}^h, \boldsymbol{\phi}^h\right) - b_s\left(\boldsymbol{\phi}^h, \mathbf{w}^h, \boldsymbol{\phi}^h\right).
\end{aligned}$$

The first two terms of the right hand side are estimated like in Section 8.1.4 leading to the bounds (8.29) and (8.30). The last term is treated differently.

Integration by parts, (5.4), and Hölder's inequality give

$$
\begin{aligned}
b_s\left(\phi^h, \mathbf{w}^h, \phi^h\right) &= \frac{1}{2} b\left(\phi^h, \mathbf{w}^h, \phi^h\right) - \frac{1}{2} b\left(\phi^h, \phi^h, \mathbf{w}^h\right) \\
&= b\left(\phi^h, \mathbf{w}^h, \phi^h\right) + \frac{1}{2}\left(\nabla \cdot \phi^h, \phi^h \cdot \mathbf{w}^h\right) \\
&\leq \left\|\nabla \mathbf{w}^h\right\|_{L^3(\Omega)} \left\|\phi^h\right\|_{L^3(\Omega)}^2 + \frac{1}{2}\left|\left(\nabla \cdot \phi^h, \phi^h \cdot \mathbf{w}^h\right)\right|.
\end{aligned}
$$

By the Sobolev embedding $H^{1/2}(\Omega) \to L^3(\Omega)$, the interpolation theorem (2.15) and Poincare's inequality, we obtain

$$
\begin{aligned}
\left\|\phi^h\right\|_{L^3(\Omega)}^2 &\leq C\left\|\phi^h\right\|_{H^{1/2}(\Omega)}^2 \leq C\left\|\phi^h\right\|_{L^2(\Omega)}\left\|\phi^h\right\|_{H^1(\Omega)} \\
&\leq C\left\|\phi^h\right\|_{L^2(\Omega)}\left\|\nabla \phi^h\right\|_{L^2(\Omega)}.
\end{aligned}
$$

Inserting this into the previous estimate and applying Korn's and Young's inequalities give

$$
\begin{aligned}
\left|b_s\left(\phi^h, \mathbf{w}^h, \phi^h\right)\right| \leq{} & \frac{\varepsilon_2}{2}\left\|\mathbb{D}\left(\phi^h\right)\right\|_{L^2(\Omega)}^2 + \frac{C}{2\varepsilon_2}\left\|\mathbb{D}\left(\mathbf{w}^h\right)\right\|_{L^3(\Omega)}^2\left\|\phi^h\right\|_{L^2(\Omega)}^2 \\
& + \frac{1}{2}\left|\left(\nabla \cdot \phi^h, \phi^h \cdot \mathbf{w}^h\right)\right|.
\end{aligned}
$$

The last term of the right hand side of this inequality is estimated by Hölder's and by Young's inequality leading to

$$
\begin{aligned}
\left|\left(\nabla \cdot \phi^h, \phi^h \cdot \mathbf{w}^h\right)\right| &\leq \left\|\nabla \cdot \phi^h\right\|_{L^2(\Omega)}\left\|\phi^h\right\|_{L^{18/7}(\Omega)}\left\|\mathbf{w}^h\right\|_{L^9(\Omega)} \\
&\leq \frac{\alpha_s}{4}\left\|\nabla \cdot \phi^h\right\|_{L^2(\Omega)}^2 + \frac{1}{\alpha_s}\left\|\phi^h\right\|_{L^{18/7}(\Omega)}^2\left\|\mathbf{w}^h\right\|_{L^9(\Omega)}^2.
\end{aligned}
$$

The Sobolev embedding theorem $W^{1,3}(\Omega) \to L^9(\Omega)$ implies together with Poincare's and Korn's inequality

$$
\left\|\mathbf{w}^h\right\|_{L^9(\Omega)}^2 \leq C\left\|\mathbf{w}^h\right\|_{W^{1,3}(\Omega)}^2 \leq C\left\|\mathbb{D}\left(\mathbf{w}^h\right)\right\|_{L^3}^2.
$$

The Sobolev embedding theorem implies also $H^{1/3}(\Omega) \to L^{18/7}(\Omega)$. With the interpolation theorem (2.15), Poincare's and Korn's inequality follow

$$
\begin{aligned}
\left\|\phi^h\right\|_{L^{18/7}(\Omega)}^2 &\leq C\left\|\phi^h\right\|_{H^{1/3}(\Omega)}^2 \leq C\left\|\phi^h\right\|_{L^2(\Omega)}^{4/3}\left\|\phi^h\right\|_{H^1(\Omega)}^{2/3} \\
&\leq C\left\|\phi^h\right\|_{L^2(\Omega)}^{4/3}\left\|\mathbb{D}\left(\phi^h\right)\right\|_{L^2(\Omega)}^{2/3}.
\end{aligned}
$$

These bounds and Young's inequality give

$$b_s\left(\boldsymbol{\phi}^h, \mathbf{w}^h, \boldsymbol{\phi}^h\right) \leq \frac{\varepsilon_2}{2}\left\|\mathbb{D}\left(\boldsymbol{\phi}^h\right)\right\|_{L^2(\Omega)}^2 + \frac{C}{2\varepsilon_2}\left\|\mathbb{D}\left(\mathbf{w}^h\right)\right\|_{L^3(\Omega)}^2\left\|\boldsymbol{\phi}^h\right\|_{L^2(\Omega)}^2$$
$$+\frac{\alpha_s}{8}\left\|\nabla\cdot\boldsymbol{\phi}^h\right\|_{L^2(\Omega)}^2 + \frac{\varepsilon_3}{6}\left\|\mathbb{D}\left(\boldsymbol{\phi}^h\right)\right\|_{L^2(\Omega)}^2$$
$$+\frac{C}{\varepsilon_3^{1/2}\alpha_s^{3/2}}\left\|\mathbb{D}\left(\mathbf{w}^h\right)\right\|_{L^3}^3\left\|\boldsymbol{\phi}^h\right\|_{L^2(\Omega)}^2.$$

This bound and (8.29) and (8.30) are substituted into (8.26) yielding the differential inequality

$$\frac{1}{2}\frac{d}{dt}\left\|\boldsymbol{\phi}^h\right\|_{L^2(\Omega)}^2 + \frac{3\alpha_s}{8}\left\|\nabla\cdot\boldsymbol{\phi}^h\right\|_{L^2(\Omega)}^2$$
$$+\left(\frac{2\nu+a_0(\delta)}{2} - \frac{\varepsilon_3}{6} - \frac{\varepsilon_2}{2}\right)\left\|\mathbb{D}\left(\boldsymbol{\phi}^h\right)\right\|_{L^2(\Omega)}^2$$
$$+\left(\frac{\underline{C}c_S\delta^2}{3} - \frac{\varepsilon_1}{3}\right)\left\|\mathbb{D}\left(\boldsymbol{\phi}^h\right)\right\|_{L^3(\Omega)}^3 + \sum_{j=1}^J\frac{\beta}{2}\left\|\boldsymbol{\phi}^h\cdot\boldsymbol{\tau}_j\right\|_{L^2(\Gamma_j)}^2$$
$$\leq\left[\frac{2}{3\left(\underline{C}c_S\right)^{1/2}\delta}\|\boldsymbol{\eta}_t\|_{V^*}^{3/2} + \alpha_s\|\nabla\cdot\boldsymbol{\eta}\|_{L^2(\Omega)}^2 + \frac{2\nu+a_0(\delta)}{2}\|\mathbb{D}(\boldsymbol{\eta})\|_{L^2(\Omega)}^2\right.$$
$$+\frac{2\overline{C}^{3/2}C_L^{3/2}c_S\delta^2}{3\underline{C}^{1/2}}\|\mathbb{D}(\boldsymbol{\eta})\|_{L^3(\Omega)}^{3/2} + \sum_{j=1}^J\frac{\beta}{2}\|\boldsymbol{\eta}\cdot\boldsymbol{\tau}_j\|_{L^2(\Gamma_j)}^2 + \frac{1}{4}\|\nabla\boldsymbol{\eta}\|_{L^3(\Omega)}^2$$
$$+\frac{C}{3\varepsilon_1^{1/2}}\left(\|\mathbf{w}\|_{L^2(\Omega)}^{3/2} + \left\|\mathbf{w}^h\right\|_{L^2(\Omega)}^{3/2}\right)\|\boldsymbol{\eta}\|_{L^6(\Omega)}^{3/2} + \frac{1}{4}\|\boldsymbol{\eta}\|_{L^6(\Omega)}^2 \tag{8.34}$$
$$\left.+\frac{1}{\alpha_s}\left\|r-q^h\right\|_{L^2(\Omega)}^2\right] + \left[\frac{1}{4}\|\mathbf{w}\|_{L^6(\Omega)}^2 + \frac{1}{4}\left\|\nabla\mathbf{w}^h\right\|_{L^3(\Omega)}^2 + \frac{C}{\varepsilon_3}\left\|\nabla\mathbf{w}^h\right\|_{L^3(\Omega)}^2\right.$$
$$\left.+C\varepsilon_3^{-1/2}\alpha_s^{-3/2}\left\|\mathbb{D}\left(\mathbf{w}^h\right)\right\|_{L^3(\Omega)}^3\right]\left\|\boldsymbol{\phi}^h\right\|_{L^2(\Omega)}^2.$$

We pick $\varepsilon_1 = \underline{C}c_S\delta^2/2$ such that

$$\frac{\underline{C}c_S\delta^2}{3} - \frac{\varepsilon_1}{3} = \frac{\underline{C}c_S\delta^2}{6}.$$

Thus, $\varepsilon_1^{-1/2} = C\delta^{-1}$. The other parameters are chosen $\varepsilon_2 = a_0(\delta)/3$, $\varepsilon_3 = a_0(\delta)$ from what

$$\frac{a_0(\delta)}{2} - \frac{\varepsilon_3}{6} - \frac{\varepsilon_2}{2} = \frac{a_0(\delta)}{6}$$

follows. We get after a multiplication with 2

$$\frac{d}{dt}\left\|\boldsymbol{\phi}^h\right\|_{L^2(\Omega)}^2 + \frac{3\alpha_s}{4}\left\|\nabla\cdot\boldsymbol{\phi}^h\right\|_{L^2(\Omega)}^2 + \left(2\nu + \frac{a_0(\delta)}{3}\right)\left\|\mathbb{D}\left(\boldsymbol{\phi}^h\right)\right\|_{L^2(\Omega)}^2$$
$$+\frac{\underline{C}c_S\delta^2}{3}\left\|\mathbb{D}\left(\boldsymbol{\phi}^h\right)\right\|_{L^3(\Omega)}^3 + \sum_{j=1}^J\beta\left\|\boldsymbol{\phi}^h\cdot\boldsymbol{\tau}_j\right\|_{L^2(\Gamma_j)}^2$$

$$
\begin{aligned}
\leq & \left[\frac{4}{3\left(\underline{C} c_S\right)^{1/2} \delta}\|\boldsymbol{\eta}_t\|_{V^*}^{3/2}+2 \alpha_s\|\nabla \cdot \boldsymbol{\eta}\|_{L^2(\Omega)}^2+\left(2 \nu+a_0(\delta)\right)\|\mathbb{D}(\boldsymbol{\eta})\|_{L^2(\Omega)}^2\right. \\
& +\frac{4 \overline{C}^{3/2} C_L^{3/2} c_S \delta^2}{3 \underline{C}^{1/2}}\|\mathbb{D}(\boldsymbol{\eta})\|_{L^3(\Omega)}^{3/2}+\sum_{j=1}^J \beta\|\boldsymbol{\eta} \cdot \boldsymbol{\tau}_j\|_{L^2(\Gamma_j)}^2+\frac{1}{2}\|\nabla \boldsymbol{\eta}\|_{L^3(\Omega)}^2 \\
& +C \delta^{-1}\left(\|\mathbf{w}\|_{L^2(\Omega)}^{3/2}+\left\|\mathbf{w}^h\right\|_{L^2(\Omega)}^{3/2}\right)\|\boldsymbol{\eta}\|_{L^6(\Omega)}^{3/2}+\frac{1}{2}\|\boldsymbol{\eta}\|_{L^6(\Omega)}^2 \\
& \left.+\frac{2}{\alpha_s}\left\|r-q^h\right\|_{L^2(\Omega)}^2\right]+\left[\frac{1}{2}\|\mathbf{w}\|_{L^6(\Omega)}^2+\frac{1}{2}\left\|\nabla \mathbf{w}^h\right\|_{L^3(\Omega)}^2\right. \\
& \left.+\frac{C}{a_0(\delta)}\left\|\nabla \mathbf{w}^h\right\|_{L^3(\Omega)}^2+C a_0(\delta)^{-1/2} \alpha_s^{-3/2}\left\|\mathbb{D}\left(\mathbf{w}^h\right)\right\|_{L^3(\Omega)}^3\right]\left\|\boldsymbol{\phi}^h\right\|_{L^2(\Omega)}^2 .
\end{aligned}
$$

Before applying Gronwall's lemma, we have to make sure that all functions are sufficiently smooth. All terms which involve only norms of $\boldsymbol{\eta}$ and derivatives of $\boldsymbol{\eta}$ are in $L^1(0,T)$. The other term in the first bracket is shown to be in $L^1(0,T)$ in Lemma 8.16. Finally, the term in the second bracket is also in $L^1(0,T)$ by Lemma 8.15. The application of Gronwall's lemma in form (8.7) gives for almost all $t \in [0,T]$

$$
\begin{aligned}
& \left\|\boldsymbol{\phi}^h(t, \mathbf{x})\right\|_{L^2(\Omega)}^2+\alpha_s\left\|\nabla \cdot \boldsymbol{\phi}^h\right\|_{L^2(0, t ; L^2(\Omega))}^2 \\
& \quad+\left(\nu+C a_0(\delta)\right)\left\|\mathbb{D}\left(\boldsymbol{\phi}^h\right)\right\|_{L^2(0, t ; L^2(\Omega))}^2+\delta^2\left\|\mathbb{D}\left(\boldsymbol{\phi}^h\right)\right\|_{L^3(0, t ; L^3(\Omega))}^3 \\
& \quad+\sum_{j=1}^J \beta\left\|\boldsymbol{\phi}^h \cdot \boldsymbol{\tau}_j\right\|_{L^2(0, t ; L^2(\Gamma_j))}^2 \\
& \leq C \exp \left(\|\lambda(t)\|_{L^1(0, t)}\right)\left\|\boldsymbol{\phi}^h(0, \mathbf{x})\right\|_{L^2(\Omega)}^2 \\
& \quad+C \exp \left(\|\lambda(t)\|_{L^1(0, t)}\right)\left[\delta^{-1}\|\boldsymbol{\eta}_t\|_{L^{3/2}(0, t ; V^*)}^{3/2}+\alpha_s\|\nabla \cdot \boldsymbol{\eta}\|_{L^2(0, t ; L^2(\Omega))}^2\right. \\
& \quad+\left(2 \nu+a_0(\delta)\right)\|\mathbb{D}(\boldsymbol{\eta})\|_{L^2(0, t ; L^2(\Omega))}^2+\delta^2 \int_0^t C_L^{3/2}\|\mathbb{D}(\boldsymbol{\eta})\|_{L^3(\Omega)}^{3/2} d t^{\prime} \\
& \quad+\sum_{j=1}^J \beta\|\boldsymbol{\eta} \cdot \boldsymbol{\tau}_j\|_{L^2(0, t ; L^2(\Gamma_j))}^2+\|\nabla \boldsymbol{\eta}\|_{L^2(0, t ; L^3(\Omega))}^2 \\
& \quad+\delta^{-1} \int_0^t\left(\|\mathbf{w}\|_{L^2(\Omega)}^{3/2}+\left\|\mathbf{w}^h\right\|_{L^2(\Omega)}^{3/2}\right)\|\boldsymbol{\eta}\|_{L^6(\Omega)}^{3/2} d t^{\prime} \\
& \quad\left.+\|\boldsymbol{\eta}\|_{L^2(0, t ; L^6(\Omega))}^2+\frac{1}{\alpha_s}\left\|r-q^h\right\|_{L^2(0, t ; L^2(\Omega))}^2\right] .
\end{aligned}
$$

Note that by the definition of C_L in (8.25), Lemma 8.5, the Cauchy-Schwarz inequality in $L^2(0,T)$ and the stability estimates (8.16) and (8.17)

$$\begin{aligned}
&\int_0^t C_L^{3/2} \|\mathbb{D}(\boldsymbol{\eta})\|_{L^3(\Omega)}^{3/2}\, dt' \\
&\le \int_0^t \left(\max\left\{\|\mathbb{D}(\mathbf{w})\|_{L^3(\Omega)}, \|\mathbb{D}(\tilde{\mathbf{w}})\|_{L^3(\Omega)}\right\}\right)^{3/2} \|\mathbb{D}(\boldsymbol{\eta})\|_{L^3(\Omega)}^{3/2}\, dt' \\
&\le C\int_0^t \|\mathbb{D}(\mathbf{w})\|_{L^3(\Omega)}^{3/2} \|\mathbb{D}(\boldsymbol{\eta})\|_{L^3(\Omega)}^{3/2}\, dt' \\
&\le C\,\|\mathbb{D}(\mathbf{w})\|_{L^3(0,t;L^3(\Omega))}^{3/2} \|\mathbb{D}(\boldsymbol{\eta})\|_{L^3(0,t;L^3(\Omega))}^{3/2} \\
&\le C(\delta)\,\|\mathbb{D}(\boldsymbol{\eta})\|_{L^3(0,t;L^3(\Omega))}^{3/2}.
\end{aligned}$$

All constants which does not depend on the problem or the discretisation are put into the generic constants C. Using the definition of $C_2(\delta)$, Lemma 8.16, and applying the essential supremum on $(0,T)$ on both sides of the inequality give

$$\begin{aligned}
&\left\|\boldsymbol{\phi}^h\right\|_{L^\infty(0,T;L^2(\Omega))}^2 + \alpha_s \left\|\nabla\cdot\boldsymbol{\phi}^h\right\|_{L^2(0,T;L^2(\Omega))}^2 \\
&\quad + (\nu + Ca_0(\delta))\left\|\mathbb{D}\left(\boldsymbol{\phi}^h\right)\right\|_{L^2(0,T;L^2(\Omega))}^2 + \delta^2\left\|\mathbb{D}\left(\boldsymbol{\phi}^h\right)\right\|_{L^3(0,T;L^3(\Omega))}^3 \\
&\quad + \sum_{j=1}^J \beta\left\|\boldsymbol{\phi}^h\cdot\boldsymbol{\tau}_j\right\|_{L^2(0,T;L^2(\Gamma_j))}^2 \\
&\le C\exp\left(\|\lambda(t)\|_{L^1(0,T)}\right)\left\|\boldsymbol{\phi}^h(0,\mathbf{x})\right\|_{L^2(\Omega)}^2 \\
&\quad + C\exp\left(\|\lambda(t)\|_{L^1(0,T)}\right)\Big[\delta^{-1}\|\boldsymbol{\eta}_t\|_{L^{3/2}(0,T;V^*)}^{3/2} + \alpha_s\|\nabla\cdot\boldsymbol{\eta}\|_{L^2(0,T;L^2(\Omega))}^2 \\
&\quad + (2\nu + a_0(\delta))\|\mathbb{D}(\boldsymbol{\eta})\|_{L^2(0,T;L^2(\Omega))}^2 + C(\delta)\|\mathbb{D}(\boldsymbol{\eta})\|_{L^3(0,T;L^3(\Omega))}^{3/2} \\
&\quad + \sum_{j=1}^J \beta\|\boldsymbol{\eta}\cdot\boldsymbol{\tau}_j\|_{L^2(0,T;L^2(\Gamma_j))}^2 + \|\nabla\boldsymbol{\eta}\|_{L^2(0,T;L^3(\Omega))}^2 \\
&\quad + C_2(\delta)\|\boldsymbol{\eta}\|_{L^{3/2}(0,T;L^6(\Omega))}^{3/2} + \|\boldsymbol{\eta}\|_{L^2(0,T;L^6(\Omega))}^2 + \frac{1}{\alpha_s}\left\|r - q^h\right\|_{L^2(0,T;L^2(\Omega))}^2\Big].
\end{aligned}$$

The triangle inequality implies

$$\begin{aligned}
&\left\|\mathbf{w}-\mathbf{w}^h\right\|_{L^\infty(0,T;L^2(\Omega))}^2 + \alpha_s\left\|\nabla\cdot\left(\mathbf{w}-\mathbf{w}^h\right)\right\|_{L^2(0,T;L^2(\Omega))}^2 \\
&\quad + (\nu + Ca_0(\delta))\left\|\mathbb{D}\left(\mathbf{w}-\mathbf{w}^h\right)\right\|_{L^2(0,T;L^2(\Omega))}^2 \\
&\quad + \delta^2\left\|\mathbb{D}\left(\mathbf{w}-\mathbf{w}^h\right)\right\|_{L^3(0,T;L^3(\Omega))}^3 + \sum_{j=1}^J \beta\left\|\left(\mathbf{w}-\mathbf{w}^h\right)\cdot\boldsymbol{\tau}_j\right\|_{L^2(0,T;L^2(\Gamma_j))}^2
\end{aligned}$$

$$\leq C\Bigg[\|\boldsymbol{\eta}\|^2_{L^\infty(0,T;L^2(\Omega))} + \alpha_s \|\nabla\cdot\boldsymbol{\eta}\|^2_{L^2(0,T;L^2(\Omega))}$$
$$+ (\nu + Ca_0(\delta)) \|\mathbb{D}(\boldsymbol{\eta})\|^2_{L^2(0,T;L^2(\Omega))} + \delta^2 \|\mathbb{D}(\boldsymbol{\eta})\|^3_{L^3(0,T;L^3(\Omega))}$$
$$+ \sum_{j=1}^{J} \beta \|\boldsymbol{\eta}\cdot\boldsymbol{\tau}_j\|^2_{L^2(0,T;L^2(\Gamma_j))} + \left\|\boldsymbol{\phi}^h\right\|^2_{L^\infty(0,T;L^2(\Omega))}$$
$$+\alpha_s \left\|\nabla\cdot\boldsymbol{\phi}^h\right\|^2_{L^2(0,T;L^2(\Omega))} + (\nu + Ca_0(\delta)) \left\|\mathbb{D}\left(\boldsymbol{\phi}^h\right)\right\|^2_{L^2(0,T;L^2(\Omega))}$$
$$+\delta^2 \left\|\mathbb{D}\left(\boldsymbol{\phi}^h\right)\right\|^3_{L^3(0,T;L^3(\Omega))} + \sum_{j=1}^{J} \beta \left\|\boldsymbol{\phi}^h\cdot\boldsymbol{\tau}_j\right\|^2_{L^2(0,T;L^2(\Gamma_j))} \Bigg].$$

With the estimate for $\boldsymbol{\phi}^h$ and since $\exp\left(\|\lambda(t)\|_{L^1(0,T)}\right) \geq 1$, we get

$$\left\|\mathbf{w}-\mathbf{w}^h\right\|^2_{L^\infty(0,T;L^2(\Omega))} + \alpha_s \left\|\nabla\cdot\left(\mathbf{w}-\mathbf{w}^h\right)\right\|^2_{L^2(0,T;L^2(\Omega))}$$
$$+ (\nu + Ca_0(\delta)) \left\|\mathbb{D}\left(\mathbf{w}-\mathbf{w}^h\right)\right\|^2_{L^2(0,T;L^2(\Omega))}$$
$$+\delta^2 \left\|\mathbb{D}\left(\mathbf{w}-\mathbf{w}^h\right)\right\|^3_{L^3(0,T;L^3(\Omega))} + \sum_{j=1}^{J} \beta \left\|\left(\mathbf{w}-\mathbf{w}^h\right)\cdot\boldsymbol{\tau}_j\right\|^2_{L^2(0,T;L^2(\Gamma_j))}$$
$$\leq C\exp\left(\|\lambda(t)\|_{L^1(0,T)}\right) \left\|\boldsymbol{\phi}^h(0,\mathbf{x})\right\|^2_{L^2(\Omega)}$$
$$+C\Bigg\{ \|\boldsymbol{\eta}\|^2_{L^\infty(0,T;L^2(\Omega))} + \delta^2 \|\mathbb{D}(\boldsymbol{\eta})\|^3_{L^3(0,T;L^3(\Omega))}$$
$$+\exp\left(\|\lambda(t)\|_{L^1(0,T)}\right) \Bigg[\delta^{-1} \|\boldsymbol{\eta}_t\|^{3/2}_{L^{3/2}(0,T;V^*)} + \alpha_s \|\nabla\cdot\boldsymbol{\eta}\|^2_{L^2(0,T;L^2(\Omega))}$$
$$+ (2\nu + a_0(\delta)) \|\mathbb{D}(\boldsymbol{\eta})\|^2_{L^2(0,T;L^2(\Omega))} + C(\delta) \|\mathbb{D}(\boldsymbol{\eta})\|^{3/2}_{L^3(0,T;L^3(\Omega))}$$
$$+ \sum_{j=1}^{J} \beta \|\boldsymbol{\eta}\cdot\boldsymbol{\tau}_j\|^2_{L^2(0,T;L^2(\Gamma_j))} + \|\nabla\boldsymbol{\eta}\|^2_{L^2(0,T;L^3(\Omega))}$$
$$+C_2(\delta) \|\boldsymbol{\eta}\|^{3/2}_{L^{3/2}(0,T;L^6(\Omega))} + \|\boldsymbol{\eta}\|^2_{L^2(0,T;L^6(\Omega))}$$
$$+\frac{1}{\alpha_s} \left\|r-q^h\right\|^2_{L^2(0,T;L^2(\Omega))} \Bigg]\Bigg\}.$$

Applying the triangle inequality

$$\left\|\boldsymbol{\phi}^h(0,\mathbf{x})\right\|^2_{L^2(\Omega)} \leq C\left(\left\|\left(\mathbf{w}-\mathbf{w}^h\right)(0,\mathbf{x})\right\|^2_{L^2(\Omega)} + \|\boldsymbol{\eta}(0,\mathbf{x})\|^2_{L^2(\Omega)}\right),$$

and taking the infimum over $\tilde{\mathbf{w}}$ completes the proof of Theorem 8.17. □

8.1.6 The Case $\nabla \mathbf{w} \in L^2(0,T;L^\infty(\Omega))$ and $a_0(\delta) \geq 0$

In this section, it is shown that an error estimate similar to that of Theorem 8.17 can be obtained also in the case $a_0(\delta) = 0$ if a higher regularity of $\mathbf{w}$ uniformly in ν is assumed, in particular

$$\nabla \mathbf{w} \in L^2(0,T;L^\infty(\Omega)).$$

Theorem 8.18. *Suppose $a_0(\delta) \geq 0, \alpha_s > 0$ and $\mathbf{w} \in L^2\left(0,T;W^{1,\infty}(\Omega)\right)$ uniformly in ν. Let*

$$a(t) := \frac{3}{4} + \|\nabla \mathbf{w}\|_{L^\infty(\Omega)} + \left(\frac{1}{4} + \frac{1}{4\alpha_s}\right) \|\mathbf{w}\|^2_{L^\infty(\Omega)} + \frac{1}{2} \|\nabla \mathbf{w}\|^2_{L^\infty(\Omega)},$$

then there is a $C_3 = C_3(\mathbf{w})$ such that

$$\|a(t)\|_{L^1(0,T)} \leq C_3(\mathbf{w}).$$

Let $C_4 = C_4(\delta)$ be such that

$$\left\|\mathbb{D}\left(\mathbf{w}^h\right)\right\|_{L^3(0,T;L^3)} \leq C_4(\delta).$$

Then, the error $\mathbf{w} - \mathbf{w}^h$ satisfies:

$$\begin{aligned}
&\left\|\mathbf{w} - \mathbf{w}^h\right\|^2_{L^\infty(0,T;L^2(\Omega))} + \delta^2 \left\|\mathbb{D}\left(\mathbf{w} - \mathbf{w}^h\right)\right\|^3_{L^3(0,T;L^3(\Omega))} \\
&\quad + \alpha_s \left\|\nabla \cdot \left(\mathbf{w} - \mathbf{w}^h\right)\right\|^2_{L^2(0,T;L^2(\Omega))} \\
&\quad + (\nu + C a_0(\delta)) \left\|\mathbb{D}\left(\mathbf{w} - \mathbf{w}^h\right)\right\|^2_{L^2(0,T;L^2(\Omega))} \\
&\quad + \sum_{j=1}^{J} \beta \left\|\left(\mathbf{w} - \mathbf{w}^h\right) \cdot \hat{\boldsymbol{\tau}}_j\right\|^2_{L^2(0,T;L^2(\Gamma_j))} \\
&\leq C \exp\left(C_3(\mathbf{w})\right) \left\|\left(\mathbf{w} - \mathbf{w}^h\right)(x,0)\right\|^2_{L^2(\Omega)} \\
&\quad + C \inf_{\tilde{\mathbf{w}} \in V^h, q^h \in Q^h} \mathcal{F}\left(\mathbf{w} - \tilde{\mathbf{w}}, r - q^h, \delta\right)
\end{aligned}$$

with

$$\begin{aligned}
&\mathcal{F}\left(\mathbf{w} - \tilde{\mathbf{w}}, r - q^h, \delta\right) \\
&= \|\mathbf{w} - \tilde{\mathbf{w}}\|^2_{L^\infty(0,T;L^2(\Omega))} + \delta^2 \|\mathbb{D}(\mathbf{w} - \tilde{\mathbf{w}})\|^3_{L^3(0,T;L^3(\Omega))} \\
&\quad + \exp\left(C_3(\mathbf{w})\right) \Bigg[\|(\mathbf{w} - \tilde{\mathbf{w}})(x,0)\|^2_{L^2(\Omega)} + \delta^{-1} \|(\mathbf{w} - \tilde{\mathbf{w}})_t\|^{3/2}_{L^{3/2}(0,T;V^*)} \\
&\quad + (2\nu + a_0(\delta)) \|\mathbb{D}(\mathbf{w} - \tilde{\mathbf{w}})\|^2_{L^2(0,T;L^2(\Omega))} \\
&\quad + \sum_{j=1}^{J} \beta \|(\mathbf{w} - \tilde{\mathbf{w}}) \cdot \hat{\boldsymbol{\tau}}_j\|^2_{L^2(0,T;L^2(\Gamma_j))} + C(\delta) \|\mathbb{D}(\mathbf{w} - \tilde{\mathbf{w}})\|^{3/2}_{L^3(0,T;L^3(\Omega))} \\
&\quad + \alpha_s^{-1} \left\|r - q^h\right\|^2_{L^2(0,T;L^2(\Omega))} + \left(\frac{1}{4} + \alpha_s\right) \|\nabla \cdot (\mathbf{w} - \tilde{\mathbf{w}})\|^2_{L^2(0,T;L^2(\Omega))}
\end{aligned}$$

$$+\|\mathbf{w}-\tilde{\mathbf{w}}\|^2_{L^2(0,T,L^2(\Omega))}$$
$$\left.+C_4(\delta)\left(\|\mathbb{D}(\mathbf{w}-\tilde{\mathbf{w}})\|^2_{L^{18/5}(0,T;L^3(\Omega))}+\|\mathbf{w}-\tilde{\mathbf{w}}\|^2_{L^6(0,T;L^6(\Omega))}\right)\right].$$

Proof. The proof is based on the differential inequality (8.26). The non-linear convective term is decomposed in the following way

$$\begin{aligned}
&\left|b_s\left(\mathbf{w},\mathbf{w},\boldsymbol{\phi}^h\right)-b_s\left(\mathbf{w}^h,\mathbf{w}^h,\boldsymbol{\phi}^h\right)\right|\\
&=\left|b_s\left(\boldsymbol{\eta}-\boldsymbol{\phi}^h,\mathbf{w},\boldsymbol{\phi}^h\right)+b_s\left(\mathbf{w}^h,\boldsymbol{\eta}-\boldsymbol{\phi}^h,\boldsymbol{\phi}^h\right)\right|\\
&=\left|b_s\left(\boldsymbol{\eta},\mathbf{w},\boldsymbol{\phi}^h\right)-b_s\left(\boldsymbol{\phi}^h,\mathbf{w},\boldsymbol{\phi}^h\right)+b_s\left(\mathbf{w}^h,\boldsymbol{\eta},\boldsymbol{\phi}^h\right)\right|
\end{aligned}$$

where $b_s\left(\mathbf{w}^h,\boldsymbol{\phi}^h,\boldsymbol{\phi}^h\right)=0$ has been used. The individual terms of the right hand side are first transformed with (5.4) and then estimated by Hölder's and by Young's inequality

$$\begin{aligned}
&\left|b_s\left(\boldsymbol{\eta},\mathbf{w},\boldsymbol{\phi}^h\right)\right|=\left|b\left(\boldsymbol{\eta},\mathbf{w},\boldsymbol{\phi}^h\right)+\frac{1}{2}\left(\nabla\cdot\boldsymbol{\eta},\boldsymbol{\phi}^h\cdot\mathbf{w}\right)\right|\\
&\le\|\nabla\mathbf{w}\|_{L^\infty(\Omega)}\|\boldsymbol{\eta}\|_{L^2(\Omega)}\left\|\boldsymbol{\phi}^h\right\|_{L^2(\Omega)}+\frac{1}{2}\|\mathbf{w}\|_{L^\infty(\Omega)}\|\nabla\cdot\boldsymbol{\eta}\|_{L^2(\Omega)}\left\|\boldsymbol{\phi}^h\right\|_{L^2(\Omega)}\\
&\le\frac{1}{2}\|\boldsymbol{\eta}\|^2_{L^2(\Omega)}+\frac{1}{2}\|\nabla\mathbf{w}\|^2_{L^\infty(\Omega)}\left\|\boldsymbol{\phi}^h\right\|^2_{L^2(\Omega)}+\frac{1}{4}\|\nabla\cdot\boldsymbol{\eta}\|^2_{L^2(\Omega)}\\
&\quad+\frac{1}{4}\|\mathbf{w}\|^2_{L^\infty(\Omega)}\left\|\boldsymbol{\phi}^h\right\|^2_{L^2(\Omega)},
\end{aligned}$$

$$\begin{aligned}
&\left|b_s\left(\boldsymbol{\phi}^h,\mathbf{w},\boldsymbol{\phi}^h\right)\right|=\left|b\left(\boldsymbol{\phi}^h,\mathbf{w},\boldsymbol{\phi}^h\right)+\frac{1}{2}\left(\nabla\cdot\boldsymbol{\phi}^h,\mathbf{w}^h\cdot\boldsymbol{\phi}^h\right)\right|\\
&\le\|\nabla\mathbf{w}\|_{L^\infty(\Omega)}\left\|\boldsymbol{\phi}^h\right\|^2_{L^2(\Omega)}+\frac{1}{2}\|\mathbf{w}\|_{L^\infty(\Omega)}\left\|\nabla\cdot\boldsymbol{\phi}^h\right\|_{L^2(\Omega)}\left\|\boldsymbol{\phi}^h\right\|_{L^2(\Omega)}\\
&\le\|\nabla\mathbf{w}\|_{L^\infty(\Omega)}\left\|\boldsymbol{\phi}^h\right\|^2_{L^2(\Omega)}+\frac{\alpha_s}{4}\left\|\nabla\cdot\boldsymbol{\phi}^h\right\|^2_{L^2(\Omega)}+\frac{1}{4\alpha_s}\|\mathbf{w}\|^2_{L^\infty(\Omega)}\left\|\boldsymbol{\phi}^h\right\|^2_{L^2(\Omega)},
\end{aligned}$$

$$\begin{aligned}
&\left|b_s\left(\mathbf{w}^h,\boldsymbol{\eta},\boldsymbol{\phi}^h\right)\right|=\left|b\left(\mathbf{w}^h,\boldsymbol{\eta},\boldsymbol{\phi}^h\right)+\frac{1}{2}\left(\nabla\cdot\mathbf{w}^h,\boldsymbol{\eta}\cdot\boldsymbol{\phi}^h\right)\right|\\
&\le\left\|\mathbf{w}^h\right\|_{L^6(\Omega)}\|\nabla\boldsymbol{\eta}\|_{L^3(\Omega)}\left\|\boldsymbol{\phi}^h\right\|_{L^2(\Omega)}+\frac{1}{2}\left\|\nabla\cdot\mathbf{w}^h\right\|_{L^3(\Omega)}\|\boldsymbol{\eta}\|_{L^6(\Omega)}\left\|\boldsymbol{\phi}^h\right\|_{L^2(\Omega)}\\
&\le\frac{1}{2}\left\|\mathbf{w}^h\right\|^2_{L^6(\Omega)}\|\nabla\boldsymbol{\eta}\|^2_{L^3(\Omega)}+C\left\|\mathbb{D}\left(\mathbf{w}^h\right)\right\|^2_{L^3(\Omega)}\|\boldsymbol{\eta}\|^2_{L^6(\Omega)}+\frac{3}{4}\left\|\boldsymbol{\phi}^h\right\|^2_{L^2(\Omega)}.
\end{aligned}$$

Korn's inequality has been used in the last line. The term $\left\|\mathbf{w}^h\right\|_{L^6(\Omega)}$ is bounded using the Gagliardo-Nirenberg inequality (2.11), Korn's inequality and the uniform boundedness of $\left\|\mathbf{w}^h\right\|_{L^2(\Omega)}$

$$\left\|\mathbf{w}^h\right\|_{L^6(\Omega)}^2 \le C\left\|\mathbf{w}^h\right\|_{L^2(\Omega)}^{2/3}\left\|\mathbb{D}\left(\mathbf{w}^h\right)\right\|_{L^3(\Omega)}^{4/3} \le C\left\|\mathbb{D}\left(\mathbf{w}^h\right)\right\|_{L^3(\Omega)}^{4/3}.$$

Combining all estimates gives

$$\begin{aligned}
&\left|b_s\left(\mathbf{w},\mathbf{w},\boldsymbol{\phi}^h\right) - b_s\left(\mathbf{w}^h,\mathbf{w}^h,\boldsymbol{\phi}^h\right)\right| \\
&\le \frac{1}{2}\|\boldsymbol{\eta}\|_{L^2(\Omega)}^2 + \frac{1}{4}\|\nabla\cdot\boldsymbol{\eta}\|_{L^2(\Omega)}^2 + C\left\|\mathbb{D}\left(\mathbf{w}^h\right)\right\|_{L^3(\Omega)}^{4/3}\|\nabla\boldsymbol{\eta}\|_{L^3(\Omega)}^2 \\
&\quad + C\left\|\mathbb{D}\left(\mathbf{w}^h\right)\right\|_{L^3(\Omega)}^2\|\boldsymbol{\eta}\|_{L^6(\Omega)}^2 + \frac{\alpha_s}{4}\left\|\nabla\cdot\boldsymbol{\phi}^h\right\|_{L^2(\Omega)}^2 \\
&\quad + \left(\frac{3}{4} + \frac{1}{2}\|\nabla\mathbf{w}\|_{L^\infty(\Omega)}^2 + \frac{1}{4}\|\mathbf{w}\|_{L^\infty(\Omega)}^2 + \|\nabla\mathbf{w}\|_{L^\infty(\Omega)} + \frac{1}{4\alpha_s}\|\mathbf{w}\|_{L^\infty(\Omega)}^2\right) \\
&\quad \times\left\|\boldsymbol{\phi}^h\right\|_{L^2(\Omega)}^2.
\end{aligned}$$

This bound is now inserted in the right hand side of (8.26) giving

$$\begin{aligned}
&\frac{1}{2}\frac{d}{dt}\left\|\boldsymbol{\phi}^h\right\|_{L^2(\Omega)}^2 + \frac{\underline{C}c_S\delta^2}{3}\left\|\mathbb{D}\left(\boldsymbol{\phi}^h\right)\right\|_{L^3(\Omega)}^3 + \frac{\alpha_s}{2}\left\|\nabla\cdot\boldsymbol{\phi}^h\right\|_{L^2(\Omega)}^2 \\
&\quad + \frac{1}{2}\left(2\nu + a_0(\delta)\right)\left\|\mathbb{D}\left(\boldsymbol{\phi}^h\right)\right\|_{L^2(\Omega)}^2 + \sum_{j=1}^{J}\frac{\beta}{2}\left\|\boldsymbol{\phi}^h\cdot\hat{\boldsymbol{\tau}}_j\right\|_{L^2(\Gamma_j)}^2 \\
&\le \Bigg[\frac{2}{3\left(\underline{C}c_S\right)^{1/2}\delta}\|\boldsymbol{\eta}_t\|_{V^*}^{3/2} + \frac{1}{2}\left(2\nu + a_0(\delta)\right)\|\mathbb{D}(\boldsymbol{\eta})\|_{L^2(\Omega)}^2 \\
&\quad + \sum_{j=1}^{J}\frac{\beta}{2}\|\boldsymbol{\eta}\cdot\hat{\boldsymbol{\tau}}_j\|_{L^2(\Gamma_j)}^2 + \frac{2}{3}\underline{C}^{-1/2}c_S\overline{C}^{3/2}C_L^{3/2}\delta^2\|\mathbb{D}(\boldsymbol{\eta})\|_{L^3(\Omega)}^{3/2} \\
&\quad + \alpha_s^{-1}\left\|r - q^h\right\|_{L^2(\Omega)}^2 + \alpha_s\|\nabla\cdot\boldsymbol{\eta}\|_{L^2(\Omega)}^2 + \frac{1}{2}\|\boldsymbol{\eta}\|_{L^2(\Omega)}^2 + \frac{1}{4}\|\nabla\cdot\boldsymbol{\eta}\|_{L^2(\Omega)}^2 \\
&\quad + C\left\|\mathbb{D}\left(\mathbf{w}^h\right)\right\|_{L^3(\Omega)}^{4/3}\|\nabla\boldsymbol{\eta}\|_{L^3(\Omega)}^2 + C\left\|\mathbb{D}\left(\mathbf{w}^h\right)\right\|_{L^3(\Omega)}^2\|\boldsymbol{\eta}\|_{L^6(\Omega)}^2\Bigg] \\
&\quad + \left[\frac{\alpha_s}{4}\left\|\nabla\cdot\boldsymbol{\phi}^h\right\|_{L^2(\Omega)}^2\right] \\
&\quad + \left(\frac{3}{4} + \|\nabla\mathbf{w}\|_{L^\infty(\Omega)} + \left(\frac{1}{4} + \frac{1}{4\alpha_s}\right)\|\mathbf{w}\|_{L^\infty(\Omega)}^2 + \frac{1}{2}\|\nabla\mathbf{w}\|_{L^\infty(\Omega)}^2\right) \\
&\quad \times\left\|\boldsymbol{\phi}^h\right\|_{L^2(\Omega)}^2.
\end{aligned}$$

To apply Gronwall's inequality, we need

$$\frac{3}{4} + \|\nabla\mathbf{w}\|_{L^\infty(\Omega)} + \left(\frac{1}{4} + \frac{1}{4\alpha_s}\right)\|\mathbf{w}\|_{L^\infty(\Omega)}^2 + \frac{1}{2}\|\nabla\mathbf{w}\|_{L^\infty(\Omega)}^2 \in L^1(0,T),$$

in other words $\mathbf{w} \in L^2\left(0,T;W^{1,\infty}\left(\Omega\right)\right)$. The term on the right hand side of this inequality containing $C_L^{3/2}$ is treated as in the proof of Theorem 8.17. In the final result of Gronwall's lemma, we must also verify that the resulting terms containing $\left\|\mathbb{D}\left(\mathbf{w}^h\right)\right\|_{L^3(\Omega)}$ are bounded uniformly in ν. To this end, apply Hölder's inequality

$$\begin{aligned}
&\int_0^T \left\|\mathbb{D}\left(\mathbf{w}^h\right)\right\|_{L^3(\Omega)}^{4/3} \left\|\mathbb{D}\left(\boldsymbol{\eta}\right)\right\|_{L^3(\Omega)}^2 dt \\
&\le \left\|\mathbb{D}\left(\mathbf{w}^h\right)\right\|_{L^{4q/3}(0,T;L^3(\Omega))}^{4/3} \left\|\mathbb{D}\left(\boldsymbol{\eta}\right)\right\|_{L^{2q'}(0,T;L^3(\Omega))}^2,
\end{aligned}$$

where $q^{-1} + {q'}^{-1} = 1$. From the stability estimates, we clearly must take q such that $4q/3 \le 3$. Accordingly, we take $q = 9/4, q' = 9/5$. This gives

$$\begin{aligned}
&\int_0^T \left\|\mathbb{D}\left(\mathbf{w}^h\right)\right\|_{L^3(\Omega)}^{4/3} \left\|\mathbb{D}\left(\boldsymbol{\eta}\right)\right\|_{L^3(\Omega)}^2 dt \\
&\le C\left\|\mathbb{D}\left(\mathbf{w}^h\right)\right\|_{L^3(0,T;L^3(\Omega))}^{4/3} \left\|\mathbb{D}\left(\boldsymbol{\eta}\right)\right\|_{L^{18/5}(0,T;L^3(\Omega))}^2 \\
&\le CC_4\left(\delta\right)\left\|\mathbb{D}\left(\boldsymbol{\eta}\right)\right\|_{L^{18/5}(0,T;L^3(\Omega))}^2.
\end{aligned}$$

Similarly, for the conjugate exponents $q = 3/2,\ q' = 3$, one obtains

$$\begin{aligned}
\int_0^T \left\|\mathbb{D}\left(\mathbf{w}^h\right)\right\|_{L^3(\Omega)}^2 \left\|\boldsymbol{\eta}\right\|_{L^6(\Omega)}^2 dt &\le \left\|\mathbb{D}\left(\mathbf{w}^h\right)\right\|_{L^{2q}(0,T;L^3(\Omega))}^2 \left\|\boldsymbol{\eta}\right\|_{L^{2q'}(0,T;L^6(\Omega))}^2 \\
&\le \left\|\mathbb{D}\left(\mathbf{w}^h\right)\right\|_{L^3(0,T;L^3(\Omega))}^2 \left\|\boldsymbol{\eta}\right\|_{L^6(0,T;L^6(\Omega))}^2 \\
&\le C_4\left(\delta\right)\left\|\boldsymbol{\eta}\right\|_{L^6(0,T;L^6(\Omega))}^2.
\end{aligned}$$

The stated error estimate now follows from Gronwall's inequality and the triangle inequality as in the proof of Theorem 8.17. □

8.1.7 Failures of the Present Analysis in Other Interesting Cases

The discrete problem (8.15) considered in this section contains three differences to the discrete problem (7.8) which is used in the computations:

1. The possibility of choosing $a_0\left(\delta\right) > 0$.
2. The possibility of choosing $\alpha_s > 0$.
3. The use of $b_s\left(\mathbf{w}^h, \mathbf{w}^h, \mathbf{v}^h\right)$ instead of the convective form $b\left(\mathbf{w}^h, \mathbf{w}^h, \mathbf{v}^h\right)$.

The three additional components in (8.15) are exploited in the analysis in this section. They are not necessary in practical computations. We want to point out which difficulties arise if they are not used in the present analysis.

If $a_0\left(\delta\right) = 0$, the analysis fails under the assumption $\nabla\mathbf{w} \in L^3\left(0,T;L^3\left(\Omega\right)\right)$ as it is described in detail in Section 8.1.4. A careful inspection of the proof of Theorem 8.17 shows that the condition $a_0\left(\delta\right) > 0$ is not used in the derivation of the differential inequality (8.34). But the next step in this proof, the choices of $\varepsilon_2, \varepsilon_3$ must be different if $a_0(\delta) = 0$. In this case, the only possibility is to choose $\varepsilon_2 = \varepsilon_3 = \mathcal{O}(\nu)$. The reminder of the proof can be done analogously to the case $a_0(\delta) > 0$. The new choices of $\varepsilon_2, \varepsilon_3$ if $a_0(\delta) = 0$ lead to constants in the right hand side of the error estimate which depend on ν. Section 8.1.6 shows that the assumption of a higher regularity of $\mathbf{w}$ uniformly in ν makes the proof of an error estimate for $a_0(\delta) = 0$ uniformly in ν possible.

The condition $\alpha_s > 0$ was used in the previous analysis to estimate the terms

$$\left(\nabla \cdot \boldsymbol{\phi}^h, \boldsymbol{\phi}^h \cdot \mathbf{w}^h\right) \quad \text{or} \quad \left(\nabla \cdot \boldsymbol{\phi}^h, \boldsymbol{\phi}^h \cdot \mathbf{w}\right) \tag{8.35}$$

arising in the transformation of $b_s\left(\boldsymbol{\phi}^h, \mathbf{w}^h, \boldsymbol{\phi}^h\right)$ or $b_s\left(\boldsymbol{\phi}^h, \mathbf{w}, \boldsymbol{\phi}^h\right)$. Possible estimates of (8.35) which can be used if α_s vanishes are

$$\begin{aligned}\left|\left(\nabla \cdot \boldsymbol{\phi}^h, \boldsymbol{\phi}^h \cdot \mathbf{w}\right)\right| &\le \left\|\nabla \cdot \boldsymbol{\phi}^h\right\|_{L^3(\Omega)} \|\mathbf{w}\|_{L^6(\Omega)} + \left\|\boldsymbol{\phi}^h\right\|_{L^2(\Omega)} \\ &\le \frac{\varepsilon}{3}\left\|\mathbb{D}\left(\boldsymbol{\phi}^h\right)\right\|_{L^3(\Omega)}^3 + \frac{2}{3\varepsilon^{1/2}}\|\mathbf{w}\|_{L^6(\Omega)}^{3/2}\left\|\boldsymbol{\phi}^h\right\|_{L^2(\Omega)}^{3/2},\end{aligned} \tag{8.36}$$

$$\begin{aligned}\left|\left(\nabla \cdot \boldsymbol{\phi}^h, \boldsymbol{\phi}^h \cdot \mathbf{w}\right)\right| &\le \left\|\nabla \cdot \boldsymbol{\phi}^h\right\|_{L^3(\Omega)} \|\mathbf{w}\|_{L^6(\Omega)} + \left\|\boldsymbol{\phi}^h\right\|_{L^2(\Omega)} \\ &\le \frac{\varepsilon}{2}\left\|\mathbb{D}\left(\boldsymbol{\phi}^h\right)\right\|_{L^3(\Omega)}^2 + \frac{1}{2\varepsilon}\|\mathbf{w}\|_{L^6(\Omega)}^{2}\left\|\boldsymbol{\phi}^h\right\|_{L^2(\Omega)}^{2}.\end{aligned} \tag{8.37}$$

Estimate (8.36) does not allow to apply Gronwall's lemma because of the power 3/2 of $\left\|\boldsymbol{\phi}^h\right\|_{L^2(\Omega)}$, see Remark 8.12. The problem in estimate (8.37) is that the term $\left\|\mathbb{D}\left(\boldsymbol{\phi}^h\right)\right\|_{L^3(\Omega)}^2$ cannot be hidden in the left hand side of (8.27) because there only the term $\left\|\mathbb{D}\left(\boldsymbol{\phi}^h\right)\right\|_{L^3(\Omega)}^3$ appears. The simple estimate

$$\left\|\mathbb{D}\left(\boldsymbol{\phi}^h\right)\right\|_{L^3(\Omega)}^2 \le 1 + \left\|\mathbb{D}\left(\boldsymbol{\phi}^h\right)\right\|_{L^3(\Omega)}^3$$

leads to a term in the right hand side of the final error estimate which does not vanish if h tends to zero. Such an estimate is worthless. Similar difficulties are encountered also if (8.35) is estimated in the more complicated way of the proof of Theorem 8.17. Under the regularity assumption $\nabla \mathbf{w} \in L^2(0,T;L^\infty(\Omega))$, one can obtain an error estimate with constants in the right hand side depending on ν along the lines of the proof of Theorem 8.18 using (8.27) and

$$\begin{aligned}\left|\left(\nabla \cdot \boldsymbol{\phi}^h, \boldsymbol{\phi}^h \cdot \mathbf{w}^h\right)\right| &\le \left\|\nabla \cdot \boldsymbol{\phi}^h\right\|_{L^2(\Omega)} \left\|\mathbf{w}^h\right\|_{L^\infty(\Omega)} + \left\|\boldsymbol{\phi}^h\right\|_{L^2(\Omega)} \\ &\le \frac{\nu}{2}\left\|\mathbb{D}\left(\boldsymbol{\phi}^h\right)\right\|_{L^2(\Omega)}^2 + \frac{1}{2\nu}\left\|\mathbf{w}^h\right\|_{L^\infty(\Omega)}^{2}\left\|\boldsymbol{\phi}^h\right\|_{L^2(\Omega)}^{2}.\end{aligned}$$

The difficulties with the convective form $b\left(\mathbf{w}^h, \mathbf{w}^h, \mathbf{v}^h\right)$ are already at the beginning of the proof in Leray's inequality, Lemma 8.8. In the proof of this inequality, $b_s\left(\mathbf{w}^h, \mathbf{w}^h, \mathbf{w}^h\right) = 0$ is used. The term $b\left(\mathbf{w}^h, \mathbf{w}^h, \mathbf{w}^h\right)$ does not posses this property and it is not clear how to estimate it. The use of the skew-symmetric form of the convective term in the analysis can be found throughout the literature, see, e.g., Heywood and Rannacher [HR82] or Marion and Temam [MT98, p. 608].

To summarize this section, in the cases $a_0(\delta) = 0$ and $\alpha_s = 0$ error estimates with constants depending on ν on the right hand side are proveable. The use of the convective form of the convective term is a general problem in the finite element analysis of Navier-Stokes type equations.

8.1.8 A Numerical Example

This section presents a numerical example to support the main statements of the error estimates

- the error in different norms is bounded independently of ν for fixed δ and h,
- for fixed δ, the order of convergence of the error in different norms is related to the interpolation error of the used pair of finite element spaces.

The numerical tests use an academic flow problem with known solution.

Example 8.19. Chorin's vortex decay problem. Chorin's vortex decay problem [Cho68] is defined in $\Omega = (0,1)^2$ and the prescribed solution has the form

$$\begin{aligned}
w_1 &= -\cos(n\pi x)\sin(n\pi y)\exp\left(-2n^2\pi^2 t/\tau\right), \\
w_2 &= \sin(n\pi x)\cos(n\pi y)\exp\left(-2n^2\pi^2 t/\tau\right), \\
r &= -\frac{1}{4}\left(\cos(2n\pi x) + \cos(2n\pi y)\right)\exp\left(-4n^2\pi^2 t/\tau\right)
\end{aligned}$$

with $\mathbf{w} = (w_1, w_2)^T$. For the relaxation time $\tau = \nu^{-1}$, this is a solution of the Navier-Stokes equations (3.5) consisting of an array of opposite signed vortices which decay exponentially as $t \to \infty$.

In the numerical tests presented here, we have used Dirichlet boundary conditions on the whole boundary, $\partial\Omega = \Gamma_0$. The right hand side $\mathbf{f}$, the initial condition $\mathbf{w}_0$ and the non-homogeneous Dirichlet boundary conditions are chosen such that $(w_1, w_2, r)^T$ is the closed form solution of (8.1).

The following model and solution parameters were used in the computations

relaxation time	$\tau = 1000$,
vortex configuration	$n = 4$,
final time	$T = 8$,
filter width	$\delta = 0.1$,
Smagorinsky constant	$c_S = 0.05$,
additional stabilisation	$a_0(\delta) = 0$.

It follows that $\mathbf{w}$ is independent of ν and Theorem 8.18 can be applied.

The computations were performed with a discrete problem which differs in two aspects from (8.15):

- the least squares constant α_s is set to be zero,
- the convective form of the convective term $b\left(\mathbf{w}^h, \mathbf{w}^h, \mathbf{v}^h\right)$ is used instead of the skew-symmetric form $b_s\left(\mathbf{w}^h, \mathbf{w}^h, \mathbf{v}^h\right)$.

These two changes lead to a form of the discrete problem which is used in applications rather than (8.15).

The fractional-step θ-scheme with an equal distant time step $\Delta t_n = 0.001$ is used as discretisation in time. The time discretisation error should be kept small by using this very small time step. In space, the Q_2/P_1^{disc} and the Q_3/P_2^{disc} finite element discretisations are applied, see Table 8.1 for the number of degrees of freedom for different mesh widths. The unit square was divided into a $h^{-1} \times h^{-1}$ mesh with $h = 1/2$ on level 0. The non-linear system in each time step is solved up to an Euclidean norm of the residual vector less than 10^{-10}.

Table 8.1. Mesh widths and degrees of freedom in space

	Q_2/P_1^{disc}			Q_3/P_2^{disc}		
mesh width	velocity	pressure	total	velocity	pressure	total
1/4	-	-	-	338	96	434
1/8	578	192	770	1250	384	1 634
1/16	2 178	768	2 946	4802	1 536	6 338
1/32	8 450	3 072	11 522	18 818	6 144	24 962
1/64	33 282	12 288	45 570	-	-	-
1/128	132 098	49 152	181 250	-	-	-

We present results for the error $\left\|\mathbf{w} - \mathbf{w}^h\right\|_{L^\infty(0,T;L^2(\Omega))}$, Tables 8.2 and 8.4, and for the error $\left\|\mathbb{D}\left(\mathbf{w} - \mathbf{w}^h\right)\right\|_{L^2(0,T;L^2(\Omega))}$, Tables 8.3 and 8.5. Let m be the interpolation order of the finite element space and assume that all terms in $\mathcal{F}\left(\mathbf{w} - \tilde{\mathbf{w}}, r - q^h, \delta\right)$ in Theorem 8.18 behave optimal for all times. Then, the order of convergence of $\mathcal{F}\left(\mathbf{w} - \tilde{\mathbf{w}}, r - q^h, \delta\right)$ is bounded, e.g., by the order of convergence of $\|\mathbb{D}(\mathbf{w} - \tilde{\mathbf{w}})\|_{L^3(0,T;L^3(\Omega))}^{3/2}$, which is $3m/2$, see Remark 7.3. That means, the expected order of convergence of the considered errors is 3/2 for the Q_2/P_1^{disc} finite element discretisation and 9/4 for the Q_3/P_2^{disc} finite element discretisation. Note, the term $\|\mathbb{D}(\mathbf{w} - \tilde{\mathbf{w}})\|_{L^3(0,T;L^3(\Omega))}^{3/2}$, which determines the interpolation order, originates already in the local Lipschitz continuity of the Smagorinsky term, (8.12).

Table 8.2. Example 8.19, $\|\mathbf{w}-\mathbf{w}^h\|_{L^\infty(0,T;L^2(\Omega))}$, Q_2/P_1^{disc} finite element discretisation, error and order of convergence with respect to h (in parentheses)

ν^{-1}	$h=1/8$	$h=1/16$	$h=1/32$	$h=1/64$	$h=1/128$
10^2	2.20176-2	2.76780-3 (2.992)	3.47796-4 (2.992)	4.35185-5 (2.999)	5.43988-6 (3.000)
10^3	3.19389-2	3.50372-3 (3.188)	4.81015-4 (2.865)	4.86864-5 (3.304)	5.50381-6 (3.145)
10^4	5.97051-2	7.01100-3 (3.090)	1.00294-3 (2.805)	1.39466-4 (2.846)	1.44706-5 (3.269)
10^5	7.67057-2	7.73782-3 (3.309)	1.09801-3 (2.817)	1.62252-4 (2.758)	1.92552-5 (3.075)
10^6	7.86394-2	7.81755-3 (3.330)	1.10830-3 (2.818)	1.64891-4 (2.749)	1.98664-5 (3.053)
10^7	7.88349-2	7.82560-3 (3.333)	1.10934-3 (2.818)	1.65161-4 (2.748)	1.99288-5 (3.051)
10^8	7.88545-2	7.82641-3 (3.333)	1.10945-3 (2.819)	1.65188-4 (2.748)	1.99371-5 (3.051)
10^9	7.88564-2	7.82649-3 (3.333)	1.10946-3 (2.819)	1.65190-4 (2.748)	1.99377-5 (3.051)
10^{10}	7.88566-2	7.82650-3 (3.333)	1.10946-3 (2.819)	1.65191-4 (2.748)	1.99380-5 (3.051)

Table 8.3. Example 8.19, $\|\mathbb{D}(\mathbf{w}-\mathbf{w}^h)\|_{L^2(0,T;L^2(\Omega))}$, Q_2/P_1^{disc} finite element discretisation, error and order of convergence with respect to h (in parentheses)

ν^{-1}	$h=1/8$	$h=1/16$	$h=1/32$	$h=1/64$	$h=1/128$
10^2	1.248279	3.13720-1 (1.992)	7.84736-2 (1.999)	1.96114-2 (2.000)	4.90234-3 (2.000)
10^3	1.569352	3.60470-1 (2.122)	8.42787-2 (2.097)	2.00913-2 (2.069)	4.93406-3 (2.026)
10^4	2.351005	4.66554-1 (2.333)	1.05387-1 (2.146)	2.34301-2 (2.169)	5.28506-3 (2.148)
10^5	2.681270	4.98844-1 (2.426)	1.14609-1 (2.122)	2.61063-2 (2.134)	5.79700-3 (2.171)
10^6	2.720373	5.02793-1 (2.436)	1.15920-1 (2.117)	2.66473-2 (2.121)	5.96091-3 (2.160)
10^7	2.724344	5.03197-1 (2.437)	1.16058-1 (2.116)	2.67093-2 (2.119)	5.98435-3 (2.158)
10^8	2.724742	5.03237-1 (2.437)	1.16072-1 (2.116)	2.67156-2 (2.119)	5.98686-3 (2.158)
10^9	2.724782	5.03241-1 (2.437)	1.16073-1 (2.116)	2.67162-2 (2.119)	5.98711-3 (2.158)
10^{10}	2.724786	5.03242-1 (2.437)	1.16073-1 (2.116)	2.67163-2 (2.119)	5.98714-3 (2.158)

Table 8.4. Example 8.19, $\left\|\mathbf{w}-\mathbf{w}^h\right\|_{L^\infty(0,T;L^2(\Omega))}$, Q_3/P_2^{disc} finite element discretisation, error and order of convergence with respect to h (in parentheses)

ν^{-1}	$h=1/4$	$h=1/8$	$h=1/16$	$h=1/32$
10^2	3.09237-2	2.14568-3 (3.849)	1.39746-4 (3.941)	8.96881-6 (3.962)
10^3	7.61050-2	2.53153-3 (4.910)	1.40003-4 (4.176)	8.85819-6 (3.982)
10^4	1.09160-1	2.79077-3 (5.290)	1.43102-4 (4.286)	8.81959-6 (4.020)
10^5	1.13716-1	2.82963-3 (5.329)	1.43815-4 (4.298)	8.81915-6 (4.027)
10^6	1.14186-1	2.83371-3 (5.333)	1.43896-4 (4.300)	8.81835-6 (4.028)
10^7	1.14234-1	2.83412-3 (5.333)	1.43904-4 (4.300)	8.81929-6 (4.028)
10^8	1.14238-1	2.83416-3 (5.333)	1.43905-4 (4.300)	8.81853-6 (4.028)
10^9	1.14239-1	2.83417-3 (5.333)	1.43905-4 (4.300)	8.81961-6 (4.028)
10^{10}	1.14239-1	2.83417-3 (5.333)	1.43905-4 (4.300)	8.81754-6 (4.029)

Table 8.5. Example 8.19, $\left\|\mathbb{D}\left(\mathbf{w}-\mathbf{w}^h\right)\right\|_{L^2(0,T;L^2(\Omega))}$, Q_3/P_2^{disc} finite element discretisation, error and order of convergence with respect to h (in parentheses)

ν^{-1}	$h=1/4$	$h=1/8$	$h=1/16$	$h=1/32$
10^2	1.15300	1.65587-1 (2.800)	2.07562-2 (3.000)	2.60050-3 (3.000)
10^3	2.87920	1.93565-1 (3.895)	2.15627-2 (3.166)	2.62325-3 (3.039)
10^4	4.79996	2.24195-1 (4.420)	2.27512-2 (3.301)	2.67749-3 (3.087)
10^5	5.07949	2.31037-1 (4.458)	2.31277-2 (3.320)	2.71302-3 (3.092)
10^6	5.10843	2.31820-1 (4.462)	2.31768-2 (3.322)	2.72525-3 (3.088)
10^7	5.11134	2.31899-1 (4.462)	2.31819-2 (3.322)	2.72679-3 (3.088)
10^8	5.11163	2.31907-1 (4.462)	2.31824-2 (3.322)	2.72674-3 (3.088)
10^9	5.11166	2.31908-1 (4.462)	2.31825-2 (3.322)	2.72698-3 (3.088)
10^{10}	5.11166	2.31908-1 (4.462)	2.31825-2 (3.322)	2.72759-3 (3.087)

It can be clearly seen that
- the error is independent of ν for a given level of refinement,
- the order of convergence depends on the interpolation order of the finite element space.

These results show some particularities which are not covered by Theorem 8.18:
- It was proved that $\sqrt{\nu}\left\|\mathbb{D}\left(\mathbf{w}-\mathbf{w}^h\right)\right\|_{L^2(0,T;L^2(\Omega))}$ is independent of ν and not the unscaled error $\left\|\mathbb{D}\left(\mathbf{w}-\mathbf{w}^h\right)\right\|_{L^2(0,T;L^2(\Omega))}$.
- The errors have a different order of convergence which is in both cases higher than expected.

Thus, the particular example which we have chosen behaves better than the theory predicts. The prescribed solution is much smoother than assumed in Theorem 8.18 and we refer the higher order of convergence to this smoothness.

A special study of this phenomenom, which has to start at (8.12), is not available.

8.2 The Taylor LES Model

An error analysis similar to that of the Smagorinsky model presented in Section 8.1 was done for the Taylor LES model with Smagorinsky subgrid scale term in a joint work with T. Iliescu and W.J. Layton [IJL02]. This section presents the main new aspects in the error analysis and states the final error estimate.

We consider the Taylor LES model with a subgrid scale term of Smagorinsky type

$$\begin{aligned} \mathbf{w}_t - \nabla\cdot\left(\left(\nu+\nu_S\left\|\nabla\mathbf{w}\right\|_F\right)\nabla\mathbf{w}\right) + (\mathbf{w}\cdot\nabla)\,\mathbf{w} & \\ +\nabla r + \nabla\cdot\frac{\delta^2}{2\gamma}\left(\nabla\mathbf{w}\nabla\mathbf{w}^T\right) &= \mathbf{f} \quad \text{in } (0,T]\times\Omega, \\ \nabla\cdot\mathbf{w} &= 0 \quad \text{in } [0,T]\times\Omega, \\ \mathbf{w} &= \mathbf{0} \quad \text{on } [0,T]\times\Gamma, \\ \mathbf{w}(0,\cdot) &= \mathbf{w}_0 \text{ in } \Omega, \\ \int_\Omega r d\mathbf{x} &= 0 \quad \text{in } (0,T], \end{aligned} \qquad (8.38)$$

where ν_S is the constant for the turbulent viscosity term. The choice of homogeneous Dirichlet boundary conditions is only for simplicity of presentation. Slip with friction boundary conditions can be incorporated in the analysis analogously to Section 8.1, see also [IJL02].

Let $V = W_0^{1,3}(\Omega)$ and $Q = L_0^2(\Omega)$. The variational formulation of (8.38) is to find $(\mathbf{w}, r) \in V\times Q$ such that for all $t\in(0,T]$ and $(\mathbf{v},q)\in V\times Q$

$$\begin{aligned} &(\mathbf{w}_t,\mathbf{v}) + \left(\left(\nu+\nu_S\left\|\nabla\mathbf{w}\right\|_F\right)\nabla\mathbf{w},\nabla\mathbf{v}\right) + b_s(\mathbf{w},\mathbf{w},\mathbf{v}) \\ &+(q,\nabla\cdot\mathbf{w}) - (r,\nabla\cdot\mathbf{v}) - \frac{\delta^2}{2\gamma}\left(\nabla\mathbf{w}\nabla\mathbf{w}^T,\nabla\mathbf{v}\right) = (\mathbf{f},\mathbf{v}) \end{aligned} \qquad (8.39)$$

and $\mathbf{w}(0,\mathbf{x}) = \mathbf{w}_0(\mathbf{x})$. This formulation does not include a stabilisation term $\alpha_s(\nabla\cdot\mathbf{w},\nabla\cdot\mathbf{v})$, $\alpha_s>0$, in contrast to the variational formulation (8.9) considered in Section 8.1. In addition, a constant artificial viscosity $\mu_0(\delta)$ is not added to the Smagorinsky subgrid scale term. It was already discussed in Section 8.1.7 that an error estimate with constants independent of ν cannot be achieved in these cases with the applied analysis.

Let $V^h\subset V$, $Q^h\subset Q$ be a pair of inf-sup stable finite element spaces. The continuous-in-time finite element method for (8.39) seeks to find $\left(\mathbf{w}^h, r^h\right)\in V^h\times Q^h$ such that for all $\left(\mathbf{v}^h,q^h\right)\in V^h\times Q^h$

$$\begin{aligned} &\left(\mathbf{w}_t^h,\mathbf{v}^h\right) + \left(\left(\nu+\nu_S\left\|\nabla\mathbf{w}^h\right\|_F\right)\nabla\mathbf{w}^h,\nabla\mathbf{v}^h\right) + b_s\left(\mathbf{w}^h,\mathbf{w}^h,\mathbf{v}^h\right) \\ &+\left(q^h,\nabla\cdot\mathbf{w}^h\right) - \left(r^h,\nabla\cdot\mathbf{v}^h\right) - \frac{\delta^2}{2\gamma}\left(\nabla\mathbf{w}^h\left(\nabla\mathbf{w}^h\right)^T,\nabla\mathbf{v}^h\right) = (\mathbf{f},\mathbf{v}) \end{aligned} \qquad (8.40)$$

and $\mathbf{w}^h(0,\mathbf{x})$ is an approximation to $\mathbf{w}_0(\mathbf{x})$.

The way of proving the error estimate is the same as given in Section 8.1.2. The new aspects for the Taylor LES model are to prove a strong monotonicity and a local Lipschitz continuity of the operator

$$\mathbf{A}(\nabla\mathbf{w}) = (\nu + \nu_S \|\nabla\mathbf{w}\|_F)\nabla\mathbf{w} - \frac{\delta^2}{2\gamma}\left(\nabla\mathbf{w}\nabla\mathbf{w}^T\right) \tag{8.41}$$

as analogue to Lemma 8.5.

Lemma 8.20. Strong monotonicity. *Let* $\nu_S = c_S\delta^2$ *and* $\nu_S \geq 2\delta^2/\gamma$. *Then, there exists a constant* $\underline{C}$ *such that for any* $\mathbf{u},\mathbf{v} \in V$

$$\begin{aligned}&\int_\Omega (\mathbf{A}(\nabla\mathbf{u}) - \mathbf{A}(\nabla\mathbf{v})) : (\nabla\mathbf{u} - \nabla\mathbf{v})\,d\mathbf{x} \\ &\geq \nu \|\nabla\mathbf{u} - \nabla\mathbf{v}\|_{L^2(\Omega)}^2 + \frac{\underline{C}}{2}c_S \|\nabla\mathbf{u} - \nabla\mathbf{v}\|_{L^3(\Omega)}^3 .\end{aligned} \tag{8.42}$$

Proof. The proof starts with the splitting

$$\begin{aligned}&(\mathbf{A}(\nabla\mathbf{u}) - \mathbf{A}(\nabla\mathbf{v})) : (\nabla\mathbf{u} - \nabla\mathbf{v}) \\ &= \left(\left[\left(\nu + \frac{\nu_S}{2}\|\nabla\mathbf{u}\|_F\right)\nabla\mathbf{u} - \frac{\delta^2}{2\gamma}\left(\nabla\mathbf{u}\nabla\mathbf{u}^T\right)\right]\right. \\ &\quad \left. - \left[\left(\nu + \frac{\nu_S}{2}\|\nabla\mathbf{v}\|_F\right)\nabla\mathbf{v} - \frac{\delta^2}{2\gamma}\left(\nabla\mathbf{v}\nabla\mathbf{v}^T\right)\right], \nabla\mathbf{u} - \nabla\mathbf{v}\right) \\ &\quad + \frac{\nu_S}{2}\left(\|\nabla\mathbf{u}\|_F \nabla\mathbf{u} - \|\nabla\mathbf{w}\|_F \nabla\mathbf{w}, \nabla\mathbf{u} - \nabla\mathbf{v}\right).\end{aligned}$$

The first term can be estimated from below by $\nu\|\nabla\mathbf{u} - \nabla\mathbf{v}\|_{L^2(\Omega)}^2$ with Lemma 6.19 since $\nu_S \geq 2\delta^2/\gamma$ which is assumption (6.34) in this lemma. Lemma 8.5 gives the estimate for the second term. The proof of this lemma does not change with the deformation tensor replaced by the gradient. □

Lemma 8.21. Local Lipschitz continuity. *Let* $\nu_S = c_S\delta^2$. *Then, there is a constant* C *such that for any* $\mathbf{u},\mathbf{v},\mathbf{w} \in V$

$$\begin{aligned}&(\mathbf{A}(\nabla\mathbf{u}) - \mathbf{A}(\nabla\mathbf{v}), \nabla\mathbf{w}) \\ &\leq C\delta^2 \max\left\{\|\nabla\mathbf{u}\|_{L^3(\Omega)}, \|\nabla\mathbf{v}\|_{L^3(\Omega)}\right\} \|\nabla(\mathbf{u} - \mathbf{v})\|_{L^3(\Omega)} \|\nabla\mathbf{w}\|_{L^3(\Omega)} \\ &\quad + \nu\|\nabla(\mathbf{u} - \mathbf{v})\|_{L^2(\Omega)} \|\nabla\mathbf{w}\|_{L^2(\Omega)} .\end{aligned}$$

Proof. It is

$$\begin{aligned}&(\mathbf{A}(\nabla\mathbf{u}) - \mathbf{A}(\nabla\mathbf{v}), \nabla\mathbf{w}) \\ &= (\nu\nabla(\mathbf{u} - \mathbf{v}), \mathbf{w}) + \nu_S\left(\|\nabla\mathbf{u}\|_F \nabla\mathbf{u} - \|\nabla\mathbf{v}\|_F \nabla\mathbf{v}, \nabla\mathbf{w}\right) \\ &\quad - \frac{\delta^2}{2\gamma}\left(\nabla\mathbf{u}\left(\nabla\mathbf{u}^T - \nabla\mathbf{v}^T\right) - (\nabla\mathbf{v} - \nabla\mathbf{u})\nabla\mathbf{v}^T, \mathbf{w}\right).\end{aligned}$$

The first term is estimated by the Cauchy-Schwarz inequality, the estimate of the second term is analogously as in the proof of Lemma 8.5 and the third term is estimated by Hölder's inequality. One obtains

$$\begin{aligned}
&(\mathbf{A}(\nabla\mathbf{u}) - \mathbf{A}(\nabla\mathbf{v}), \nabla\mathbf{w}) \\
&\leq \nu \|\nabla(\mathbf{u}-\mathbf{v})\|_{L^2(\Omega)} \|\mathbf{w}\|_{L^2(\Omega)} \\
&\quad + \left(C\delta^2 \max\left\{ \|\nabla\mathbf{u}\|_{L^3(\Omega)}, \|\nabla\mathbf{v}\|_{L^3(\Omega)} \right\} + \frac{\delta^2}{2\gamma} \left(\|\nabla\mathbf{u}\|_{L^3(\Omega)} + \|\nabla\mathbf{v}\|_{L^3(\Omega)} \right) \right) \\
&\quad \times \|\nabla(\mathbf{u}-\mathbf{v})\|_{L^3(\Omega)} \|\nabla\mathbf{w}\|_{L^3(\Omega)},
\end{aligned}$$

which proves the lemma. □

The final finite element error estimate proved in Iliescu et al. [IJL02] has the following form.

Theorem 8.22. *Let* $(\mathbf{w}, r)$ *be the solution of (8.39) and* $(\mathbf{w}^h, r^h)$ *be the solution of (8.40). Let the finite element spaces fulfil the inf-sup condition*

$$\inf_{\lambda^h \in Q^h} \sup_{\mathbf{v}^h \in V^h} \frac{(\lambda^h, \nabla \cdot \mathbf{v}^h)}{\|\lambda^h\|_{L^2(\Omega)} \|\nabla\mathbf{v}^h\|_{L^2(\Omega)}} \geq C > 0,$$

let $c_S \geq 2/\gamma$ *and let* $a^h(t)$ *be*

$$a^h(t) := \left\|\mathbf{w}^h(t)\right\|_{L^2(\Omega)}^{1/2} \left\|\nabla\mathbf{w}^h(t)\right\|_{L^2(\Omega)}^{1/2}.$$

Under the assumption

$$\nabla\mathbf{w} \in L^4(0,T;L^3(\Omega)),$$

the error $\mathbf{w} - \mathbf{w}^h$ *satisfies*

$$\begin{aligned}
&\left\|\mathbf{w}-\mathbf{w}^h\right\|_{L^\infty(0,T;L^2(\Omega))}^2 + \nu \left\|\nabla\left(\mathbf{w}-\mathbf{w}^h\right)\right\|_{L^2(0,T;L^2(\Omega))}^2 \\
&\quad + c_S\delta^2 \left\|\nabla\left(\mathbf{w}-\mathbf{w}^h\right)\right\|_{L^3(0,T;L^3(\Omega))}^3 \\
&\leq CC^*(T) \left\|\mathbf{w}(\mathbf{x},0) - \mathbf{w}^h(\mathbf{x},0)\right\|_{L^2(\Omega)}^2 \\
&\quad + C \inf_{\tilde{\mathbf{w}} \in V^h, q^h \in Q^h} \mathcal{F}\left(\mathbf{w}-\tilde{\mathbf{w}}, r-q^h, \nu, \delta, c_S, T\right)
\end{aligned}$$

with

$$C^*(T) = \exp\left(\int_0^T \left(1 + C\nu^{-3} \|\nabla\mathbf{w}\|_{L^2(\Omega)}^4\right) dt \right)$$

and

$$
\begin{aligned}
&\mathcal{F}\left(\mathbf{w}-\tilde{\mathbf{w}}, r-q^{h}, \nu, \delta, c_{S}, T\right) \\
&= C\Bigg\{ \|\mathbf{w}-\tilde{\mathbf{w}}\|_{L^{\infty}(0,T;L^{2}(\Omega))}^{2} + c_{S}\delta^{2}\left\|\nabla\left(\mathbf{w}-\tilde{\mathbf{w}}\right)\right\|_{L^{3}(0,T;L^{3}(\Omega))}^{3} \\
&\quad + C^{*}\left(T\right)\Big[\left\|\left(\mathbf{w}-\tilde{\mathbf{w}}\right)\left(\mathbf{x},0\right)\right\|_{L^{2}(\Omega)}^{2} + C\left(\delta, c_{S}, T\right)\left\|\nabla\left(\mathbf{w}-\tilde{\mathbf{w}}\right)\right\|_{L^{3}(0,T;L^{3}(\Omega))}^{3/2} \\
&\quad + \nu^{-1}\left(\left\|a^{h}\left(t\right)\right\|_{L^{4}(0,T)}^{2} + \left\|\nabla\mathbf{w}\right\|_{L^{4}(0,T;L^{2}(\Omega))}^{2}\right)\left\|\nabla\left(\mathbf{w}-\tilde{\mathbf{w}}\right)\right\|_{L^{4}(0,T;L^{2}(\Omega))}^{2} \\
&\quad + \nu\left\|\nabla\left(\mathbf{w}-\tilde{\mathbf{w}}\right)\right\|_{L^{2}(0,T;L^{2}(\Omega))}^{2} + \nu^{-1}\left\|\left(\mathbf{w}-\tilde{\mathbf{w}}\right)_{t}\right\|_{L^{2}(0,T;H^{-1}(\Omega))}^{2} \\
&\quad + \nu^{-1}\left\|r-q^{h}\right\|_{L^{2}(0,T;L^{2}(\Omega))}^{2}\Big]\Bigg\}.
\end{aligned}
$$

9

The Solution of the Linear Systems

The discretisation and linearisation of the LES models, described in Chapter 7, lead to linear systems of the abstract form

$$Ax = b \tag{9.1}$$

with non-symmetric matrix A. These systems have to be solved in each step of the fixed point iteration (7.7) for each sub time step. The solution of these large number of systems of form (9.1) is the most time consuming part of the computations. That's why, the solver applied to (9.1) is one of the most important components for the efficiency of the numerical simulations with the LES models.

Our approach is to use an iterative scheme with preconditioner to solve (9.1) approximately. We will describe in this chapter a simple iterative scheme, the fixed point iteration in Section 9.1, and a more sophisticated iterative scheme, the flexible GMRES method in Section 9.2. The efficiency of these iterative schemes is mainly determined by the preconditioner. We use as preconditioner another iterative scheme, namely a coupled multigrid method. This multigrid method is the core of the solver and will be described in Section 9.3. Section 9.4 contains the description of the solver for the auxiliary problem in the rational LES model with auxiliary problem and in the Iliescu-Layton subgrid scale model (4.31).

All algorithms described in this chapter have been implemented in the code MooNMD.

9.1 The Fixed Point Iteration for the Solution of Linear Systems

The fixed point iteration is the simplest iterative scheme for solving (9.1). Let M be a preconditioner, i.e. M^{-1} is an approximation to A^{-1}. System (9.1) can be transformed into the fixed point form

$$x = x + M^{-1}(b - Ax).$$

Given an iterate x^k, the next iterate x^{k+1} in the fixed point iteration is computed by

$$x^{k+1} = x^k + M^{-1}\left(b - Ax^k\right).$$

Algorithm 9.1. Fixed point iteration for the solution of (9.1). Given A, M, b and an initial iterate x^0. The fixed point iteration for solving (9.1) has the following form:

```
 1. d := b - Ax^0
 2. res0 := res := ||d||_2
 3. if res <= resmin
 4.     maxit := minit
 5. for (i := 1; i <= maxit; i++)
 6.     solve Mv = d
 7.     x^i := x^{i-1} + v
 8.     d := b - Ax^i
 9.     res := ||d||_2
10.     if i < minit continue
11.     if res <= resmin break
12.     if res <= res0 · red_factor break
13.     if res >= res0 · div_factor exit
14. endfor
15. return
```

If M^{-1} stands for a multigrid cycle, this algorithm is just a standard multigrid iteration written as fixed point iteration.

Remark 9.2. Parameters of the fixed point iteration. The user prescribed parameters in the algorithm have the following meanings:

- `minit` : minimal number of iterations,
- `maxit` : maximal number of iterations,
- `resmin` : absolute value for the Euclidean norm of the residual for which the iteration stops,
- `red_factor` : factor for the reduction of the initial Euclidean norm of the residual after which the iteration stops,
- `div_factor` : factor for the increase of the initial Euclidean norm of the residual after which the iteration is considered as divergent and the program terminates without result.

□

In line 3, it is tested if the stopping criterion with respect to the minimal residual is already fulfilled before the iteration. In this case, exactly `minit` iterations are performed. One iteration step consists of the solution of the preconditioner equation (line 6), the update of the iterate (line 7), the computation of the new residual vector (line 8) and its Euclidean norm (line 9).

After this, the available stopping criteria are checked if at least `minit` iterations were already performed. A possible damping of the update in line 7 is incorporated into the solution of the preconditioner equation.

Remark 9.3. Additional memory requirements of the fixed point iteration. Let $A \in \mathbb{R}^{n \times n}$. The solution `v` of the preconditioner system can be stored in the right hand side array `d` in the implementation of Algorithm 9.1. Thus, the fixed point iteration requires one additional array of size n. □

9.2 Flexible GMRES (FGMRES) With Restart

The second iterative scheme which we use belongs to the class of Krylov subspace methods. There are two important requirements for our choice of the method. First, it must be able to solve systems with a non-symmetric matrix A. Second, it must work with flexible preconditioners. Analogously to the fixed point iteration, we like to use as preconditioner a multigrid method, i.e. an iterative scheme. Thus, M can vary from step to step within the Krylov subspace iteration, e.g., depending on stopping criteria within the multigrid method. In the numerical tests presented in this monograph, we have used the right preconditioned flexible GMRES algorithm with restart, which was developed by Saad, see [Saa93] or [Saa96, p. 256 ff.].

Algorithm 9.4. Flexible GMRES for the solution of (9.1). Given A, M, b and x^0.

```
V^0 := b − Ax^0
res := res0 := β := ||V^0||_2
if (res ≤ resmin)
    maxit := minit
j := 1
while (j ≤ maxit)
    V^0 := V^0/β
    s := 0
    s(0) := β
    for (i := 0; i < restart AND j ≤ maxit; i++, j++)
      Y^i := 0
      if (prec_maxit > 1)
        dnorm0 := ||V^i||_2 · prec_red_factor
      for (k := 0; k < prec_maxit; k++)
        solve MY = V^i − AY^i
        Y^i := Y^i + Y
        defect := V^i − AY^i
        if (k < prec_maxit)
          if (||defect||_2 < dnorm0) break
      endfor
```

```
21.         for (k := 0; k ≤ i; k + +)
22.            H (k, i) := V^k · defect
23.            defect := defect − H (k, i) V^k
24.         endfor
25.         H (i + 1, i) := ||defect||_2
26.         V^{i+1} := defect
27.         V^{i+1} := V^{i+1}/H (i + 1, i)
28.         convert H to a diagonal matrix by Givens rotations,
            update vector s
29.         res := |s (i + 1)|
30.         if ((j ≥ minit) AND
                   ((res ≤ resmin) OR (res ≤ res0 · red_factor)))
31.            x^0 := x^0 + YH^{-1}s
32.            return
33.         endif
34.      endfor
35.      x^0 := x^0 + YH^{-1}s
36.      V^0 := b − Ax^0
37.      β := ||V^0||_2
38. endwhile
39. return
```

Remark 9.5. Parameters of FGMRES. The parameters minit, maxit, resmin and red_factor have the same meaning as in the fixed point iteration, see Remark 9.2. The task of the other parameters is:

- restart : number of FGMRES iterations after which a restart will be performed. A large restart may improve the performance of FGMRES considerably but also increases the memory requirements, see Remark 9.6.
- prec_maxit : maximal iterations for the preconditioner, at least one. This parameter allows to apply more than one multigrid cycle for preconditioning.
- prec_red_factor : factor for the reduction of the Euclidean norm of the residual vector (before the preconditioner iteration) after which the preconditioner iteration stops.

□

The GMRES algorithm minimizes in step k the Euclidean norm of the residual vector in the Krylov space span$\{\mathtt{V}^0, \ldots, \mathtt{A}^{k-1}\mathtt{V}^0\}$. Thus, the norm of the residual vector cannot increase during the iteration. For more mathematical properties of the GMRES algorithm, we refer to Saad and Schultz [SS86] or Saad [Saa96]. We will explain now the individual steps of Algorithm 9.4 in more detail.

The flexible GMRES starts by computing the residual (line 1). In line 3, it is tested if the stopping criterion is fulfilled before the iteration. In this case, exactly minit iterations are performed.

The actual iteration, with at most `maxit` steps is given from line 6 to 38. There is an inner iteration loop with at most `restart` steps from line 10 to 34.

The application of the preconditioner is given in lines 11 to 20. The system $\mathtt{AY^i} = \mathtt{V^i}$ has to be solved, which is done iteratively. To this end, this system is transformed into the fixed point form

$$\mathtt{Y^i} = \mathtt{Y^i} + \mathtt{M}^{-1}\left(\mathtt{V^i} - \mathtt{AY^i}\right),$$

where M stands for the multigrid preconditioner. Given an iterate $\mathtt{Y^{i,k}}$, the succeeding iterate $\mathtt{Y^{i,k+1}}$ is computed by

$$\mathtt{Y^{i,k+1}} = \mathtt{Y^{i,k}} + \mathtt{M}^{-1}\left(\mathtt{V^i} - \mathtt{AY^{i,k}}\right), \quad 0 \le k < \mathtt{prec\_maxit}.$$

The result of this iteration depends of course on the initial iterate $\mathtt{Y^{i,0}}$, which is set in line 11. We use in all computations the zero vector as initial iterate. We think that this choice is not optimal since the Euclidean norm of the right hand side $\mathtt{V^i}$ is one, and thus, the solution $\mathtt{Y^i}$ of $\mathtt{AY^i} = \mathtt{V^i}$ cannot be expected to be close to the zero vector. But we are not aware of any studies how to choose the initial iterate. We use the zero vector just for simplicity and expect that a clever choice of the initial iterate will enhance the performance of the method. At most `prec_maxit` iterations with the preconditioner are performed. If not the last preconditioner iteration is reached, it is checked in line 19 if the preconditioner iteration can be stopped due to a sufficient reduction of the norm of the residual.

In lines 21 to 28, the matrices and vectors which are used by FGMRES are updated. The i-th column of the matrix H is computed in lines 21 to 24. This matrix becomes a so-called upper Hessenberg matrix, i.e. the lower triangle without the diagonal just below the main diagonal is zero. The vector `defect` is updated in line 23 such that it becomes orthogonal to the vectors $\mathtt{V^0}, \ldots, \mathtt{V^i}$. The final result of these updates defines $\mathtt{V^{i+1}}$ which is normalised in line 27. The computation of the update of the FGMRES iteration requires the solution of the system

$$Hy = s. \tag{9.2}$$

Since H is an upper Hessenberg matrix, (9.2) can be transformed to a triangular system efficiently with Givens rotations. Also the right hand side s of (9.2) must be transformed. This is done in line 28. The immediate transformation gives the valuable information on the norm of the residual of the current FGMRES iterate although the iterate itself is not directly available. The norm of the residual is the absolute value of $\mathtt{s\,(i+1)}$, line 29.

In line 30, it is checked if one of the stopping criteria is fulfilled. If this is the case, the update of the initial iterate $\mathtt{x^0}$ is computed by solving (9.2) and multiplying the solution with the matrix Y, line 31.

The inner loop with at most `restart` iterations finishes in line 34. If the stopping criteria are not fulfilled within `restart` iterations, the current iterate

$\mathbf{x}^0$ is updated (line 35), a new FGMRES iteration is prepared (lines 36 and 37) and started.

Remark 9.6. Additional memory requirements of FGMRES. Let $A \in \mathbb{R}^{n\times n}$ and $\mathtt{restart} \le \mathtt{maxit}$. Then, the additional arrays in Algorithm 9.4 have the following dimensions: $\mathtt{V}, \mathtt{Y} : \mathtt{n} \cdot (\mathtt{restart} + 1)$; $\mathtt{defect} : \mathtt{n}$, $\mathtt{H} : \mathtt{restart} \cdot (\mathtt{restart} + 1)$; $\mathtt{s} : \mathtt{restart} + 1$. In addition, two arrays of size `restart` are needed to perform the Givens rotations in line 28. □

Remark 9.7. Advantages of FGMRES. The FGMRES iteration, Algorithm 9.4, is more expensive than the fixed point iteration, Algorithm 9.1, and the memory requirements are much higher. The advantages which are expected by FGMRES are the followings:

- The rate of convergence of FGMRES should be better than with the fixed point iteration using the same multigrid preconditioner. Or, the rates should be similar if a cheaper multigrid preconditioner is used in FGMRES. Since the application of the multigrid method is the most time consuming part of the iteration, the computations with FGMRES should be faster.
- Numerical experiences show that the behaviour of the fixed point iteration is sometimes very sensitive with respect to some parameters in the multigrid preconditioner. The FGMRES method proves to be much more robust with respect to these parameters, which simplifies their choice by the user.

The expected advantages of using FGMRES, higher efficiency and better robustness, could be observed in a numerical study of the 3d steady state Navier-Stokes equations, see John [Joh02a]. A numerical study with time dependent equations is not yet available. □

Remark 9.8. Motivation for choosing FGMRES. There are also other Krylov subspace methods, like CGS and BiCGStab, which can be used for solving (9.1). Within the fixed point iteration (7.7), it is sufficient to solve (9.1) only approximately and we want to apply only a very small number of iterations. We have decided to use FGMRES because the norm of the residual cannot increase. This is in contrast to other Krylov subspace methods which might show an unpredictable behaviour at the beginning of the iteration. Of course, there are linear systems where a non-increase of the norm of the residual does not imply that the norm of the error does not increase. However, we do not expect this to be the standard case in our applications. □

9.3 The Coupled Multigrid Method

The preconditioner is the most important component of the linear system solvers described in the previous sections. It determines essentially the rate of convergence and, in general, the solution of the preconditioner system, line 6 in the fixed point iteration and line 15 in the FGMRES method, is the most time consuming part of the iteration. In all numerical tests, we have used a so-called coupled multigrid method as preconditioner which will be described in this section. For an introduction to multigrid methods, we refer to the books by Hackbusch [Hac85] or Briggs et al. [BHM00].

The multigrid method is applied to a linear saddle point problem of the abstract form

$$\mathcal{A}\begin{pmatrix} w \\ r \end{pmatrix} = \begin{pmatrix} A & B \\ C & 0 \end{pmatrix}\begin{pmatrix} w \\ r \end{pmatrix} = \begin{pmatrix} f \\ g \end{pmatrix}, \tag{9.3}$$

see (7.9). The (strongly) coupled multigrid method solves (9.3) for both types of unknowns, the velocity w and the pressure r, together. This is in contrast to so-called weakly coupled multigrid methods which solve (9.3) with a Schur complement approach for the pressure using separate multigrid solvers for the velocity and for the pressure, e.g., see John [Joh99]. In this study, the strongly coupled multigrid approach has been proved to be more efficient than the weakly coupled one for the steady state and the time dependent Navier-Stokes equations. These experiences are the main reason why we prefer the strongly coupled approach.

A multigrid method is defined by

- the grid hierarchy,
- the grid transfer operators (function prolongation, defect restriction and function restriction),
- the smoother on finer levels,
- the coarse grid solver.

As coarse grid solver, we apply the same iterative scheme which is used as smoother on the finer levels. The grid transfer operators, the smoother on finer levels and possible grid hierarchies will be described in more detail in the following.

9.3.1 The Transfer Between the Levels of the Multigrid Hierarchy

The Function Prolongation

The multiple discretisation multilevel method described in Section 9.3.3 requires the transfer between non-nested finite element spaces. This will be accomplished by a transfer operator proposed and analysed by Schieweck [Sch00], which allows the transfer between arbitrary finite element spaces. For simplicity, we describe the transfer for scalar finite element spaces and we will concentrate on the aspects which are important for implementing the transfer operator.

We consider the transfer (prolongation) from a finite element space V_{l-1}^h to a finite element space V_l^h. Let $\mathcal{T}_{l-1}$ and $\mathcal{T}_l$ be the corresponding triangulations of the domain Ω such that $\mathcal{T}_l$ originates either from a refinement of $\mathcal{T}_{l-1}$ or $\mathcal{T}_{l-1} = \mathcal{T}_l$. The second case is relevant in the multiple discretisation multilevel method for $l = L+1$, see Figure 9.7 below.

Let Σ_l^h be a discontinuous finite element space defined on $\mathcal{T}_l$

$$\Sigma_l^h = \left\{ w \in L^2(\Omega) \,:\, w|_K \in S_l^h(K) \;\forall K \in \mathcal{T}_l \right\}.$$

The choice of the local spaces $S_l^h(K)$ depends on V_{l-1}^h and V_l^h. It has to be done such that the inclusion

$$V_{l-1}^h + V_l^h \subset \Sigma_l^h \tag{9.4}$$

holds. From the practical point of view, the spaces $S_l^h(K)$ are not needed for implementing the transfer operator. From the theoretical point of view, it can be proved that appropriate spaces $S_l^h(K)$ always exist for triangulations consisting of simplices, see John et al. [JKMT02].

The transfer operator is based on the concept of nodal functionals. For each mesh cell $K \in \mathcal{T}_l$ and for the finite element space Σ_l^h there exist a local finite element basis $\left\{\psi_{l,j}^h|_K\right\}$ and a dual basis $\left\{N_{l,j}^K\right\}$ of local nodal functionals such that

$$N_{l,j}^K\left(\psi_{l,i}^h|_K\right) = \delta_{ij}, \quad 0 \le i,j \le \dim\left(S_l^h(K)\right),$$

where δ_{ij} is the Kronecker delta.

Remark 9.9. Examples of local nodal functionals. For finite element spaces like $P_k, Q_k, k \ge 0$, the local nodal functionals are point values

$$N_{l,j}^K\left(w^h\right) = w^h|_K(\mathbf{x}),$$

where $\mathbf{x} \in \overline{K}$ is a given point. The local nodal functionals for the non-conforming P_1^{nc} and the two dimensional mean value oriented Q_1^{rot} finite element spaces are given by integral mean values on the faces of K

$$N_{l,j}^K\left(w^h\right) = \frac{1}{|\partial K_j|}\int_{\partial K_j} w^h|_K ds,$$

where ∂K_j is a face of K with $(d-1)$ dimensional measure $|\partial K_j|$. For the finite element spaces P_k^{disc}, $k \ge 1$, consisting of discontinuous functions, the local nodal functionals are given by integral mean values on K. These mean values are defined with integral mean values on $\hat{K}$ and the reference map F_K. The mean values on $\hat{K}$ are given by integrals of products of w^h with weighting functions from $P_k\left(\hat{K}\right)$. □

Let $\left\{\varphi_{l,j}^h\right\}$ be a finite element basis of V_l^h. The indices j are called nodes or degrees of freedom. The set of all nodes of V_l^h is denoted by $I_l\left(V_l^h\right)$. The set of local nodes with respect to the mesh cell K is given by

$$I_l\left(K, V_l^h\right) = \left\{ i \in I_l\left(V_l^h\right) \,:\, \mathrm{supp}\left(\varphi_{l,i}^h\right) \cap K \ne \emptyset \right\}, \tag{9.5}$$

where supp $\left(\varphi_{l,i}^h\right)$ is the support of $\varphi_{l,i}^h$. Furthermore, we define for any node $j \in I_l\left(V_l^h\right)$

$$\mathcal{T}_{l,j} = \left\{K \in \mathcal{T}_l \,:\, j \in I_l\left(K, \Sigma_l^h\right)\right\},$$

the set of all mesh cells which are connected to the node j, see Figure 9.1. Then, the global nodal functional which is associated with a node $j \in I_l\left(V_l^h\right)$ and whose argument is a function $w^h \in \Sigma_l^h$ is defined by the arithmetic mean of local nodal functionals

$$N_{l,j}\left(w^h\right) = \frac{1}{\operatorname{card}\left(\mathcal{T}_{l,j}\right)} \sum_{K\in\mathcal{T}_{l,j}} N_{l,j}^K\left(w^h|_K\right), \quad w^h \in \Sigma_l^h,$$

where $\operatorname{card}\left(\mathcal{T}_{l,j}\right)$ denotes the number of mesh cells in $\mathcal{T}_{l,j}$. The transfer operator for the prolongation is defined with the help of the global nodal functionals:

$$P_{l-1}^l : \Sigma_l^h \to V_l^h \qquad P_{l-1}^l\left(w^h\right) = \sum_{i=1}^{\dim\left(V_l^h\right)} N_{l,i}\left(w^h\right)\varphi_{l,i}^h. \tag{9.6}$$

From the inclusion (9.4) follows that this operator is defined especially for functions from V_{l-1}^h.

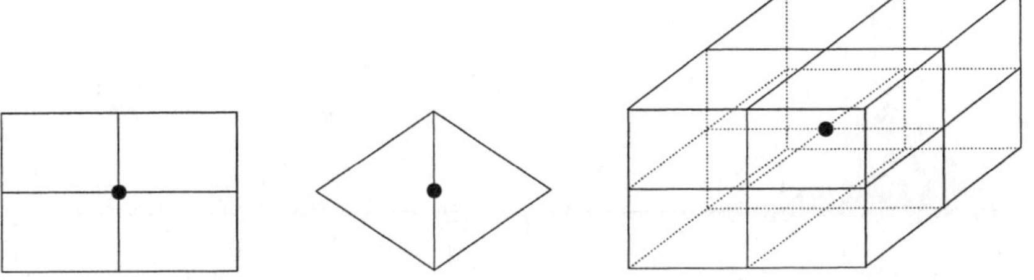

Fig. 9.1. Examples for domains $\mathcal{T}_{l,j}$, the ball indicates the position of the node j

Let $\left\{\varphi_{l-1,i}^h\right\}$ be a finite element basis of V_{l-1}^h and

$$w_{l-1}^h = \sum_{i=1}^{\dim\left(V_{l-1}^h\right)} w_{l-1,i}\varphi_{l-1,i}^h \in V_{l-1}^h.$$

For evaluating the coefficient of $\varphi_{l,i}^h$ for the prolongation of the function w_{l-1}^h, one has to compute

$$\begin{aligned}
N_{l,i}\left(w_{l-1}^h\right) &= \frac{1}{\operatorname{card}\left(\mathcal{T}_{l,i}\right)} \sum_{K\in\mathcal{T}_{l,i}} N_{l,i}^K\left(w_{l-1}^h|_K\right) \\
&= \frac{1}{\operatorname{card}\left(\mathcal{T}_{l,i}\right)} \sum_{K\in\mathcal{T}_{l,i}} \sum_{j=1}^{\dim\left(V_{l-1}^h\right)} w_{l-1,j} N_{l,i}^K\left(\varphi_{l-1,j}^h|_K\right).
\end{aligned}$$

Remark 9.10. The computation of $N_{l,i}^K\left(\varphi_{l-1,j}^h|_K\right)$. We will give some concrete examples how to compute the local nodal functionals. For simplicity of presentation, we restrict ourselves to two dimensional finite elements. Let $\mathcal{P}(K) \in \mathcal{T}_{l-1}$ be the parent mesh cell of $K \in \mathcal{T}_l$. The local degrees of freedom of $\mathcal{P}(K)$ are represented by balls in the following figures and the local degrees of freedom of K by squares.

• *Red refined triangles,* $P_1(\mathcal{P}(K)) \to P_1(K)$. We consider the two situations given in Figure 9.2. For this finite element, the local nodal functionals are point values, i.e. $N_{l,i}^K\left(\varphi_{l-1,j}^h|_K\right)$ is the value of $\varphi_{l-1,j}^h$ at the position of the local degree of freedom i in K. One obtains the following values

	Figure 9.2, left			Figure 9.2, right		
	$\varphi_{l-1,0}^h\|_K$	$\varphi_{l-1,1}^h\|_K$	$\varphi_{l-1,2}^h\|_K$	$\varphi_{l-1,0}^h\|_K$	$\varphi_{l-1,1}^h\|_K$	$\varphi_{l-1,2}^h\|_K$
$N_{l,0}^K$	1	0	0	0.5	0.5	0
$N_{l,1}^K$	0.5	0.5	0	0	0.5	0.5
$N_{l,2}^K$	0.5	0	0.5	0.5	0	0.5

It turns out for the prolongation that this is just the standard inclusion.

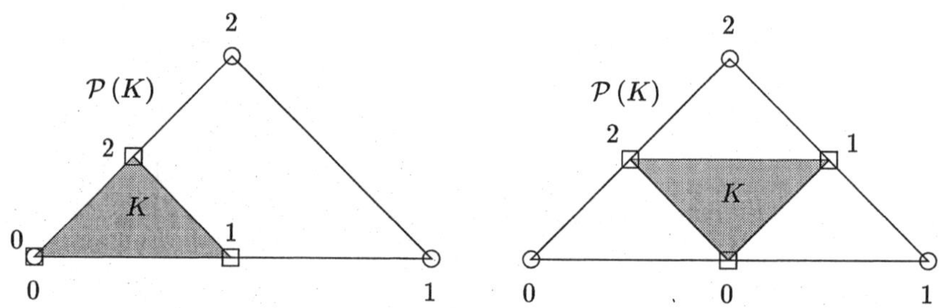

Fig. 9.2. Red refined triangles, $P_1(\mathcal{P}(K)) \to P_1(K)$

• *Red refined triangles,* $P_1^{\mathrm{nc}}(\mathcal{P}(K)) \to P_1^{\mathrm{nc}}(K)$. We consider again two situations, see Figure 9.3. In this case, the local nodal functionals are given by integral mean values at the edges of K, e.g.,

$$N_{l,0}^K\left(\varphi_{l-1,j}^h|_K\right) = \frac{1}{\|\mathbf{x}_0 - \mathbf{x}_1\|_2}\int_{\mathbf{x}_0}^{\mathbf{x}_1} \varphi_{l-1,j}^h|_K d\mathbf{s}.$$

One obtains for the local nodal functionals

	Figure 9.3, left			Figure 9.3, right		
	$\varphi_{l-1,0}^h\|_K$	$\varphi_{l-1,1}^h\|_K$	$\varphi_{l-1,2}^h\|_K$	$\varphi_{l-1,0}^h\|_K$	$\varphi_{l-1,1}^h\|_K$	$\varphi_{l-1,2}^h\|_K$
$N_{l,0}^K$	1	-0.5	0.5	0.5	0.5	0
$N_{l,1}^K$	0.5	0	0.5	0	0.5	0.5
$N_{l,2}^K$	0.5	-0.5	1	0.5	0	0.5

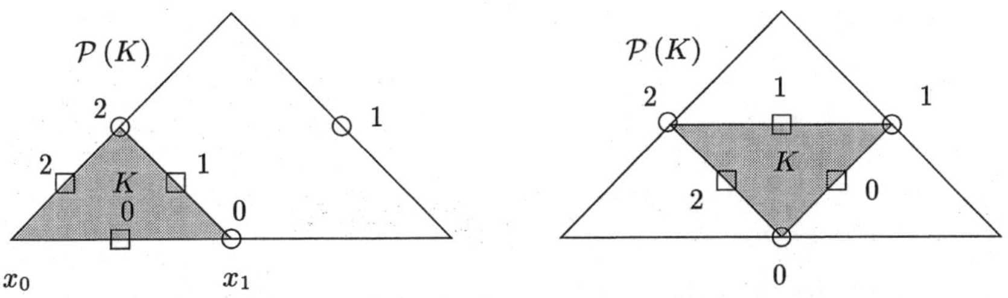

Fig. 9.3. Red refined triangles, $P_1^{\mathrm{nc}}(\mathcal{P}(K)) \to P_1^{\mathrm{nc}}(K)$

Applying these local nodal functionals in the prolongation operator (9.6), one gets a standard averaging operator, see Figure 9.4. In this figure, the square denotes the degree of freedom of V_l^h whose value has to be computed and the balls stand for the nodes of V_{l-1}^h. The numbers give the weights which have to be applied to the coefficients of the function from V_{l-1}^h corresponding to these nodes. It is easy to see that the prolongated value in the left picture of Figure 9.4 is just the average in this point of the values of the finite element function of V_{l-1}^h restricted to the triangles in $\mathcal{T}_{l-1}$.

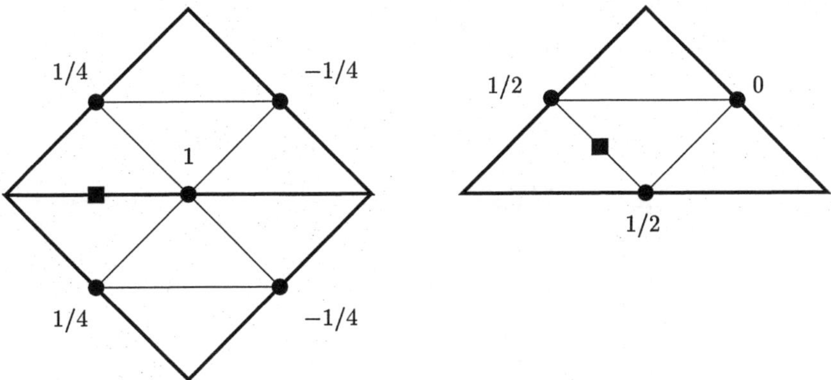

Fig. 9.4. Red refined triangles, the weights in the prolongation for $P_1^{\mathrm{nc}}(\mathcal{P}(K)) \to P_1^{\mathrm{nc}}(K)$

- *No refined affine mapped quadrilaterals,* $Q_1^{\mathrm{rot}}(\mathcal{P}(K)) \to Q_2(K)$. In this case, we have $\mathcal{P}(K) = K$. We start by a change of the basis in $Q_1^{\mathrm{rot}}\left(\hat{K}\right)$. The original basis is given in (7.12). The new basis looks as follows

$$
\begin{aligned}
Q_1^{\mathrm{rot}}\left(\hat{K}\right) &= \mathrm{span}\left\{\hat{\varphi}_{l-1,0}^h, \hat{\varphi}_{l-1,1}^h, \hat{\varphi}_{l-1,2}^h, \hat{\varphi}_{l-1,3}^h\right\} \\
&= \mathrm{span}\left\{-\frac{3}{8}\left(\hat{x}_1^2-\hat{x}_2^2\right)-\frac{1}{2}\hat{x}_2+\frac{1}{4}, \frac{3}{8}\left(\hat{x}_1^2-\hat{x}_2^2\right)+\frac{1}{2}\hat{x}_1+\frac{1}{4},\right. \\
&\quad \left. -\frac{3}{8}\left(\hat{x}_1^2-\hat{x}_2^2\right)+\frac{1}{2}\hat{x}_2+\frac{1}{4}, \frac{3}{8}\left(\hat{x}_1^2-\hat{x}_2^2\right)-\frac{1}{2}\hat{x}_1+\frac{1}{4}\right\}.
\end{aligned}
$$

It is straightforward to check that

$$
\frac{1}{\left|\hat{E}_i\right|}\int_{\hat{E}_i} \hat{\varphi}_{l-1,j}^h ds = \delta_{ij},
$$

where the edges $\hat{E}_i$ of $\hat{K}$ are numbered counter clockwise, starting with the bottom edge. Let K be an arbitrary mesh cell with an affine reference transformation and let $(\hat{x}_1, \hat{x}_2) \in \overline{\hat{K}}$ be transformed to $(x_1, x_2) \in \overline{K}$. Then it holds for the transformation of the basis functions that $\hat{\varphi}^h(\hat{x}_1, \hat{x}_2) = \varphi^h(x_1, x_2)$. Thus, the values of the transformed basis functions can be easily computed by values of the reference basis functions in $\overline{\hat{K}}$. The affine reference transformation leads to a situation as presented in Figure 9.5. The local nodal functionals of $Q_2(K)$ are defined as point values of the local basis functions of $Q_1^{\mathrm{rot}}(K)$. As pointed out, the evaluation of these point values can be done in $\hat{K}$ which gives, independent of K,

	$\varphi_{l-1,0}^h\|_K$	$\varphi_{l-1,1}^h\|_K$	$\varphi_{l-1,2}^h\|_K$	$\varphi_{l-1,3}^h\|_K$
$N_{l,0}^K$	3/4	-1/4	-1/4	3/4
$N_{l,1}^K$	9/8	-1/8	1/8	-1/8
$N_{l,2}^K$	3/4	3/4	-1/4	-1/4
$N_{l,3}^K$	-1/8	1/8	-1/8	9/8
$N_{l,4}^K$	1/4	1/4	1/4	1/4
$N_{l,5}^K$	-1/8	9/8	-1/8	1/8
$N_{l,6}^K$	-1/4	-1/4	3/4	3/4
$N_{l,7}^K$	1/8	-1/8	9/8	-1/8
$N_{l,8}^K$	-1/4	3/4	3/4	-1/4

If the reference map F_K is bilinear, we use for simplicity also the values of the local nodal functionals given in this table.

These examples show that the values of the local nodal functionals $N_{l,i}^K\left(\varphi_{l-1,j}^h|_K\right)$ are, in general, the same for a large number of mesh cells. The values can be computed in a preprocessing step and stored in a data base. In computing the prolongation, only local matrix-vector products have to be performed with these values. This strategy is used to accelarete the computation of the prolongation (9.6). □

An algorithm for computing the prolongation (9.6) looks as follows.

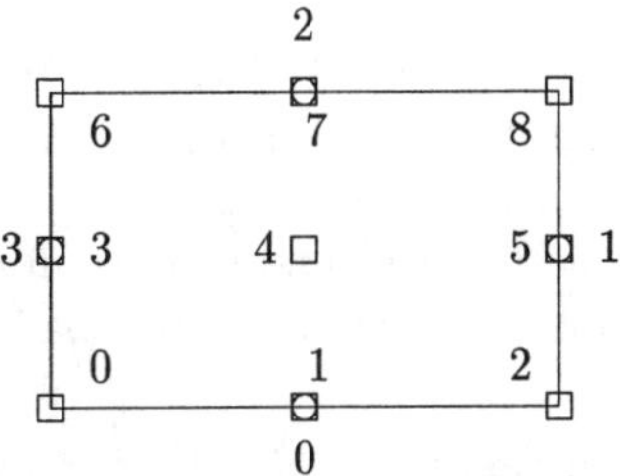

Fig. 9.5. No refined affine mapped quadrilaterals, $Q_1^{\text{rot}}(\mathcal{P}(K)) \to Q_2(K)$

Algorithm 9.11. Prolongation. Given the coefficient vector $\mathtt{w}_{l-1}$ of the finite element function $w_{l-1}^h \in V_{l-1}^h$.

```
w_l = 0
card = 0
for K ∈ T_l
    for i ∈ I_l(K, V_l^h)
        for (j := 0; j < dim(V_{l-1}^h); j++)
            if supp (φ_{l-1,j})^h |_K ∩ K = ∅
                continue
            w_l(i) := w_l(i) + w_{l-1}(j) N_{l,i}^K (φ_{l-1,j}^h |_K)
            card(i) := card(i) + 1
        endfor
    endfor
endfor
for (i := 0; i < dim(V_l^h); i++)
    w_l(i) := w_l(i)/card(i)
endfor
```

The Defect Restriction

The definition of the operator for the defect restriction $R_l^{*,l-1} : \left(V_l^h\right)^* \to \left(V_{l-1}^h\right)^*$ uses the prolongation operator given in (9.6). Let $d_l \in \left(V_l^h\right)^*$ be a given defect functional, its restriction to $\left(V_{l-1}^h\right)^*$ is defined by

$$\int_\Omega R_l^{*,l-1}(d_l)\, \varphi_{l-1}^h d\mathbf{x} = \int_\Omega d_l P_{l-1}^l \left(\varphi_{l-1}^h\right) d\mathbf{x} \quad \forall \varphi_{l-1}^h \in V_{l-1}^h.$$

This is the standard definition and since the prolongation operator turns out to be standard in many situations, see Remark 9.10, the same holds for the defect restriction operator.

The Function Restriction

In the multigrid approaches for solving the linear saddle point problem (9.3), the matrix of this problem has to be assembled also on the coarse levels.

Therefore, the finite element functions $\mathbf{w}^h_{\text{old}}$ and $\mathbf{w}^h_{k-1}$ must be available on the coarse grids. A restriction operator $R_l^{l-1} : V_l^h \to V_{l-1}^h$ which maps a finite element function from the finite element space connected to level l in the multilevel hierarchy to a finite element function connected to level $l-1$ is necessary. We use a function restriction which is based on local L^2 projections and averaging.

The bases of V_l^h, V_{l-1}^h are again denoted by $\left\{\varphi_{l,j}^h\right\}$, $\left\{\varphi_{l-1,i}^h\right\}$. Let $\mathbf{w}_l^h \in V_l^h$ with

$$\mathbf{w}_l^h = \sum_{i=1}^{\dim\left(V_l^h\right)} w_{l,i}\varphi_{l,i}^h$$

be given. The goal is to compute a function

$$R_l^{l-1}\left(\mathbf{w}_l^h\right) = \sum_{i=1}^{\dim\left(V_{l-1}^h\right)} w_{l-1,i}\varphi_{l-1,i}^h.$$

We consider a mesh cell K on the geometric grid which is connected with V_{l-1}^h and assume that K possesses an affine reference transformation. Local values of the unknown coefficients $w_{l-1,i}$ are determined by the local L^2 projection

$$\begin{aligned} &\sum_{i=1}^{\operatorname{card}\left(I_l\left(K,V_l^h\right)\right)} w_{l,i}|_K \left(\varphi_{l,i}^h, \varphi_{l-1,j}^h\right)_K \\ = &\sum_{i=1}^{\operatorname{card}\left(I_{l-1}\left(K,V_{l-1}^h\right)\right)} w_{l-1,i}|_K \left(\varphi_{l-1,i}^h, \varphi_{l-1,j}^h\right)_K \end{aligned}$$

for all $j \in I_{l-1}\left(K, V_{l-1}^h\right)$, where $I_l\left(K, V_l^h\right)$ is defined in (9.5). The transformation to the reference cell $\hat{K}$ gives

$$\begin{aligned} &\sum_{i=1}^{\operatorname{card}\left(I_l\left(K,V_l^h\right)\right)} w_{l,i}|_K \int_{\hat{K}} \hat{\varphi}_{l,i}^h \hat{\varphi}_{l-1,j}^h \left|\det J_K\left(\hat{\mathbf{x}}\right)\right| d\hat{\mathbf{x}} \\ = &\sum_{i=1}^{\operatorname{card}\left(I_{l-1}\left(K,V_{l-1}^h\right)\right)} w_{l-1,i}|_K \int_{\hat{K}} \hat{\varphi}_{l-1,i}^h \hat{\varphi}_{l-1,j}^h \left|\det J_K\left(\hat{\mathbf{x}}\right)\right| d\hat{\mathbf{x}} \end{aligned}$$

for all $j \in I_{l-1}\left(K, V_{l-1}^h\right)$. Since $\left|\det J_K\left(\hat{\mathbf{x}}\right)\right|$ is constant, this relation simplifies to

$$\begin{aligned} &\sum_{i=1}^{\operatorname{card}\left(I_l\left(K,V_l^h\right)\right)} w_{l,i}|_K \int_{\hat{K}} \hat{\varphi}_{l,i}^h \hat{\varphi}_{l-1,j}^h d\hat{\mathbf{x}} \\ = &\sum_{i=1}^{\operatorname{card}\left(I_{l-1}\left(K,V_{l-1}^h\right)\right)} w_{l-1,i}|_K \int_{\hat{K}} \hat{\varphi}_{l-1,i}^h \hat{\varphi}_{l-1,j}^h d\hat{\mathbf{x}} \end{aligned} \tag{9.7}$$

for all $j \in I_{l-1}\left(K, V_{l-1}^h\right)$. This is a linear system of equations of the form

$$Gw_l|_K = Mw_{l-1}|_K.$$

Thus, the local values of the unknown coefficients are given by

$$w_{l-1}|_K = M^{-1}Gw_l|_K = Rw_l|_K. \tag{9.8}$$

The matrix R is independent of K. That means, for all other mesh cells whose basis on the reference mesh cell has the same form as for K, one also needs the matrix R. This will be the case very often. E.g., if the grids are uniformly refined and the same finite element space is used on every level, the matrix R is needed for each mesh cell on each level ! This matrix R will be computed once and then stored in a data base. Then, only a local matrix-vector product has to be computed in (9.8) which leads to a very fast algorithm. The final restriction is computed by an averaging

$$w_{l-1,i} = \frac{1}{\text{card}\left(\mathcal{T}_{l-1,i}\right)} \sum_{K \in \mathcal{T}_{l-1,i}} w_{l-1,i}|_K.$$

For mesh cells with a non-affine reference transformation, we use for simplicity also (9.7) such that they are handled in the same way as mesh cells with an affine transformation. One can consider this approach also as a function restriction which is a local L^2 projection on the reference mesh cell and which is an approximation of a L^2 projection on the original mesh cell.

9.3.2 The Vanka Smoothers

This section contains a detailed description of the smoothers which are used. In addition, we illustrate the increase of the numerical costs of these smoothers in three dimensions compared to two dimensions.

The coupled multigrid method is used with local smoothers, so-called Vanka-type smoothers, Vanka [Van86]. Vanka-type smoothers can be considered as block Gauss-Seidel methods. Let $\mathcal{V}^h$ and $\mathcal{Q}^h$ be the set of velocity and pressure degrees of freedom, respectively. These sets are decomposed into

$$\mathcal{V}^h = \cup_{j=1}^J \mathcal{V}_j^h, \quad \mathcal{Q}^h = \cup_{j=1}^J \mathcal{Q}_j^h. \tag{9.9}$$

The subsets are not required to be disjoint.

Let $\mathcal{A}_j$ be the block of the matrix $\mathcal{A}$ which is connected with the degrees of freedom of $\mathcal{W}_j^h = \mathcal{V}_j^h \cup \mathcal{Q}_j^h$, i.e. the intersection of the rows and columns of $\mathcal{A}$ with the global indices belonging to $\mathcal{W}_j^h$,

$$\mathcal{A}_j = \begin{pmatrix} A_j & B_j \\ C_j & 0 \end{pmatrix} \in \mathbb{R}^{\dim\left(\mathcal{W}_j^h\right) \times \dim\left(\mathcal{W}_j^h\right)}.$$

In addition, we define

$$\mathcal{D}_j = \begin{pmatrix} \operatorname{diag}(A_j) & B_j \\ C_j & 0 \end{pmatrix} \in \mathbb{R}^{\dim(\mathcal{W}_j^h)\times\dim(\mathcal{W}_j^h)}.$$

Similarly, we denote by $(\cdot)_j$ the restriction of a vector on the rows corresponding to the degrees of freedom in $\mathcal{W}_j^h$. Each smoothing step with a Vanka-type smoother consists in a loop over all sets $\mathcal{W}_j^h$, where for each $\mathcal{W}_j^h$ a local system of equations connected with the degrees of freedom in this set is solved. The local solutions are updated in a Gauss-Seidel manner. The diagonal Vanka smoother computes the velocity and pressure values connected to $\mathcal{W}_j^h$ by

$$\begin{pmatrix} w \\ r \end{pmatrix}_j := \begin{pmatrix} w \\ r \end{pmatrix}_j + \mathcal{D}_j^{-1}\left(\begin{pmatrix} f \\ g \end{pmatrix} - \mathcal{A}\begin{pmatrix} w \\ r \end{pmatrix}\right)_j.$$

The full Vanka smoother computes new velocity and pressure values by

$$\begin{pmatrix} w \\ r \end{pmatrix}_j := \begin{pmatrix} w \\ r \end{pmatrix}_j + \mathcal{A}_j^{-1}\left(\begin{pmatrix} f \\ g \end{pmatrix} - \mathcal{A}\begin{pmatrix} w \\ r \end{pmatrix}\right)_j.$$

The general strategy for choosing the sets $\mathcal{V}_j^h$ and $\mathcal{Q}_j^h$ is as follows. First, pick some pressure degrees of freedom which define $\mathcal{Q}_j^h$. Second, $\mathcal{V}_j^h$ is formed by all velocity degrees of freedom which are connected with the pressure degrees of freedom from $\mathcal{Q}_j^h$ by non-zero entries in the matrix C.

We have applied two types of Vanka smoothers with respect to this strategy. The first one is called mesh cell oriented, because $\mathcal{Q}_j^h$ is defined by all pressure degrees of freedom which are connected to the mesh cell j. For this type of Vanka smoother, J coincides with the number of mesh cells. The mesh cell oriented Vanka smoother was applied only for discretisations with discontinuous pressure approximation, i.e. $Q^h \in \{P_0, Q_0, P_1^{\text{disc}}, P_2^{\text{disc}}\}$. For such discretisations, $\mathcal{V}_j^h$ consists of all velocity degrees of freedom which are connected to the mesh cell j. This property is not given for discretisations with continuous pressure approximation. For these discretisations, we use a decomposition in which $\mathcal{Q}_j^h$ is defined by a single pressure degree of freedom, $\dim \mathcal{Q}_j^h = 1$. This smoother is called pressure node oriented Vanka smoother. For this smoother, the number of subsets J in decomposition (9.9) is equal to the number of pressure degrees of freedom.

For a continuous pressure approximation, a pressure degree of freedom on a given mesh cell K is in general connected to velocity degrees of freedom on other mesh cells. In this case it is also possible to generate the local systems with all pressure degrees of freedom on K in order to perform a mesh cell oriented Vanka smoother. But in comparison to discontinuous pressure approximations, the overhead of searching all connections and allocating the local matrix and right hand side is much higher.

Remark 9.12. The size of the local systems. Using discontinuous discrete pressure, the local matrix of the mesh cell oriented Vanka smoother can be generated on the current mesh cell. The size of the local systems is known a priori and it is given for the different discretisations in Table 9.1. It can be seen that the local systems for the higher order discretisations in 3d are relatively large.

Table 9.1. Degrees of freedom for the local systems of the mesh cell oriented Vanka smoother (velocity: each component)

	2d			3d		
	velocity	pressure	total	velocity	pressure	total
Q_1^{rot}/Q_0	4	1	9	6	1	19
Q_2/P_1^{disc}	9	3	21	27	4	85
Q_3/P_2^{disc}	16	6	38	64	10	202
P_1^{nc}/P_0	3	1	7	4	1	13

The size of the local systems for the pressure node Vanka smoother applied in discretisations with continuous pressure approximation depends on the particular pressure degree of freedom and on the given grid. In addition, the size of the local systems cannot be bounded a priori if adaptive grid refinement is used since it depends on the maximal number of neighbour cells of K. A neighbour is a mesh cell K_1 with $\overline{K} \cap \overline{K_1} \neq \emptyset$ and the maximal number of neighbour cells of a mesh cell K can increase on adaptively refined grids. To illustrate the size of the local systems, we give it for concrete degrees of freedom in typical situations, Figure 9.6, in Table 9.2. It can be observed that these size are considerably larger than for the mesh cell oriented Vanka smoother for discretisations with discontinuous pressure, Table 9.1. In addition, the number of local systems which must be solved in each smoothing step is in general a multiple compared to the mesh cell oriented Vanka smoother since the number of pressure degrees of freedom is in general larger than the number of mesh cells.

Table 9.2. Degrees of freedom for the local systems of the pressure node oriented Vanka smoother (velocity: each component)

	2d			3d		
	velocity	pressure	total	velocity	pressure	total
Q_2/Q_1	25	1	51	125	1	376
Q_3/Q_2	49	1	99	343	1	1030
P_2/P_1	19	1	39	65	1	196
P_3/P_2	37	1	75	175	1	526

The mesh cell oriented and the pressure node oriented Vanka smoother are equivalent for piecewise constant discrete pressure if the mesh cells and the pressure nodes are numbered in the same way. □

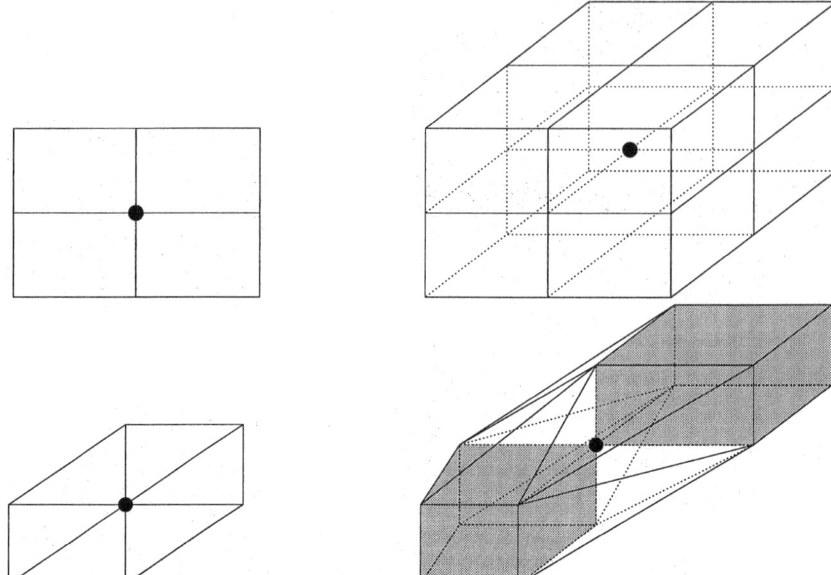

Fig. 9.6. Degree of freedom for which the size of the local systems in the pressure node oriented Vanka smoother is given in Table 9.2. (bottom right: 6 tetrahedra in two directions (coloured) and 2 tetrahedra in six directions, i.e. 24 tetrahedra are connected with this pressure degree of freedom)

Remark 9.13. Solving the local systems. The local system of the pressure node oriented diagonal Vanka smoother has the form

$$\begin{pmatrix} A & b \\ c & 0 \end{pmatrix} \begin{pmatrix} w \\ r \end{pmatrix} = \begin{pmatrix} a_{11} & 0 & \cdots & 0 & b_1 \\ 0 & a_{22} & \cdots & 0 & b_2 \\ \vdots & \vdots & \ddots & \vdots & \vdots \\ 0 & 0 & \cdots & a_{nn} & b_n \\ c_1 & c_2 & \cdots & c_n & 0 \end{pmatrix} \begin{pmatrix} w_1 \\ w_2 \\ \vdots \\ w_n \\ r \end{pmatrix} = \begin{pmatrix} f_1 \\ f_2 \\ \vdots \\ f_n \\ g \end{pmatrix} = \begin{pmatrix} f \\ g \end{pmatrix}.$$

Thus, the solution of this system can be easily achieved by

$$r = \frac{g - cA^{-1}f}{cA^{-1}b}, \quad w = A^{-1}(f - rb)$$

where

$$cA^{-1}f = \sum_{l=1}^{n} \frac{c_l f_l}{a_{ll}}, \quad cA^{-1}b = \sum_{l=1}^{n} \frac{c_l b_l}{a_{ll}}, \quad w_k = \frac{f_l - rb_l}{a_{ll}} \quad l = 1, \ldots, n.$$

In this case, the solution of the local systems is obtained very fast. Such a simple way of solving the local systems is not possible for the full Vanka smoothers and the mesh cell oriented diagonal Vanka smoother for non-constant pressure approximations. The size of the arising local systems may be rather large, especially for higher order finite element discretisations in 3d, see Tables 9.1 and 9.2. We have applied in the computations two approaches for their solution:

- direct solution using the Gaussian elimination with column pivoting,
- approximate solution using the GMRES method which stops after having reduced the Euclidean norm of the residual by a prescribed factor. In our numerical tests, the factor 10 is prescribed.

A numerical study at the steady state Navier-Stokes equations, John [Joh02a], shows that the approximate solution of the local systems may improve the efficiency of the solver considerably but also may destroy the robustness of the solver. As result of this study, it is proposed to use Gaussian elimination for the mesh cell oriented Vanka smoother and discretisations with discontinuous pressure approximation. For discretisations with continuous pressure approximation in three dimensions, the approximative solution of the local systems is recommended. However, the stopping criterion for the GMRES iteration in the solution of the local systems is a sensible parameter which balances robustness and efficiency. In both approaches, the solution of the large number of local systems is the most time consuming part of the whole algorithm.

The use of the full Vanka smoothers becomes necessary if too much information from the off diagonals are neglected by the diagonal Vanka smoothers. A numerical study at the 2d Navier-Stokes equations, John and Tobiska [JT00], shows that the diagonal Vanka smoother may be more efficient (measured in computing times) than the full Vanka smoother for time dependent problems and lowest order non-conforming discretisations with upwind stabilisation. □

Remark 9.14. Numbering of the degrees of freedom in the Vanka-type smoothers. Since the Vanka-type smoothers are block Gauss-Seidel methods, their performance depends on the ordering of the degrees of freedom. We did not study this dependence in our tests. If an example possesses a dominant flow direction, like the flow in a channel, we ordered the degrees of freedom, with possibly few exceptions, from the inflow to the outflow (downwind numbering). □

Remark 9.15. Damping of the smoother iterate. Sometimes it becomes necessary to damp the smoother iterate. Let (w_l, r_l) be the current iterate on the multigrid level l and $(\delta w_l, \delta r_l)$ be the update computed by one iteration of the smoother. Then, the new iterate is computed by $(w_l, r_l) + \omega_l\,(\delta w_l, \delta r_l)$. The damping parameter can be chosen differently on all levels of the multigrid hierarchy. If not mentioned otherwise, we did not apply a damping, i.e. $\omega_l = 1$ for all l. □

Remark 9.16. Experiences with the coupled multigrid approach compared to other solvers. The behaviour of coupled multigrid methods with Vanka-type smoothers applied in the solution of the Navier-Stokes equations has been studied numerically for non-conforming finite element spaces of first order in a number of papers John [Joh99], John and Tobiska [JT00], John et al. [JMM+00]. In these studies, they behaved superior to several other classes of solvers. □

9.3.3 The Standard Multigrid Method and the Multiple Discretisation Multilevel Method

We have used two different multigrid approaches for the solution of (9.3), see Figure 9.7.

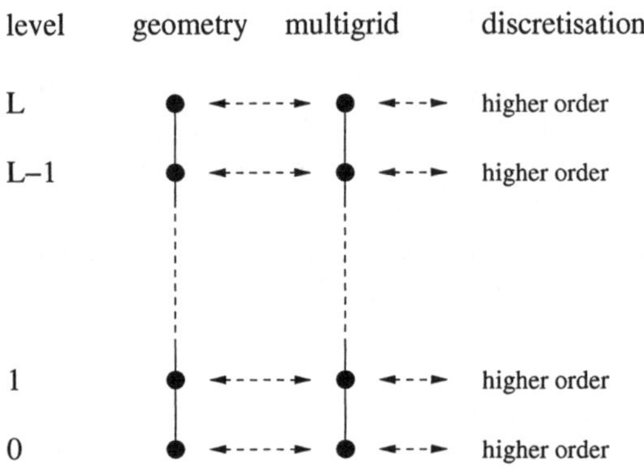

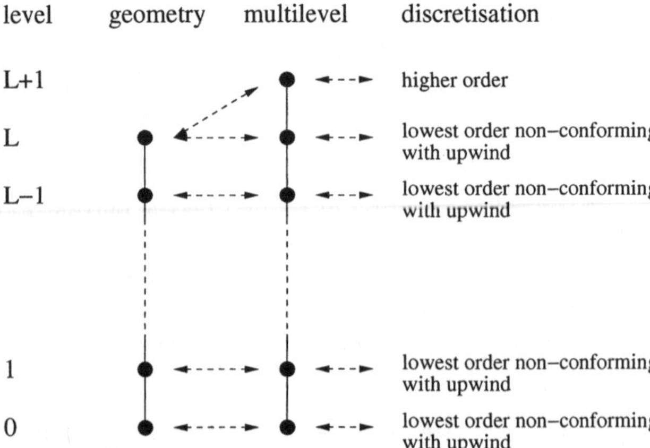

Fig. 9.7. The standard and the multiple discretisation multilevel approach for higher order discretisations

In the standard multigrid approach, the number of geometric grid levels and the number of levels in the multigrid hierarchy coincide. The same discretisation is used on each multigrid level. In numerical studies of benchmark problems for the steady state Navier-Stokes equations in 2d, John and Matthies [JM01], and in 3d, John [Joh02a], difficulties are reported with the standard multigrid approach for solving linear saddle point problems arising in some higher order finite element discretisations. Higher order finite element discretisations have given very accurate results for the benchmark reference values. In contrast, lowest order non-conforming discretisation with upwind stabilisation were rather inaccurate but the standard multigrid approach has been proved as a very efficient solver, see also John and Tobiska [JT00]. This situation led to the idea of constructing a multilevel method for higher order finite element discretisations which is based on a multilevel method with stable lowest order non-conforming finite element discretisations. We will call this approach multiple discretisation multilevel method. There are of course many possibilities to construct such a multilevel method. We use in our tests the approach depicted in Figure 9.7. In this approach, the multilevel hierarchy possesses one level more than the geometric grid hierarchy. On the finest geometric grid, level L, two discretisations are applied. One of them, which forms the finest level of the multilevel hierarchy, is the discretisation which we are interested in, e.g., a higher order discretisation. The second discretisation on the geometric level L is a lowest order non-conforming discretisation with upwind stabilisation. On all coarser geometric levels, also a stabilised lowest order non-conforming discretisation is applied.

A convergence analysis of the multiple discretisation multilevel approach applied to the Stokes equations can be found in John et al. [JKMT02]. The analysis is done for smoothers of Braess-Sarazin-type, Braess and Sarazin [BS97].

9.3.4 Schematic Overview and Parameters

We will give an overview of the multigrid method and explain the parameters which are used for its control.

The multigrid method is performed by the recursive call of the following routine.

Algorithm 9.17. Multigrid method for the solution of (9.3).

```
multigrid(level)
   if (level == 0) //coarsest level
     compute norm of the initial residual res0 = res
     while (res ≥ coarse_red_factor · res0)
       apply coarse_smoother and compute update
       compute new iterate by adding the old iterate and
       the update damped with smooth_damp_factor
       compute norm of the residual res
```

```
8.          if coarse_maxit reached then break
9.        endwhile
10.     else //finer levels
11.       for (j = 0; j < pre_smooth; j + +)
12.         apply smoother and compute update
13.         compute new iterate by adding the old iterate and
            the update damped with smooth_damp_factor
14.       endfor
15.       compute defect
16.       restrict defect to level − 1
17.       for (i = 0; i < recursion(i); i + +) multigrid(level-1)
18.       prolongate update from level − 1
19.       compute new iterate by adding the old iterate and
          the update damped with prolo_damp_factor
20.       for (j = 0; j < post_smooth; j + +)
21.         apply smoother and compute update
22.         compute new iterate by adding the old iterate and
            the update damped with smooth_damp_factor
23.       endfor
24.     endif
25. return
```

Remark 9.18. Parameters of the multigrid method. The multigrid method can be controlled with the following parameters:

- `mg_type` : type of the multigrid method (standard or multiple discretisation),
- `mg_cycle` : type of the multigrid cycle. This parameter defines the array `recursion` in line 17, e.g., `recursion(i)` $= 1$ for the V-cycle and `recursion(i)` $= 2$ for the W-cycle. To obtain the F-cycle, this array has to be modified during the cycle appropriately.
- `smoother` : type of the smoother (diagonal or full, mesh cell or pressure node oriented),
- `pre_smooth` : number of pre-smoothing steps on the finer levels, line 11,
- `post_smooth` : number of post-smoothing steps on the finer levels, line 20,
- `smooth_damp_factor_fine` : damping factor for the smoothing iteration on the finest level, this parameter replaces on the finest level `smooth_damp_factor` in lines 13 and 22,
- `smooth_damp_factor` : damping factor for the smoothing iteration on all coarser levels, lines 13 and 22,
- `prolo_damp_factor_fine` : damping factor for the update on the finest level, this parameter replaces on the finest level `prolo_damp_factor` in line 19,
- `prolo_damp_factor` : damping factor for the update on all coarser levels, line 19,

- `coarse_smoother` : smoother on the coarsest grid,
- `coarse_maxit` : maximal number of iterations on the coarsest grid,
- `coarse_red_factor` : factor for the reduction of the Euclidean norm of the initial residual after which the iteration on the coarsest grid stops.

The choice of different values for some parameters on the finest and on all other levels is sometimes helpful to enhance the performance of the multiple discretisations multilevel method. □

9.4 The Solution of the Auxiliary Problem in the Rational LES Model

The second order rational LES model with auxiliary problem, Section 4.2.2, requires the solution of

$$-\frac{\delta^2}{4\gamma}\Delta\mathbb{X} + \mathbb{X} = \nabla\mathbf{w}\nabla\mathbf{w}^T \tag{9.10}$$

with homogeneous Neumann boundary conditions. The discretisation of (9.10) is described in Section 7.7. There arise d^2 systems of linear equations but due to the symmetry of $\nabla\mathbf{w}\nabla\mathbf{w}^T$ not more than $d\,(d+1)\,/2$ posses a different right hand side. These $d\,(d+1)\,/2$ systems are solved of course simultaneously. For simplicity, we will describe the solution for one right hand side and for a system of the form

$$Ax = b. \tag{9.11}$$

The matrix A in (9.11) is symmetric and positive definite. These properties suggest to apply the conjugate gradient method by Hestenes and Stiefel [HS52] with a preconditioner M as solver for (9.11). We use the following implementation of this method.

Algorithm 9.19. Preconditioned conjugate gradient method for the solution of (9.11). Given A, M, b and an initial iterate x^0. The preconditioned conjugate gradient method for solving (9.11) has the following form:

```
r := Ax^0 − b
res0 := res := ||r||_2
if res ≤ resmin
    maxit := minit
q := 0
κ := 1
for (i := 1; i ≤ maxit; i++)
    z := 0
    solve Mz = r
    ρ := r^T z
        q := z + (ρ/κ) q
```

```
    κ := ρ
    z := Aq
    α := ρ/(q^T z)
    x^{i+1} := x^i + αq
    r := r − αz
    res := ||r||_2
    if i < minit continue
    if res ≤ resmin break
    if res ≤ res0 · red_factor break
endfor
return
```

We found in computational experiments, which will not reported here, that it is not necessary to use a very sophisticated preconditioner in the conjugate gradient method. In all computations, we used one SSOR step as preconditioner.

Algorithm 9.20. Symmetric successive overrelaxation for the solution of (9.11). Given $A \in \mathbb{R}^{n\times n}, b \in \mathbb{R}^n$, $x^0 \in \mathbb{R}^n$ and $\omega \in (0,2)$. One SSOR step applied to (9.11) has the following form:

for $(i := 1; i \le n; i++)$

$$x_i^{1/2} := x_i^0 + \frac{\omega}{a_{ii}}\left(b_i - \sum_{j=1}^{i-1} a_{ij}x_j^{1/2} - \sum_{j=i}^{n} a_{ij}x_j^0\right)$$

for $(i := n; i \ge 1; i--)$

$$x_i^1 := x_i^{1/2} + \frac{\omega}{a_{ii}}\left(b_i - \sum_{j=1}^{i} a_{ij}x_j^{1/2} - \sum_{j=i+1}^{n} a_{ij}x_j^1\right)$$

We have used in all computations $\omega = 1.5$. Since (9.10) has to be solved only once in each sub time step, see Section 7.2, the computational costs are small compared to performing the fixed point iteration (7.7).

Remark 9.21. Initial iterate and stopping criteria. Because the computational costs of solving (9.11) for all right hand sides are a very small part of the total costs, we solve these equations always accurately. If not mentioned otherwise, the preconditioned conjugate gradient method is stopped if the Euclidean norm of the residual, computed with all $d(d+1)/2$ components, is less than 10^{10} and at least one iteration was performed. The initial iterate for the preconditioned conjugate gradient iteration is the solution of the previous sub time step. If **w** has not changed much from the previous sub time step, the right hand side of the auxiliary system does not change much, too. In this case, an initial iterate which is close to the solution is easily available. □

Remark 9.22. Additional memory requirements. Let n be the dimension of the finite element space for one component of the velocity. The preconditioned conjugate gradient method requires the additional arrays q, r and z. Since

$d(d+1)/2$ systems of size n are solved together, each of the additional arrays has the dimension $nd(d+1)/2$. □

Remark 9.23. The Iliescu-Layton subgrid scale model (4.31). If the Iliescu-Layton subgrid scale model (4.31) is used, a system of form (9.10) with the right hand side replaced by $\mathbf{w}$ has to be solved. The discretisation leads to d linear systems of equations of form (9.11) with the same matrix A which are solved simultaneously. The solution of these systems is done exactly in the same way as described in this section for the auxiliary problem. □

10

A Numerical Study of a Necessary Condition for the Acceptability of LES Models

In this section, we present a numerical study at a simple test problem which investigates if the LES models considered in this monograph fulfil or violate a condition which is in our opinion necessary for the acceptability of LES models.

Let $\mathbf{u}$ be the velocity of a flow in a domain Ω with total kinetic energy

$$E_{\text{kin}}(\mathbf{u}) = \frac{1}{2}\int_{\Omega} \mathbf{u}^T \mathbf{u} d\mathbf{x} < \infty \quad \text{for } t \in [0, T].$$

Extending $\mathbf{u}$ by $\mathbf{0}$ outside Ω, it follows by Young's inequality for convolutions (2.18) that $\overline{\mathbf{u}}$ also possesses bounded total kinetic energy. The same has to be expected from the velocity $\mathbf{w}$ obtained with a LES model since $\mathbf{w}$ should be an approximation of $\overline{\mathbf{u}}$. In this chapter, it will be studied if the velocities $\mathbf{w}^h$, computed with the different LES models, possess bounded total kinetic energy $E_{\text{kin}}(\mathbf{w}^h)$ for each discrete time step $t_n \in [0, T]$. If a LES model fails to satisfy this condition, it can be concluded with certainty that this model is not suited to model turbulent flows and we will not consider this model in further numerical studies. On the other hand, if a model fulfils this condition does not mean automatically that it is a good model. Such models require further investigations to understand their advantages and drawbacks.

10.1 The Flow Through a Channel

We consider the flow through the channel presented in Figure 10.1. On Γ_{in}, the inflow boundary condition

$$\mathbf{w}(t, 0, y, z) = \begin{pmatrix} 4y(1-y) \\ 0 \\ 0 \end{pmatrix},$$

at the top and bottom wall no slip boundary conditions and at the lateral walls free slip boundary conditions are prescribed. The flow leaves the channel at

Γ_{out} by outflow boundary conditions. The initial velocity is given by

$$\mathbf{w}(0,x,y,z) = \begin{pmatrix} 4y(1-y) \\ 0 \\ 0 \end{pmatrix} + c_{\text{noise}} \begin{pmatrix} -4\pi \sin(4\pi y) \\ -3\pi \sin(3\pi z) \\ 3\pi \cos(3\pi x) \end{pmatrix} \tag{10.1}$$

and the right hand side is chosen to be $\mathbf{f} = \mathbf{0}$. Without noise in the initial condition, $c_{\text{noise}} = 0$,

$$\mathbf{w}(t,x,y,z) = \begin{pmatrix} 4y(1-y) \\ 0 \\ 0 \end{pmatrix}, \quad r(t,x,y,z) = -8\nu(x-10)$$

is a solution of the Navier-Stokes equations (3.5) with $\mathbf{f} = \mathbf{0}$ whose total kinetic energy is $2.66\overline{6}$ for each time t. We present results of computations for the viscosity $\nu^{-1} = 10^5$ and the scaling factor $c_{\text{noise}} = 0.01$. The Reynolds number of the flow, based on the average inflow W and the height of the channel L is

$$Re = \frac{WL}{\nu} = \frac{2}{3}10^5 \approx 66667.$$

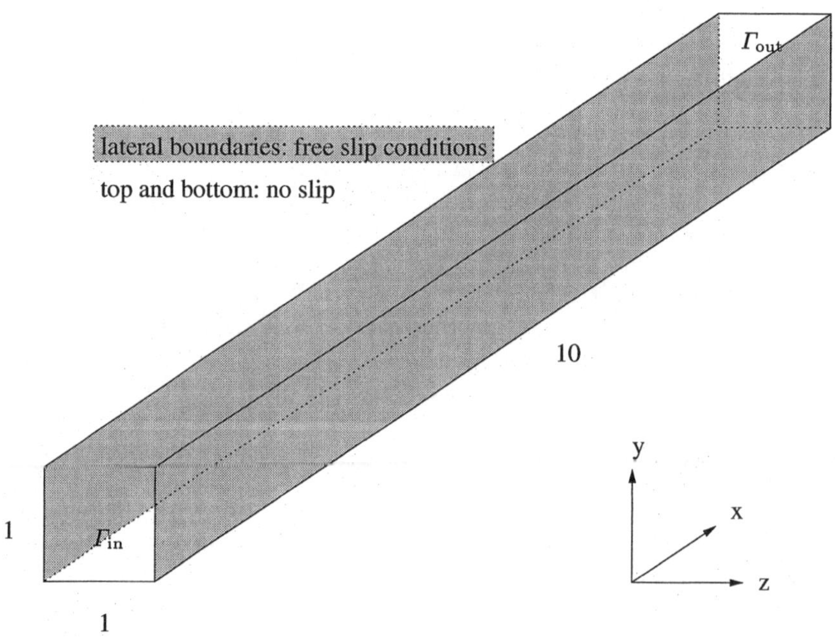

Fig. 10.1. The channel

We will present computations on hexahedral and tetrahedral grids. The initial hexahedral grid, level 0, consists of 80 cubes of size 0.5. For the initial tetrahedral grid, these cubes are decomposed each into six tetrahedra such that the initial tetrahedral grid consists of 480 tetrahedra. All computations are carried out on level 1. The number of degrees of freedom for different discretisations in space on this level is given in Table 10.1. The mesh cells on level 1 of the hexahedral grid have the diameter $h = 0.25\sqrt{3} \approx 0.433$. In the tetrahedral mesh, the mesh cells are of different sizes. The minimal diameter on level 1 is $h_{K,min} \approx 0.354$ and the maximal diameter is $h_{K,max} \approx 0.559$. We use rather coarse meshes in the computations to reflect the typical situation in turbulent flow computations that there are too less degrees of freedom compared to the size of the Reynolds number. The filter width is chosen to be $\delta = 0.5$.

Table 10.1. Degrees of freedom on level 1

finite element space	velocity	pressure	total
Q_2/Q_1	19 683	1 025	20 708
$Q_2/P_1^{\rm disc}$	19 683	2 560	22 243
$Q_3/P_2^{\rm disc}$	61 347	6 400	67 747
P_2/P_1	19 683	1 025	20 708

The time discretisations are applied with an equidistant time step $\Delta t_n = 0.01$. The fixed point iteration in each time step was stopped for an Euclidean norm of the residual vector less than 10^{-6}.

Remark 10.1. Computations with other kinds of noises and without noise in the initial condition. We performed numerical studies also with other values of the scaling factor $c_{\rm noise}$ in the initial condition (10.1) and with other types of initial noises, e.g., initial noises which satisfy no slip boundary conditions at $y = 0$ and $y = 1$. The LES models were tested even without noise in the initial condition. We always obtained qualitatively the same results as those which are presented in this chapter. Only such quantities like blow-up times were different. □

10.2 The Failure of the Taylor LES Model

In this section, we present numerical results obtained with the Taylor LES model with Smagorinsky subgrid scale model and two values of the scaling factor c_S. The choice $c_S = 0.01$ is done accordingly to Remarks 4.4 and 4.12. For this value, it turns out that the Taylor LES model fails to compute solutions with bounded total kinetic energy. To support the analytical results of Sections 6.2 and 8.2, which state the existence and uniqueness of a weak solution and give error estimates for finite element approximations if c_S is large enough, we performed computations also with $c_S = 1/3$.

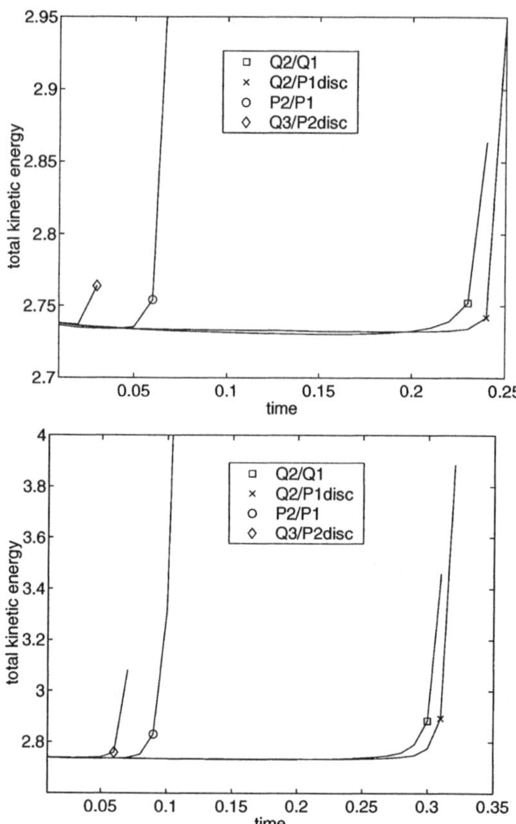

Fig. 10.2. The blow-up of the total kinetic energy of the solution computed with the Taylor LES model and Smagorinsky subgrid scale term with $c_S = 0.01$, various finite element discretisations, the fractional-step θ-scheme (top), Crank-Nicolson time discretisation (bottom)

The total kinetic energy of the computed solutions for the case of a small scaling factor in the Smagorinsky subgrid scale model, $c_S = 0.01$, is presented in Figure 10.2. The results of the upper picture are obtained with the fractional-step θ-scheme (variant 1) and the results of the lower picture with the Crank-Nicolson time discretisation. We present results for various pairs of finite element spaces. It can be seen that the total kinetic energy of the computed solution blows up independent of the discretisation in space and time. The blow up occurs fastest for the highest order finite element discretisation which was used. Applying other discretisations with third order velocity and second order pressure, P_3/P_2 or Q_3/Q_2, we could not solve the non-linear system in the initial time step. The non-linear fixed point iteration diverged.

Altogether, the Taylor LES model with Smagorinsky subgrid scale term and a standard choice of c_S did not compute solutions with bounded total kinetic energy. It is even much more instable than the Galerkin finite element discretisation of the Navier-Stokes equations. The total kinetic energy of the Galerkin Q_2/P_1^{disc} finite element solution and the fractional-step θ-scheme blows up at 4.22 s, which is much later than for the Taylor LES model. This model is not suited to model turbulent flows. The reason for the failure of the Taylor LES model is the completely wrong approximation of the Fourier transform of the Gaussian filter for high wave numbers, see Section 4.2.1.

Remark 10.2. The failure of the Taylor LES model in other computations. The failure of the Taylor LES model was observed for the 2d driven cavity problem by Coletti [Col97] and for the 2d and 3d driven cavity problem in Iliescu et al. [IJL+03]. Numerical studies by Cantekin et al. [CWL94] and in [Col97] show even problems with the Taylor LES model applied in laminar flows. In [IJL+03], the Taylor LES model was studied for different mesh sizes, lengths of the time step and initial conditions. The total kinetic energy of the computed solution blew up for nearly all configurations. Since there is the clear conclusion in [IJL+03] that the behaviour of the Taylor LES model is not improved by varying h, Δt_n or $\mathbf{w}_0$, we do not present studies with respect to these parameters.

In Chapter 11, flows will be considered where the region of interest is away from the boundary of the domain. In simulating these flows with the Taylor LES model with Smagorinsky subgrid scale term and $c_S = 0.01$, a fast blow up of the solution could be observed, too. Thus, the failure of the Taylor LES model is not caused by effects coming from the boundary of the domain. □

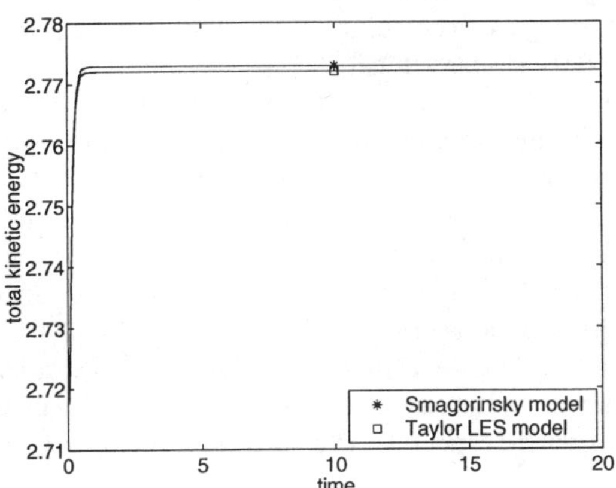

Fig. 10.3. The total kinetic energy of the solution computed with the Taylor LES model and Smagorinsky subgrid scale term with $c_S = 1/3$, fractional-step θ-scheme time discretisation and Q_2/P_1^{disc} finite element discretisation

The analysis of the Taylor LES model with Smagorinsky subgrid scale model, Sections 6.2 and 8.2, proves that a unique weak solution exists and can be approximated by a finite element method if the Smagorinsky constant c_S is large enough. However, see Remark 6.20 for a discussion on the contradiction between the derivation of the Taylor LES model and choosing c_S to be large. A guideline for the size of c_S fitting into the analysis gives Lemma 8.20: $c_S = 2/\gamma = 1/3$. Results with this value of c_S are presented here to support the analytical results, see Figure 10.3. We used in this computation the fractional-step θ-scheme time discretisation and the Q_2/P_1^{disc} finite element discretisation in space. As expected by the analysis, the computed solution possesses bounded total kinetic energy.

The Taylor LES model will not be considered further in the numerical studies.

10.3 The Rational LES Model

If we speak in the following of *the rational LES model*, we mean always both types of this model. The two variants of the rational LES model with convolution are described in Remark 7.11. The computations presented in this section are performed with the fractional-step θ-scheme time discretisation (variant 1) and the Q_2/P_1^{disc} finite element discretisation in space.

10.3.1 Computations With the Smagorinsky Subgrid Scale Model

Figure 10.4 presents results for the two types of the rational LES model with Smagorinsky subgrid scale term with the scaling factor $c_S = 0.01$. In addition, the total kinetic energy of the solution obtained with the pure Smagorinsky model with $c_S = 0.01$ is also shown.

It can be observed that all computed solutions posses bounded total kinetic energy. After having damped out the initial noise, the absolute values of the total kinetic energy are similar for the rational LES model and the Smagorinsky model, see lower picture in Figure 10.4. Finally, the total kinetic energy obtained with the Smagorinsky model is constant in time which suggests that this model leads to a steady state solution. In contrast, a time dependence of the solutions obtained with the rational LES model can be observed. Considering both variants of the rational LES model with convolution, the first variant behaves more similar to the rational LES model with auxiliary problem with respect to the total kinetic energy, see lower picture in Figure 10.4.

Figure 10.5 presents the time averaged total kinetic energy for different values of the scaling factor c_S in the Smagorinsky model. For a large scaling factor, $c_S = 0.1$, there is almost no difference between the two types of the rational LES model with Smagorinsky subgrid scale term and the Smagorinsky model. The difference is in general larger for smaller values of c_S. This is natural since the influence of the Smagorinsky subgrid scale term decays. If c_S is sufficiently small, $c_S = 10^{-3}$, then the Smagorinsky model produces

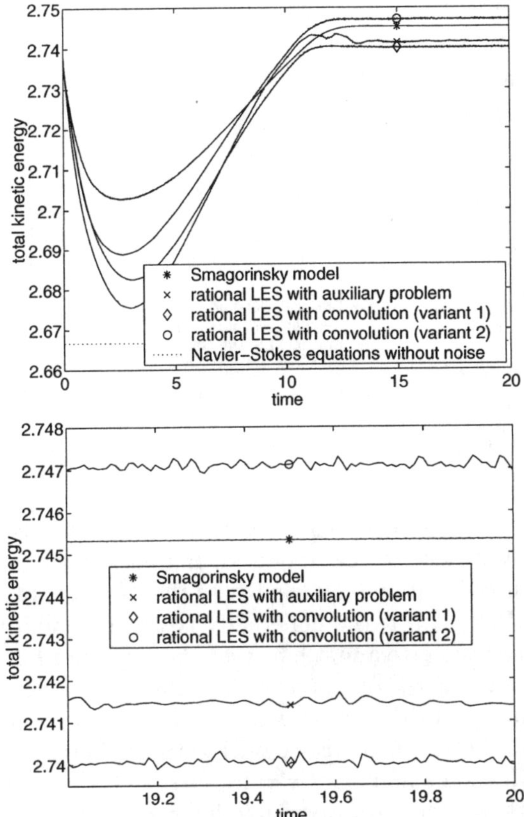

Fig. 10.4. The total kinetic energy computed with the rational LES model with Smagorinsky subgrid scale term and the Smagorinsky model, $c_S = 0.01$, fractional-step θ-scheme time discretisation and Q_2/P_1^{disc} finite element discretisation

still a solution with bounded total kinetic energy whereas both types of the rational LES model lead to a blow-up of the solution. One of the drawbacks of the Smagorinsky model is that it is overly diffusive, see Remark 4.3. The term coming from the derivation of the rational LES model seems to counteract the diffusivity of the Smagorinsky model to some extend.

Remark 10.3. Numerical studies with the rational LES model with Smagorinsky subgrid scale model in the literature. A numerical study of the rational LES model at the two and three dimensional driven cavity flow for the Reynolds number $Re = 10000$ can be found in Iliescu et al. [IJL$^+$03]. The results of this study with respect to the total kinetic energy are similar like in this monograph. In addition, especially in the two dimensional driven cavity problem, large differences of the solutions obtained with the Smagorinsky model and the rational LES model are observed. The solution obtained with the

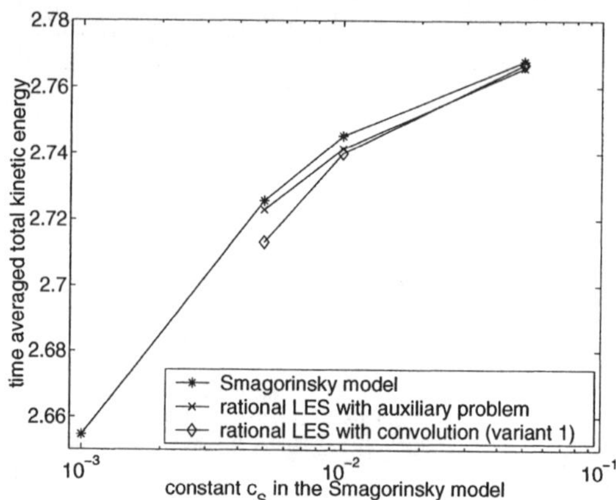

Fig. 10.5. Time averaged total kinetic energy for different values of c_S, $t \in [19, 20]$

Smagorinsky model possesses a main eddy which is considerably too small and its centre is too close to the upper lid. Because of the diffusivity of this model, the computed solution looks similar to the solution of the driven cavity Stokes problem. In contrast, the main eddy is well captured by the rational LES model. These tests give a first hint that the rational LES model is better suited to model turbulent flows than the Smagorinsky model.

Fischer and Iliescu [FI01] compared the Smagorinsky model with van Driest damping and the rational LES model with auxiliary problem for a 3d channel flow of Reynolds number $Re = 180$ based on the wall shear velocity. Both models were applied with a variable filter width tending to zero at the walls. Based on the evaluation of a number of flow statistics, they concluded that the rational LES model performed in general better near the wall and the Smagorinsky model with van Driest damping was better in the centre of the channel. Altogether, both models were comparable in accuracy. □

10.3.2 Computations With the Iliescu-Layton Subgrid Scale Model

The total kinetic energy of the solutions obtained with the rational LES model with the Iliescu-Layton subgrid scale term (4.31) is presented in Figure 10.6. The computations were performed with the scaling factor $c_S = 0.17$. This scaling factor has been proposed by Layton and Lewandowski [LL02].

The rational LES model with the Iliescu-Layton subgrid scale term computes solutions with bounded total kinetic energy. But there are considerable differences in comparison to the Smagorinsky model as subgrid scale term. Using the Iliescu-Layton subgrid scale model, there is no initial decrease of the total kinetic energy and the fully developed flow possesses considerably larger total kinetic energy.

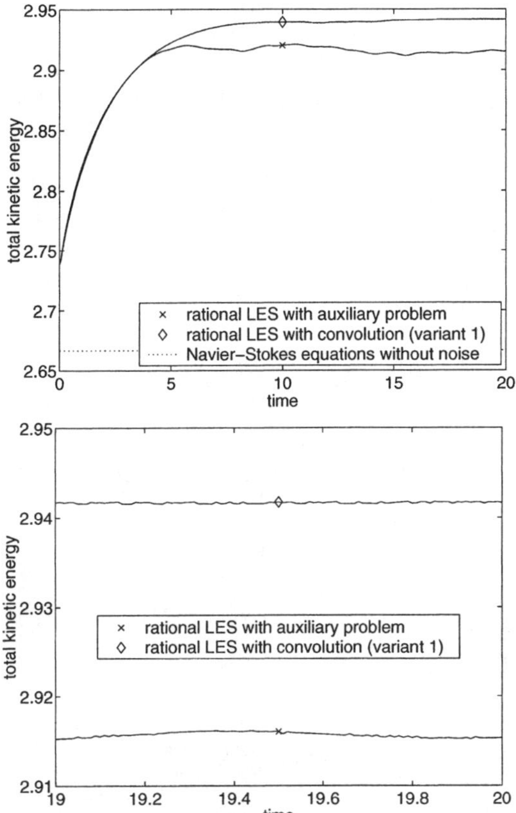

Fig. 10.6. The total kinetic energy computed with the rational LES model with Iliescu-Layton subgrid scale model (4.31), $c_S = 0.17$, fractional-step θ-scheme time discretisation and Q_2/P_1^{disc} finite element discretisation

10.3.3 Computations Without Model for the Subgrid Scale Term

The application of the rational LES model with a very small constant c_S in the Smagorinsky subgrid scale model or even without model for the subgrid scale term $\overline{\mathbf{u}'\mathbf{u}'^T}$ leads to a blow up of the solutions, see Figure 10.7. This effect shows the necessity of using an appropriate model for $\overline{\mathbf{u}'\mathbf{u}'^T}$.

10.4 Summary

This section summarises the most important results obtained in the numerical study presented in this chapter:

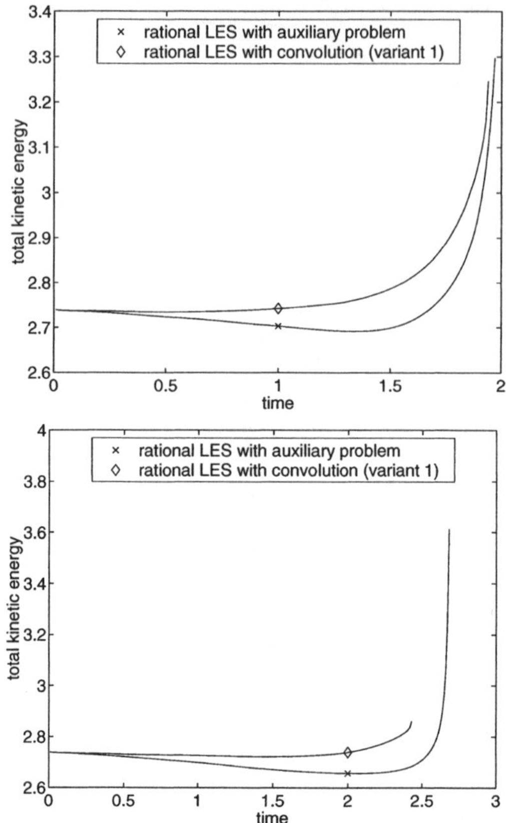

Fig. 10.7. The blow-up of the total kinetic energy, rational LES models without model for the subgrid scale term (top) and with Smagorinsky subgrid scale model with $c_S = 0.001$ (bottom), fractional-step θ-scheme time discretisation and Q_2/P_1^{disc} finite element discretisation

- The rational LES model with Smagorinsky subgrid scale model (4.3) and with Iliescu-Layton subgrid scale model (4.31) computes solutions with bounded total kinetic energy if the scaling factor c_S is sufficiently large.
- After having smoothed the initial noise, the total kinetic energy behaves different for the Smagorinsky and the Iliescu-Layton subgrid scale model.
- The rational LES model without model for the subgrid scale term blows up.
- The Taylor LES model with Smagorinsky subgrid scale model and a standard choice of the scaling factor c_S blows up.

The numerical tests show that the Taylor LES model is not suited for modelling turbulent flows. The rational LES model deserves and needs further investigations.

11

A Numerical Study of the Approximation of Space Averaged Flow Fields by the Considered LES Models

Numerical studies of LES models which can be found in the literature try, in general, to simulate a turbulent flow as good as possible and to compare the numerical results with statistics of the flow field known from experiments or DNS data. However, the studies presented in this chapter address a different question. The main purpose of LES models is to provide an accurate approximation of $(\overline{\mathbf{u}}, \overline{p})$. A natural and very important question is : *How good is the approximation of* $(\overline{\mathbf{u}}, \overline{p})$ *by the flow field computed with LES models ?* This is a fundamental question for each LES model. The performance of numerical tests studying this question requires reliable data for $(\overline{\mathbf{u}}, \overline{p})$ in time and space. Section 11.1 studies the above formulated question for the LES models considered in this monograph in a situation where reliable data for $(\overline{\mathbf{u}}, \overline{p})$ can be computed, namely a 2d mixing layer problem at $Re = 10000$. Section 11.2 presents numerical tests for a mixing layer problem in three dimensions and $Re = 714$. A comparison with filtered DNS data cannot be presented since it was not possible to perform a DNS in the three dimensional example.

Let Ω be a domain with boundaries. The rational LES model which we are using in the computations was derived by applying two simplifications in this case:

- The commutation error, Definition 3.5, is neglected.
- The models for the large scale advective term and the cross terms are derived in $\mathbb{R}^d$, see Chapter 4. In the case $\Omega \neq \mathbb{R}^d$, the application of the rational LES model is described in Remarks 4.7 and 4.9. These applications are not supported by analytical considerations.

Thus, it can be expected that in a vicinity of the boundary additional errors occur besides the modelling error. It is also well known that the Smagorinsky model behaves unsatisfactorily in near boundary regions. Since we will investigate in this chapter the principal properties of the rational LES model in the form which is studied hitherto in this monograph, i.e. with constant filter width δ and without special treatment of subregions, we will consider flows whose regions of interest are away from the boundary of the domain.

11.1 A Mixing Layer Problem in Two Dimensions

The mixing layer problem in 2d is a popular test case in large eddy simulation. Numerical studies can be found, e.g., in Lesieur et al. [LSLC88], Boersma et al. [BKNW97], Nägele and Wittum [NW03] or Griebel and Koster [GK00].

The aim of this numerical study is a comparison of solutions obtained with the considered LES models on a coarse grid with the filtered solution of the Navier-Stokes equations on a fine grid. This filtered solution of the Navier-Stokes equations is the correct quantity to compare with since the LES models try to predict the behaviour of $(\overline{\mathbf{u}}, \overline{p})$ and not of $(\mathbf{u}, p)$.

11.1.1 The Definition of the Problem and the Setup of the Numerical Tests

The Definition of the Problem

The problem is defined in $\Omega = (-1,1)^2$. Free slip boundary conditions are applied at $y = -1$ and $y = 1$. At $x = 1$ and $x = -1$, periodic boundary conditions are prescribed. With the periodic boundary condition, a channel of infinite length is simulated. The initial velocity is given by

$$\mathbf{w}_0 = \begin{pmatrix} W_\infty \tanh\left(\dfrac{2y}{\sigma_0}\right) \\ 0 \end{pmatrix} + c_{\text{noise}} W_\infty \begin{pmatrix} \dfrac{\partial \psi}{\partial y} \\ -\dfrac{\partial \psi}{\partial x} \end{pmatrix} \qquad (11.1)$$

with

$$\psi = \exp\left(-\left(2y/\sigma_0\right)^2\right)\left(\cos\left(8\pi x\right) + \cos\left(20\pi x\right)\right).$$

An illustration of the first component of the initial velocity field is presented in Figure 11.1.

The mixing layer problem in two dimensions is well analysed, e.g., see Lesieur [Les97, Section 3.3.1]. The problem is known to be inviscidly unstable. Slight perturbations in the initial condition are amplified by so-called Kelvin-Helmholtz instabilities. The most amplified mode corresponds to the wave length $\lambda_a = 7\sigma_0$, see Michalke [Mic64]. For a domain having the extension l_x in x-direction with $l_x = n\lambda_a$, $n \in \mathbb{N}$, the number of primary vortices which are expected to develop is equal to n, see [Les97, p. 312].

We will present computations with four primary vortices, i.e. $n = 4$. Since $l_x = 2$, it follows that we have to choose $\sigma_0 = 1/14$. The other parameters in the computations are chosen to be

- $W_\infty = 1$,
- scaling factor $c_{\text{noise}} = 0.001$,
- viscosity $\nu^{-1} = 140000$.

The Reynolds number of this flow, based on σ_0, W_∞ and ν is

$$Re = \frac{\sigma_0 W_\infty}{\nu} = 10000.$$

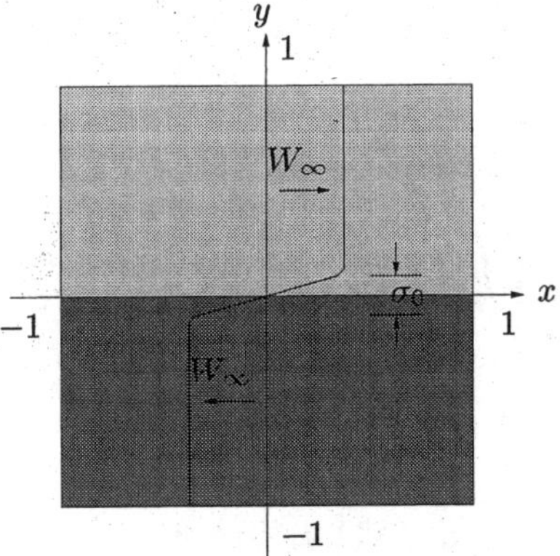

Fig. 11.1. First component of the initial velocity for the mixing layer problem in 2d (without noise)

The Vorticity Thickness

For the evaluation of the computational results, we consider the vorticity $\boldsymbol{\omega}$ of the flow. The vorticity is the curl of the velocity $\mathbf{w} = (w_1, w_2, w_3)^T$

$$\boldsymbol{\omega} = \nabla \times \mathbf{w} = \begin{pmatrix} (w_3)_y - (w_2)_z \\ (w_1)_z - (w_3)_y \\ (w_2)_x - (w_1)_y \end{pmatrix}. \tag{11.2}$$

Thus, for a two dimensional flow, only the last component $\omega_3 = (w_2)_x - (w_1)_y$ does not vanish. All plots of the vorticity in this section show ω_3. The vorticity thickness $\sigma(t)$ is defined by

$$\sigma(t) = \frac{2W_\infty}{\sup\limits_{y\in[-1,1]} |\langle\omega_3\rangle(t,y)|}, \tag{11.3}$$

where $\langle\omega_3\rangle(t,y)$ is the integral mean in periodic direction

$$\langle\omega_3\rangle(t,y) = \frac{\int_{-1}^{1} \omega_3(t,x,y)\,dx}{\int_{-1}^{1} dx} = \frac{1}{2}\int_{-1}^{1} \omega_3(t,x,y)\,dx.$$

This definition, taken from Nägele and Wittum [NW03], differs slightly from the definition by Lesieur et al. [LSLC88] who used

$$\frac{2W_\infty}{\sup\limits_{y\in[-1,1]} \dfrac{d\langle w_1\rangle(t,y)}{dy}}. \tag{11.4}$$

Assuming w_1 to be sufficiently smooth and $(w_2)_x$ to be small, one obtains

$$\begin{aligned}\frac{d\langle w_1\rangle(t,y)}{dy} &= \frac{1}{2}\frac{d}{dy}\int_{-1}^{1} w_1(t,x,y)\,dx = \frac{1}{2}\int_{-1}^{1}(w_1)_y(t,x,y)\,dx \\ &\approx \frac{1}{2}\int_{-1}^{1}(w_1)_y(t,x,y)-(w_2)_x(t,x,y)\,dx \\ &= -\frac{1}{2}\int_{-1}^{1}\omega_3(t,x,y)\,dx.\end{aligned}$$

The last integral turned out to be negative in the computations for all considered times. Thus (11.3) and (11.4) give similar results.

The parameter σ_0 is the vorticity thickness of the unperturbed initial velocity. Considering (11.1) for $c_{\text{noise}}=0$ and using the fact that the hyperbolic cosine function takes its minimum for $y=0$, one obtains

$$\sup_{y\in[-1,1]}|\langle\omega_3\rangle(0,y)| = \sup_{y\in[-1,1]}\left|\frac{W_\infty}{\sigma_0\cosh^2\left(\frac{2y}{\sigma_0}\right)}\int_{-1}^{1}dx\right| = \frac{2W_\infty}{\sigma_0}.$$

It follows from (11.3) that $\sigma(0)=\sigma_0$.

In the computations, the term $2W_\infty/|\langle\omega_3\rangle(t,y)|$ can be computed only for a finite number of values y. We compute this term on all grid lines which are parallel to the x-axis. From the values computed in this way, the maximum is taken to obtain $\sigma(t)$. In the evaluation of the computations, we consider the vorticity thickness relative to σ_0: $\sigma(t)/\sigma_0$.

The Discretisation in Time

We present results which are obtained with the fractional-step θ-scheme as time discretisation. A time unit $\bar{t}=\sigma_0/W_\infty$ is defined and an equal distant time step of length $\Delta t_n = 0.1\bar{t} = 0.1/14 \approx 7.1428e-3$ is used. The final time is set to be $T=200\bar{t}\approx 14.285$.

In computations with the Galerkin finite element discretisation of the Navier-Stokes equations on level 8, the Crank-Nicolson time discretisation turned out to be unstable using the already rather fine time step given above. During the pairing of the four primary eddies (see below), the solution of the arising discrete systems was not possible.

The Discretisation in Space

The Q_2/P_1^{disc} finite element discretisation is used in the computations. The initial computational grid (level 0), consists of four squares of edge length one. This grid is refined uniformly and the number of degrees of freedom on finer levels is presented in Table 11.1. In the computations with the LES models, the non-linear discrete problem in each time step was solved up to an Euclidean norm of the residual vector less than 10^{-12}. With a less accurate stopping criterion, e.g., 10^{-8}, we obtained in some cases considerably different results. In the computations with the Galerkin finite element discretisation of the Navier-Stokes equations on level 8, the stopping criterion 10^{-8} was used. Note, that the dimension of the residual vector on level 8 is much larger than on level 5. Thus, the required accuracy per component is comparable for both levels.

Table 11.1. Degrees of freedom and mesh width on different levels

level	h	velocity	pressure	total
5	$\sqrt{2}/32$	33 024	12 288	45 312
7	$\sqrt{2}/128$	525 312	196 608	721 920
8	$\sqrt{2}/256$	2 099 200	786 432	2 885 632

Parameters for the LES Models

The filter width δ was chosen in all tests to be $\delta = h$, where h denotes as usual the diameter of the mesh cells, see Table 11.1. This choice makes sense for the second order velocity finite element space since each averaging circle with a degree of freedom as centre and with radius δ contains several neighbour degrees of freedom, see Figure 11.2.

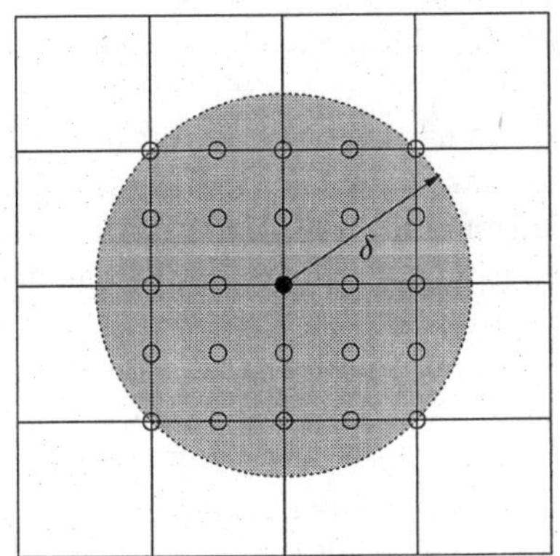

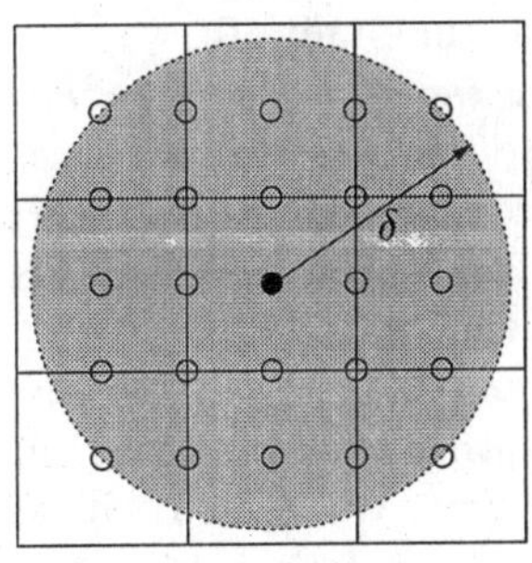

Fig. 11.2. Averaging circle for second order velocity with a degree of freedom as centre and radius $\delta = h$

In addition to using the Smagorinsky term as subgrid scale model in the rational LES model, we will also present numerical tests for using one of the models of Iliescu and Layton [IL98], namely model (4.31).

In the rational LES model, either an auxiliary problem has to be solved or a convolution has to be computed. The auxiliary problem is equipped with periodic boundary conditions at $x = -1$ and $x = 1$. At the other boundaries, we use as usual homogeneous Neumann boundary conditions. For the computation of the convolution, the functions are extended periodically in x direction at $x = -1$ and $x = 1$. At the other boundaries, we extend the functions by zero.

Evolution of the Flow

The evolution of the flow can be described with the help of Figure 11.3.[1] These pictures are the result of a computation using the Galerkin finite element method on level 8. We will refer to this computation as direct numerical simulation (DNS). The evolution of the relative vorticity thickness of the DNS solution behaves very similar to the evolution of the relative vorticity thickness of the filtered DNS solution which is presented, e.g., in Figures 11.14 and 11.19.

- *Development of the four primary eddies.* Starting with the initial noise, the four primary vortices develop. They can be seen clearly after 30 time units. The vorticity thickness is approximately doubled in comparison to the initial vorticity thickness σ_0.
- *Pairing of the four primary eddies.* It can be clearly seen that one of these pairings starts earlier than the other one. At time unit 80, the different developments of both pairings are obviously. The resulting eddies are called secondary eddies. The pairing of the four primary eddies is connected with another doubling of the relative vorticity thickness whose value is around 4 at time unit 80.
- *Pairing of the two secondary eddies.* The pairing of the four primary eddies into pairs of two is succeeded immediately by the pairing of these two secondary eddies into one eddy, see time unit 100. The relative vorticity thickness reaches values of more than 6. The pairing is in principal finished at time unit 140.
- *Rotation of the final eddy.* After time unit 140, the final eddy rotates at a rather fixed position. Since this eddy has an elliptic shape, the relative vorticity thickness oscillates and it takes values between 4 and 6.

The aim of the LES models is to predict the behaviour of $(\overline{\mathbf{u}}, \overline{p})$. An approximation of $(\overline{\mathbf{u}}, \overline{p})$ can be obtained by filtering the solution $\left(\mathbf{u}^h, p^h\right)$ of the DNS with the same filter width which was used in the computations with the LES models, $\delta = \sqrt{2}/32$. This approximation is denoted by $\left(\overline{\mathbf{u}^h}, \overline{p^h}\right)$.

[1] The vorticity isolines in this section were plotted using the software package GRAPE.

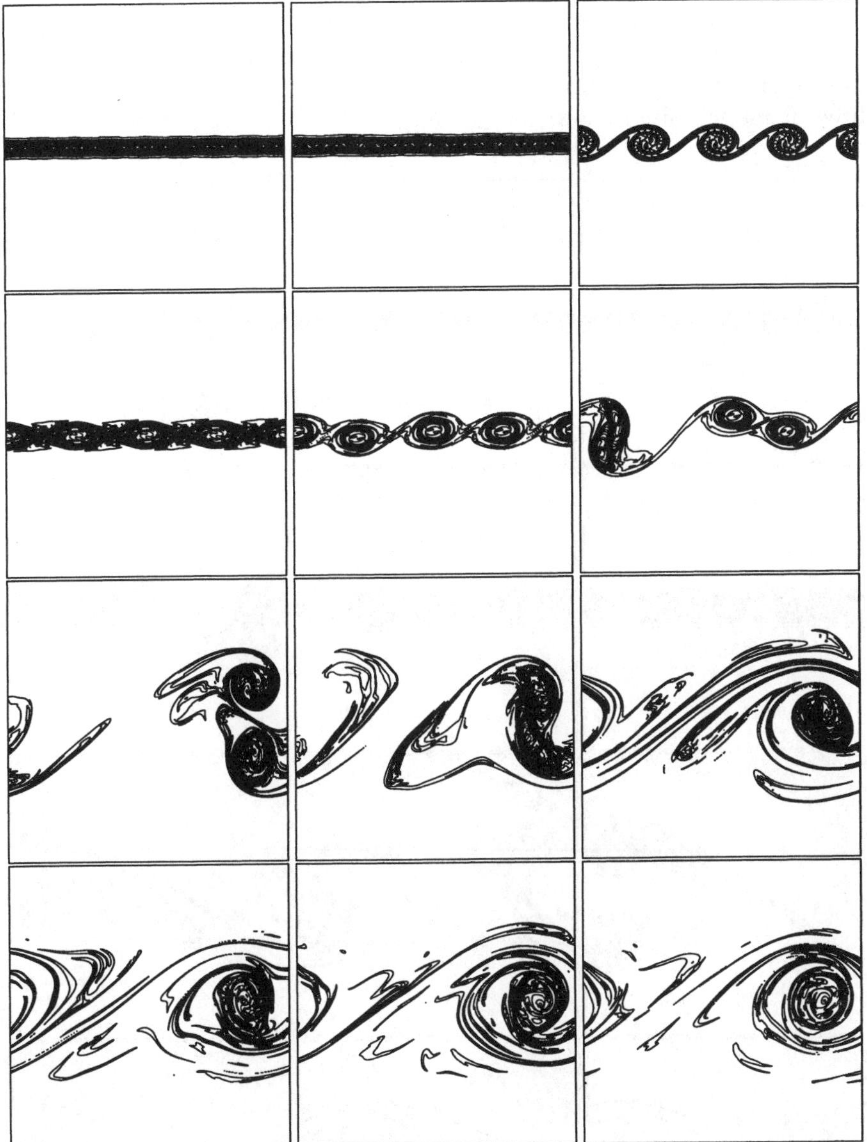

Fig. 11.3. Galerkin FEM, vorticity on level 8, at time units 0, 20, 30, 50, 70, 80, 100, 120, 140, 160, 180, 200 (left to right, top to bottom)

We will compare the Smagorinsky model and the rational LES model with the reference solution $\left(\overline{\mathbf{u}^h}, \overline{p^h}\right)$. The evolution of the vorticity of $\overline{\mathbf{u}^h}$ is presented in Figure 11.4.

The Smagorinsky model and both types of the rational LES model are applied on level 5. The Galerkin finite element solution blows up on this level after approximately 20 time units. Thus, this level is too coarse to admit a direct numerical simulation.

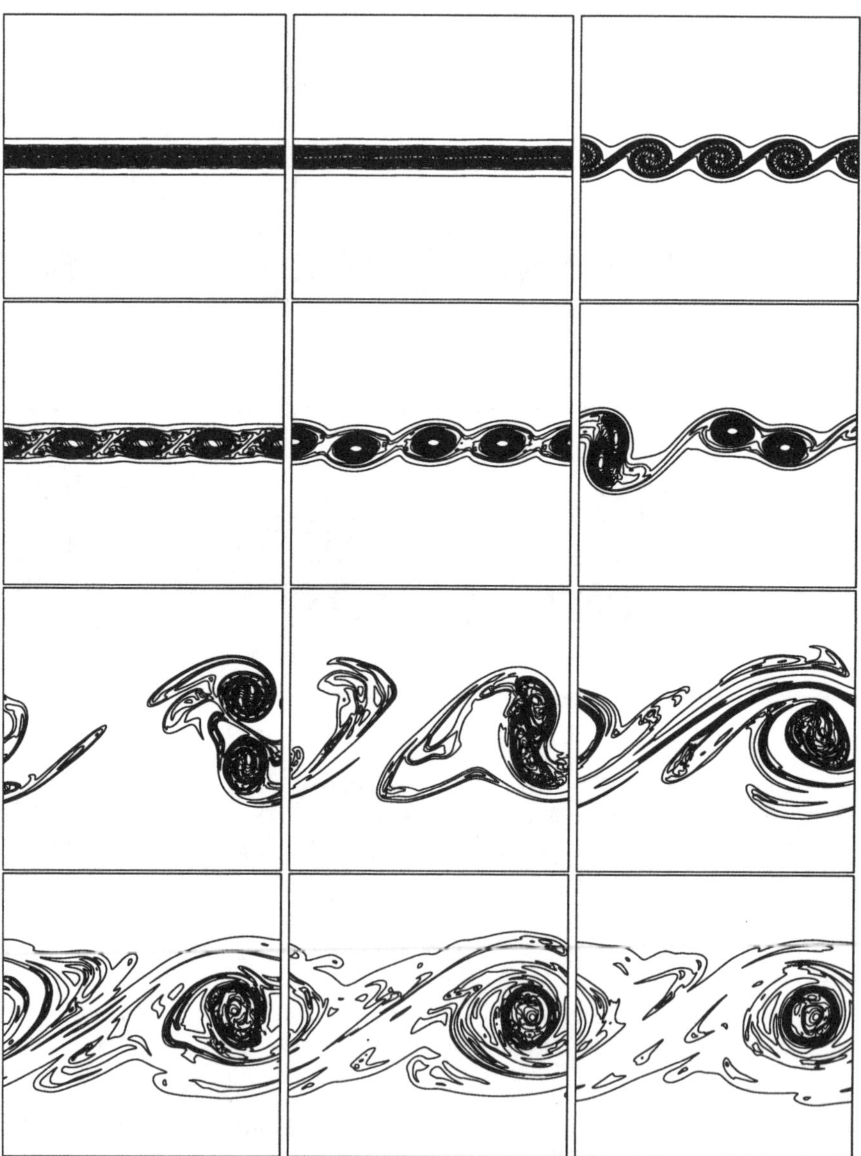

Fig. 11.4. Vorticity of $\overline{\mathbf{u}^h}$, level 8, at time units 0, 20, 30, 50, 70, 80, 100, 120, 140, 160, 180, 200 (left to right, top to bottom)

11.1.2 The Smagorinsky Model and the Rational LES Model With Smagorinsky Subgrid Scale Term

The rational LES model with Smagorinsky subgrid scale term and the Smagorinsky model were applied with the scaling factors $c_S = 0.01$ and $c_S = 0.005$.

The Scaling Factor $c_S = 0.01$

The development of the vorticity of the computed solutions with the scaling factor $c_S = 0.01$ is presented in Figures 11.5 - 11.7. It can be seen, also in some features of the computed flows shown in Figures 11.8 - 11.10, that there are only small differences of the solutions obtained with the Smagorinsky model and the rational LES model in the time interval $[0, 80\bar{t}]$.

- *Development of the four primary eddies.* The development of the four primary eddies is computed very badly. In the reference solution, Figure 11.4, these eddies are clearly seen at time unit 30 whereas these eddies can hardly be recognised in Figures 11.5 - 11.7. There is also a clear difference in the development of the relative vorticity thickness, Figure 11.8. For the reference solution, one can observe a rapid increase from 1 to 2 between time unit 25 and 30. For the Smagorinsky model and the rational LES model with Smagorinsky subgrid scale term and $c_S = 0.01$, the increase from 1 to 2 is very slowly and it takes place from time unit 10 to 80. The four primary eddies are very flat in comparison to the reference solution. After time unit 50, they seem to disappear again.
- *Pairing of the four primary eddies.* This pairing starts after time unit 80 and it is connected with an increase of the relative vorticity thickness from 2 to 4. The beginning of the pairing is too late. This pairing is finished at time unit 100. A non-simultaneous pairing cannot be observed.
- *Pairing of the two secondary eddies.* In the rational LES model, this pairing occurs immediately after the previous pairing. This is like in the reference solution. Since the pairing of the four primary eddies happened too late, also the pairing of the two secondary eddies takes place too late, at around time unit 130. The relative vorticity thickness reaches values of around 12. In the Smagorinsky model, there is some delay between the pairing of the four primary eddies and the pairing of the two secondary eddies. The pairing of the two secondary eddies occurs at time unit 140. This phase of the flow is computed better by the rational LES model.
- *Rotation of the final eddy.* The final eddy of the solutions computed with the LES models rotates with a somewhat lower speed than the final eddy of the reference solution. This can be seen at the lower frequency of the oscillation of the relative vorticity thickness in Figure 11.8. The values of the relative vorticity thickness are larger than the values of the reference solution which indicates that the final eddy computed with the LES mod-

els is larger. The position of the final eddy is computed well with all LES models.

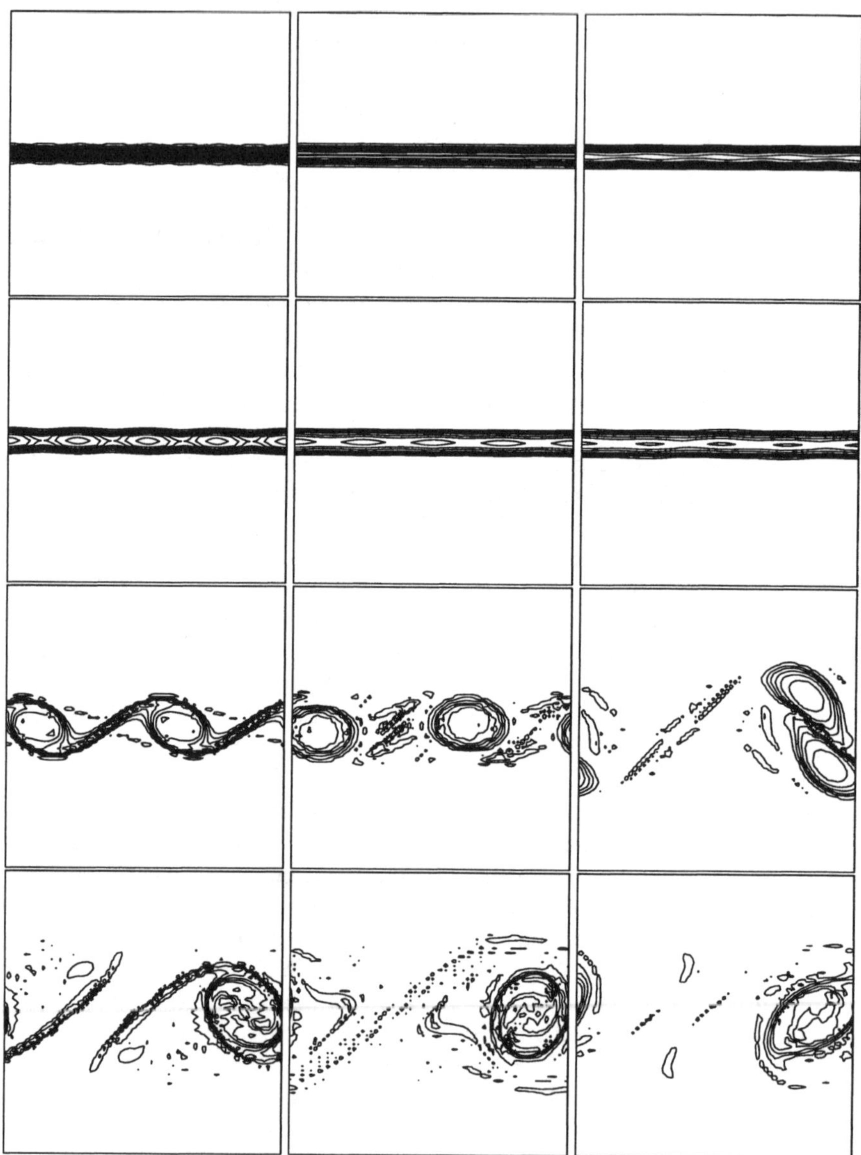

Fig. 11.5. Smagorinsky model (4.3), $c_S = 0.01$, vorticity on level 5, at time units 0, 20, 30, 50, 70, 80, 100, 120, 140, 160, 180, 200 (left to right, top to bottom)

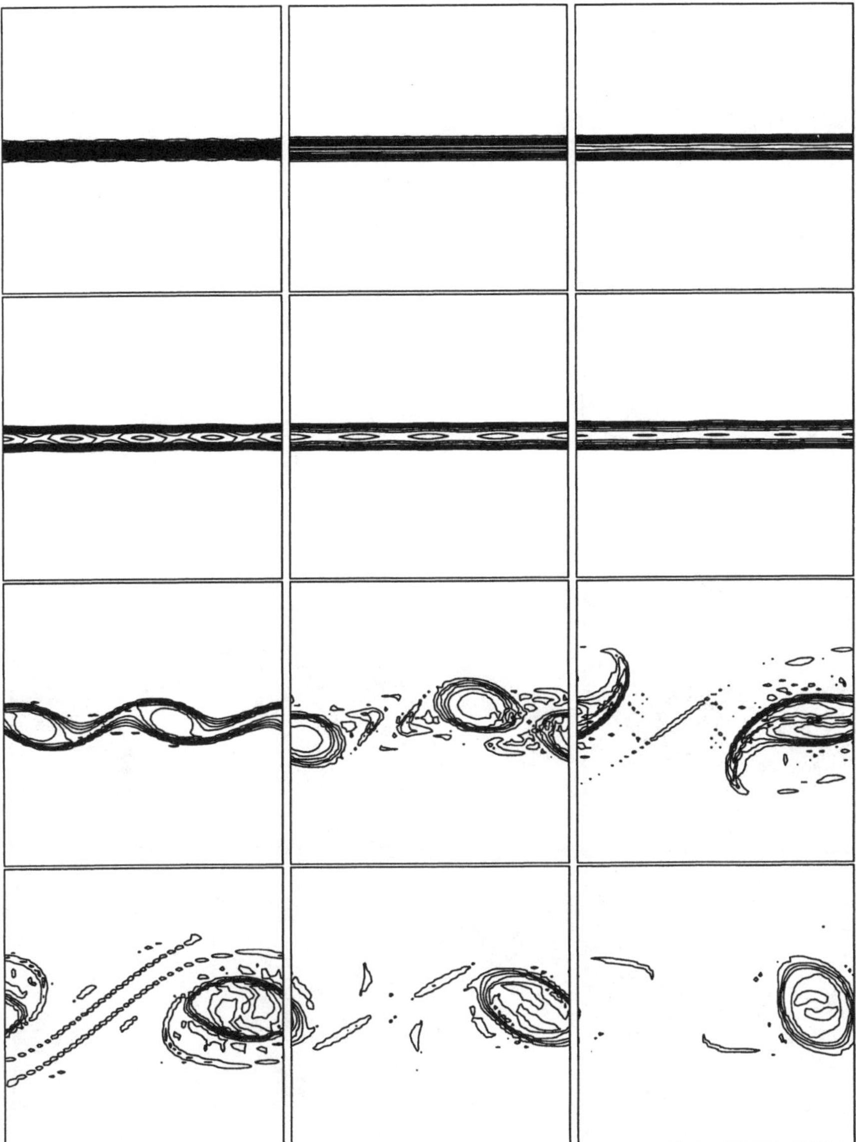

Fig. 11.6. Rational LES model with auxiliary problem, Smagorinsky model (4.3) as subgrid scale model with $c_S = 0.01$, vorticity on level 5, at time units 0, 20, 30, 50, 70, 80, 100, 120, 140, 160, 180, 200 (left to right, top to bottom)

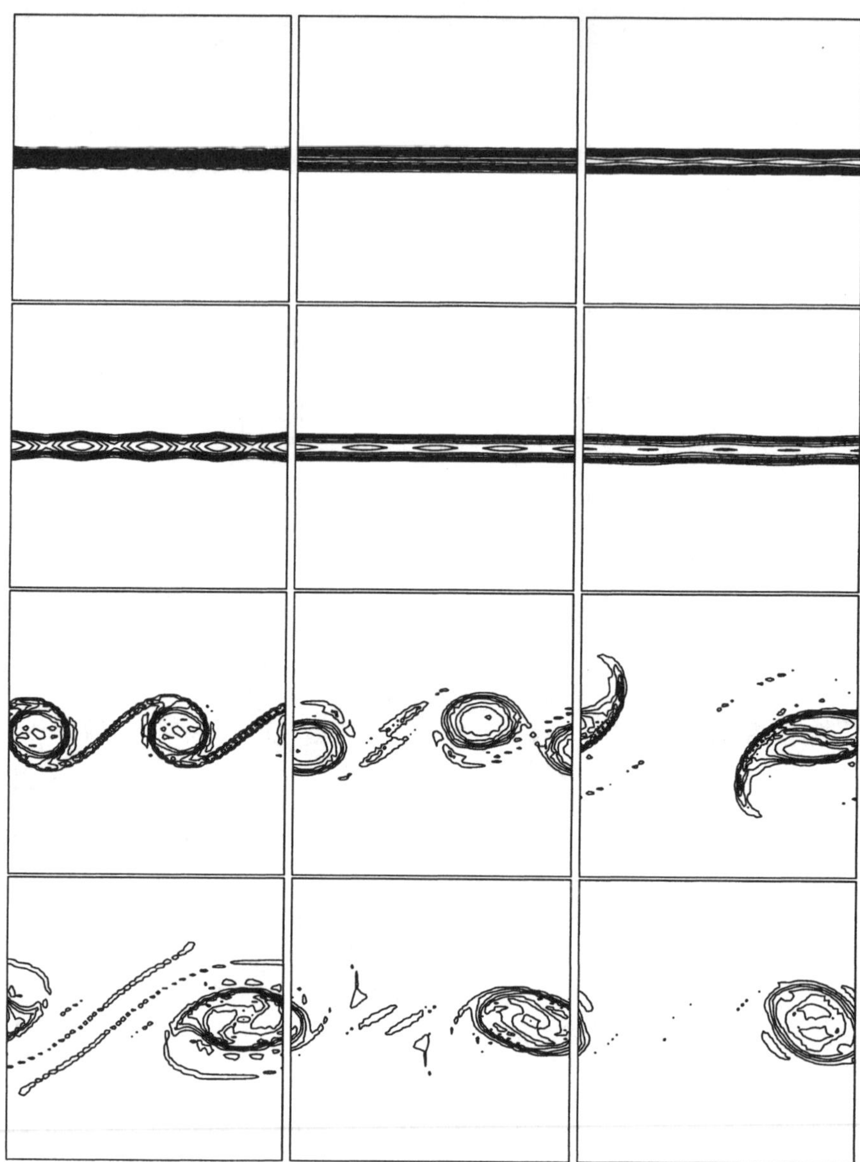

Fig. 11.7. Rational LES model with convolution, Smagorinsky model (4.3) as subgrid scale model with $c_S = 0.01$, vorticity on level 5, at time units 0, 20, 30, 50, 70, 80, 100, 120, 140, 160, 180, 200 (left to right, top to bottom)

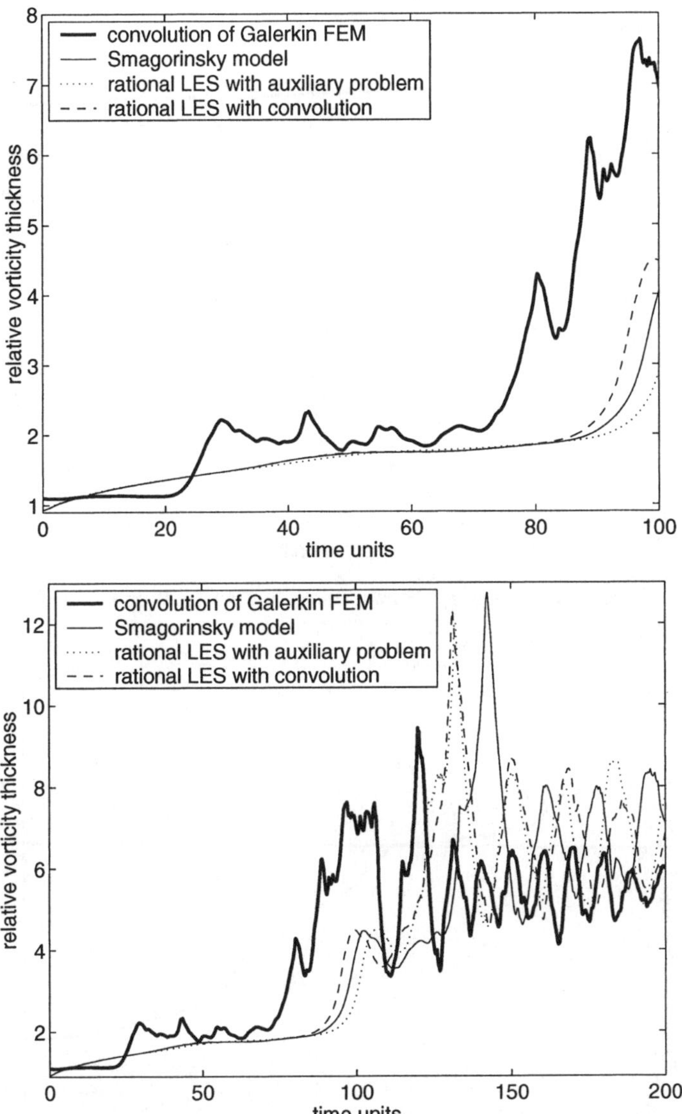

Fig. 11.8. Relative vorticity thickness to σ_0, Smagorinsky model (4.3) as subgrid scale model with $c_S = 0.01$, level 5

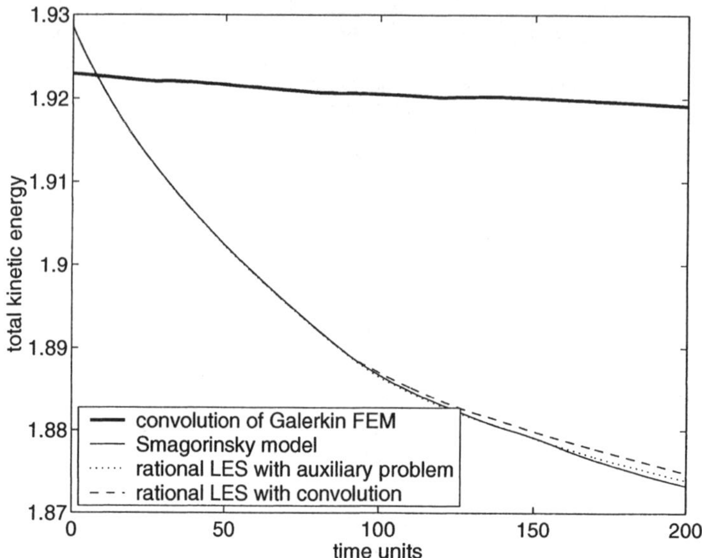

Fig. 11.9. Total kinetic energy, Smagorinsky model (4.3) as subgrid scale model with $c_S = 0.01$, level 5

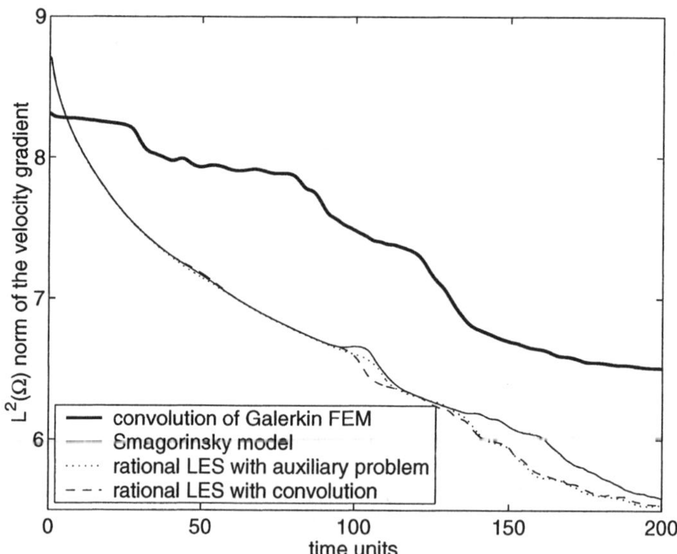

Fig. 11.10. $L^2(\Omega)$ norm of the gradient of the velocity, Smagorinsky model (4.3) as subgrid scale model with $c_S = 0.01$, level 5

The Scaling Factor $c_S = 0.005$

The results for the scaling factor $c_S = 0.005$ are presented in Figures 11.11 - 11.13. It can be seen that the Smagorinsky model and the rational LES model with convolution behave rather similar whereas the rational LES model with auxiliary problem shows a quite strange behaviour after the pairing of the four primary eddies.

We will compare here also the computed results obtained with $c_S = 0.005$ and $c_S = 0.01$. It will be mentioned only if one result is better or worse than the other one. If there is no statement of comparison, the results are considered to be similar good or bad.

- *Development of the four primary eddies.* The four primary eddies can be seen clearly at time unit 30. In the rational LES model with auxiliary problem, they are present already at time unit 20. The rapid increase of the relative vorticity thickness from 1 to 2 can be seen for all computations, Figure 11.14. The pairing in the rational LES model with auxiliary problem starts a bit too early and in the other two LES models, it starts somewhat too late. This phase of the flow is computed much better than with the scaling factor $c_S = 0.01$.
- *Pairing of the four primary eddies.* The pairing of the four primary eddies is computed completely wrong by the rational LES model with auxiliary problem, see Figure 11.12. The four primary eddies are disappeared in time units 70 and 80 and a rather strange flow pattern can be seen. The results obtained with the Smagorinsky model and the rational LES model with convolution are much better. In these models, the pairing of the four primary eddies starts, somewhat too late, at time unit 85. The peaks in the relative vorticity thickness reach values of more than 5, which is somewhat too high. For both LES models, the two secondary eddies in time unit 100 look rather the same. Thus, a non-simultaneous pairing did not happen.
- *Pairing of the two secondary eddies.* The two secondary eddies are present in all computations at time unit 100. Then, there is some delay before their pairing starts. In the rational LES model with convolution, it starts at time unit 130, in the Smagorinsky model a little bit later and in the rational LES model with auxiliary problem at around time unit 150. Since there was no delay between both pairings, this phase of the flow was predicted better for the scaling factor $c_S = 0.01$ and the rational LES model.
- *Rotation of the final eddy.* The rotation of the final eddy happens in all computations with a speed which is too small and a relative vorticity thickness which is too large. But, compared to the results obtained with $c_S = 0.01$, the values and the frequency of the relative vorticity thickness are closer to the reference solution. The position of the final eddy is computed best with the Smagorinsky model. This phase of the flow is computed somewhat better using the scaling factor $c_S = 0.005$.

The solutions are much more noisy for $c_S = 0.005$ which results in a larger total kinetic energy and $L^2(\Omega)$ norm of the gradient of the velocity. These quantities are underestimated for $c_S = 0.01$ and their prediction is somewhat better for $c_S = 0.005$.

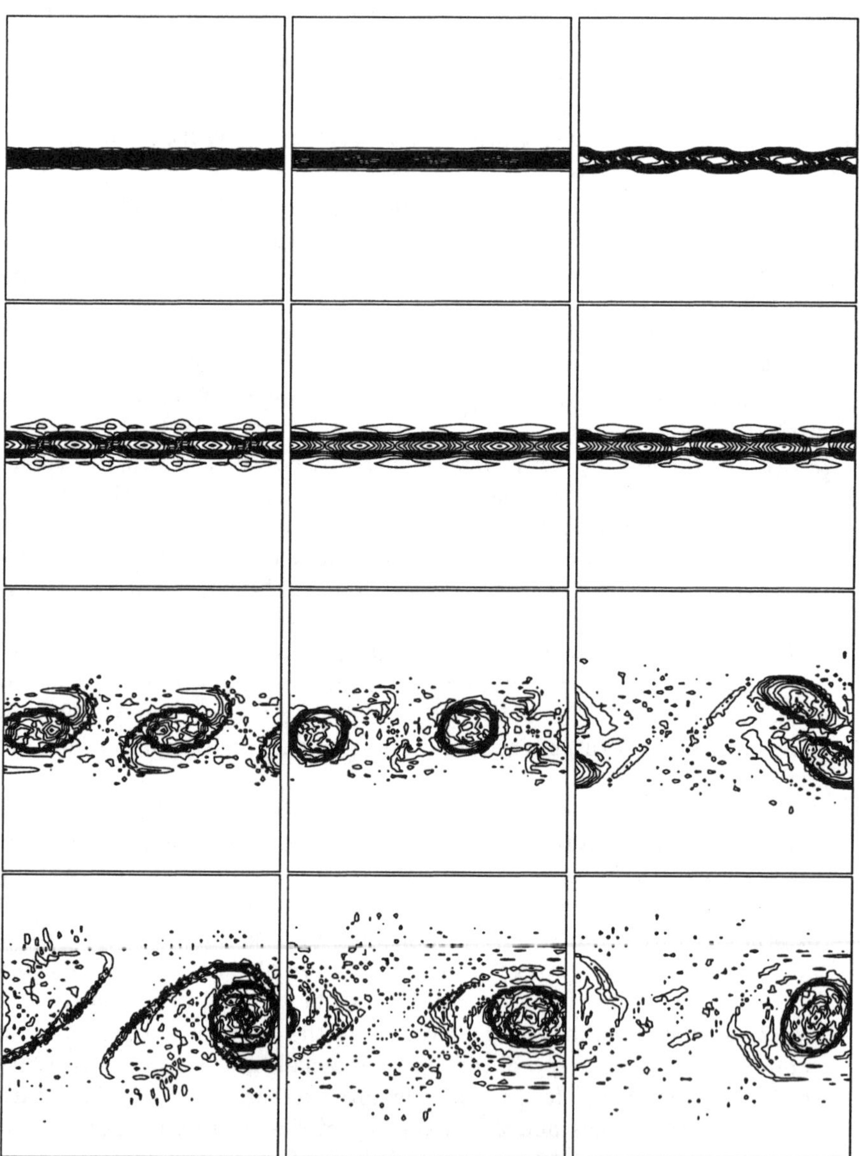

Fig. 11.11. Smagorinsky model (4.3), $c_S = 0.005$, vorticity on level 5, at time units 0, 20, 30, 50, 70, 80, 100, 120, 140, 160, 180, 200 (left to right, top to bottom)

Considering the whole time interval, the use of $c_S = 0.005$ instead of $c_S = 0.01$ improves the results obtained with the Smagorinsky model and the rational LES model with convolution. The rational LES model with auxiliary problem behaves completely wrong for $c_S = 0.005$ after the first phase of the flow.

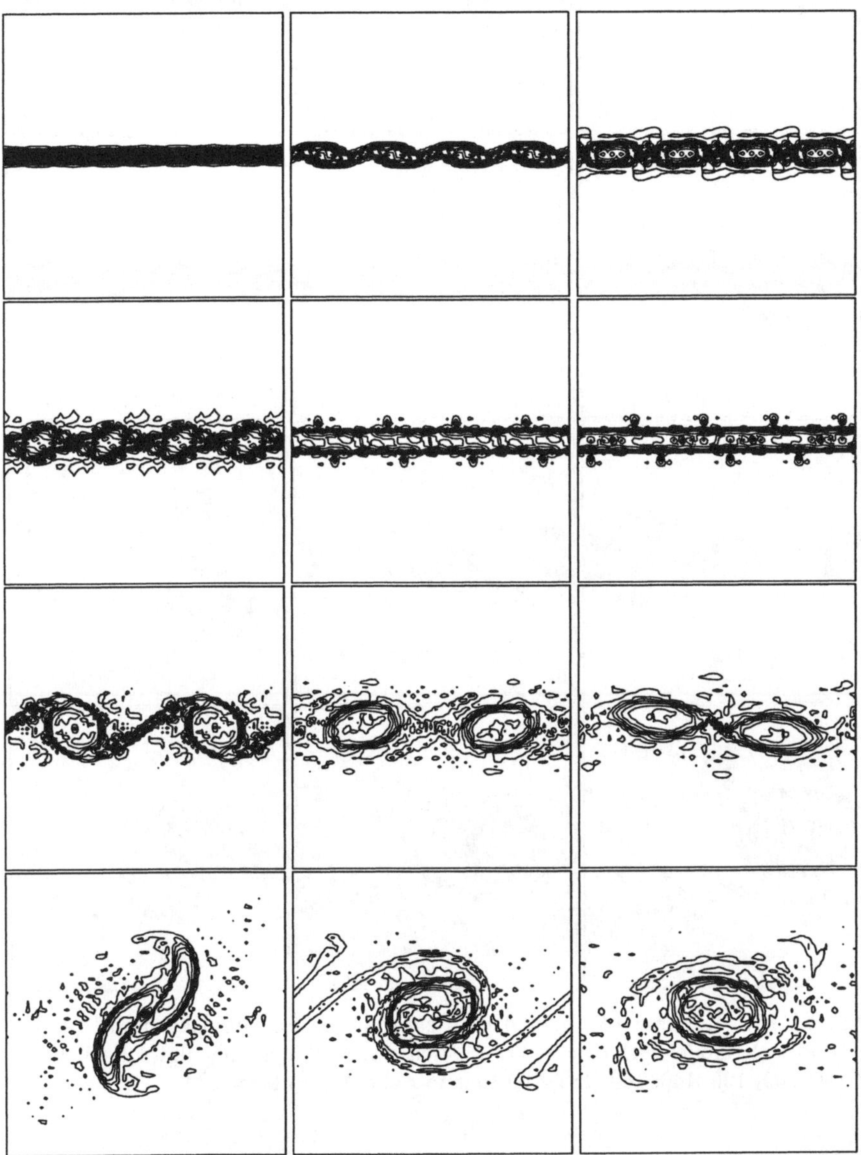

Fig. 11.12. Rational LES model with auxiliary problem, Smagorinsky model (4.3) as subgrid scale model with $c_S = 0.005$, vorticity on level 5, at time units 0, 20, 30, 50, 70, 80, 100, 120, 140, 160, 180, 200 (left to right, top to bottom)

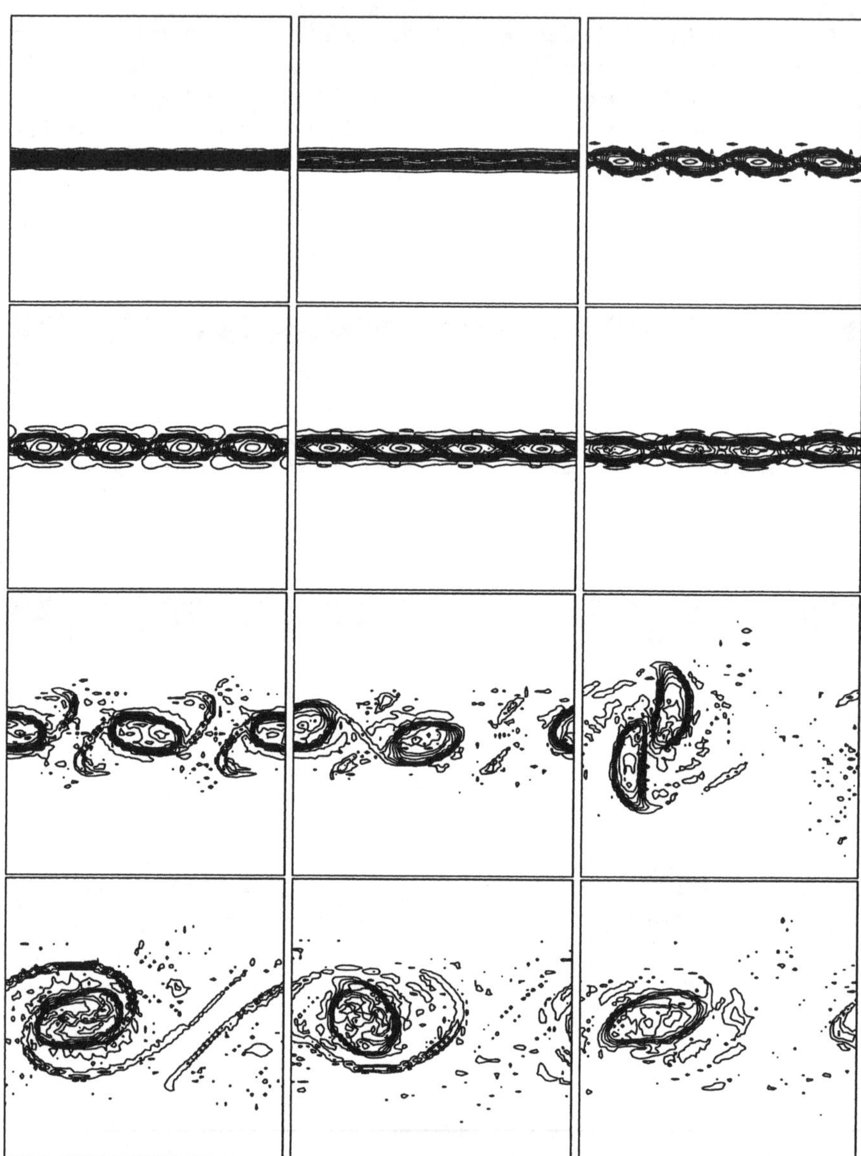

Fig. 11.13. Rational LES model with convolution, Smagorinsky model (4.3) as subgrid scale model with $c_S = 0.005$, vorticity on level 5, at time units 0, 20, 30, 50, 70, 80, 100, 120, 140, 160, 180, 200 (left to right, top to bottom)

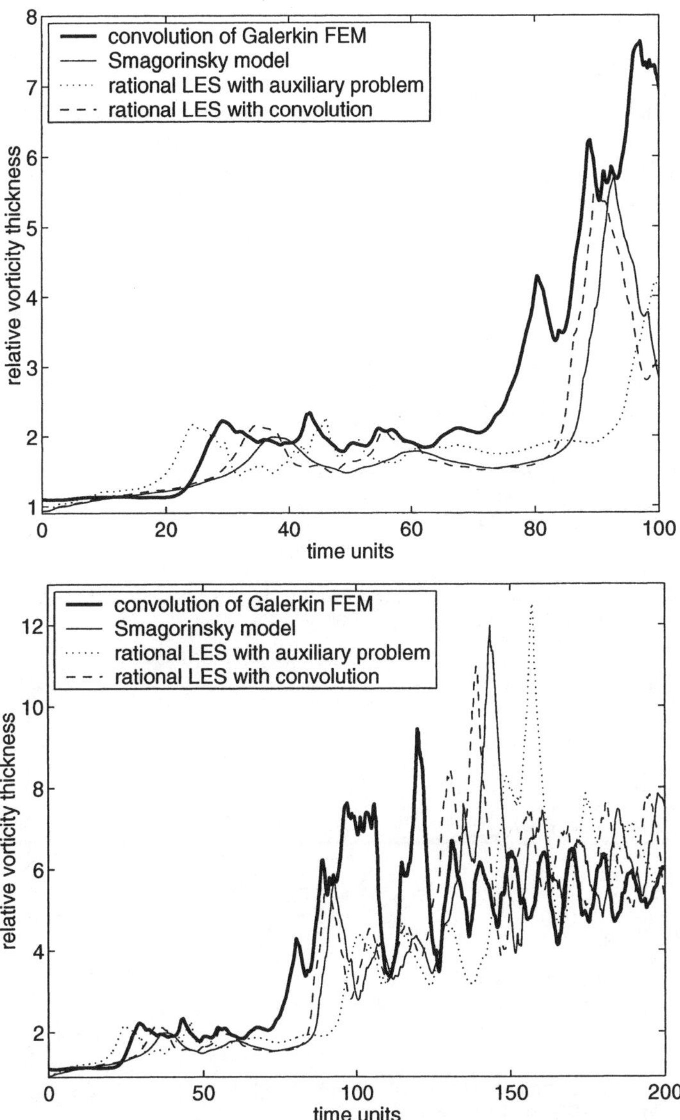

Fig. 11.14. Relative vorticity thickness to σ_0, Smagorinsky model (4.3) as subgrid scale model with $c_S = 0.005$, level 5

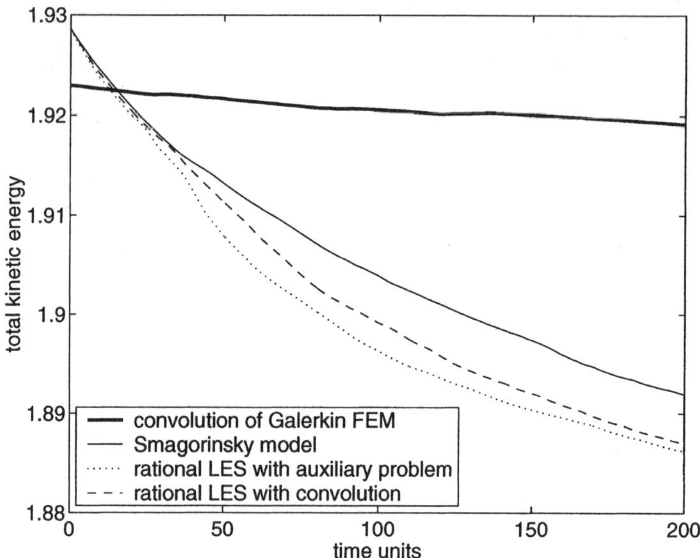

Fig. 11.15. Total kinetic energy, Smagorinsky model (4.3) as subgrid scale model with $c_S = 0.005$, level 5

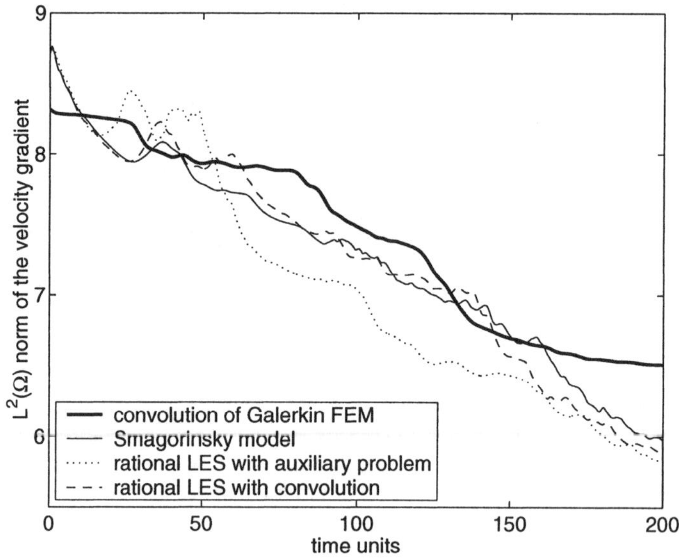

Fig. 11.16. $L^2(\Omega)$ norm of the gradient of the velocity, Smagorinsky model (4.3) as subgrid scale model with $c_S = 0.005$, level 5

11.1.3 The Rational LES Model With Iliescu-Layton Subgrid Scale Term

The rational LES model with Iliescu-Layton subgrid scale term (4.31) was applied with the scaling factors $c_S = 0.5$ and $c_S = 0.17$. The use of the scaling factor $c_S = 0.17$ is proposed by Layton and Lewandowski [LL02].

The Scaling Factor $c_S = 0.5$

The development of the vorticity using $c_S = 0.5$ is presented in Figures 11.17 and 11.18.

- *Development of the four primary eddies.* The primary eddies are clearly visible already at time unit 20. This is somewhat too early. The development starts at around time unit 15 and a rather steep increase of the relative vorticity thickness from 1 to 2 can be observed, Figure 11.19. A similar steep increase can be seen also for the reference solution.
- *Pairing of the four primary eddies.* This pairing happens somewhat too early, at time unit 65, for the rational LES model with auxiliary problem. For the rational LES model with convolution, the time of this pairing is nearly the same as in the reference solution. The sharp peaks in the relative vorticity thickness can be seen very well. A non-simultaneous pairing of the four primary eddies cannot be observed.
- *Pairing of the two secondary eddies.* This pairing occurs, in contrast to the reference solution, not directly after the pairing of the four primary eddies. There is some delay and the pairing of the secondary eddies takes place too late, at around time unit 140 for the rational LES model with auxiliary problem and at around time unit 150 for the rational LES model with convolution.
- *Rotation of the final eddy.* The speed of the rotation of the final eddy is too small which is reflected in a too small frequency of the oscillation of the relative vorticity thickness. The values of the relative vorticity thickness are too large, which means that the final eddy is larger than in the reference solution. The position of the final eddy is computed somewhat better using the rational LES model with convolution.

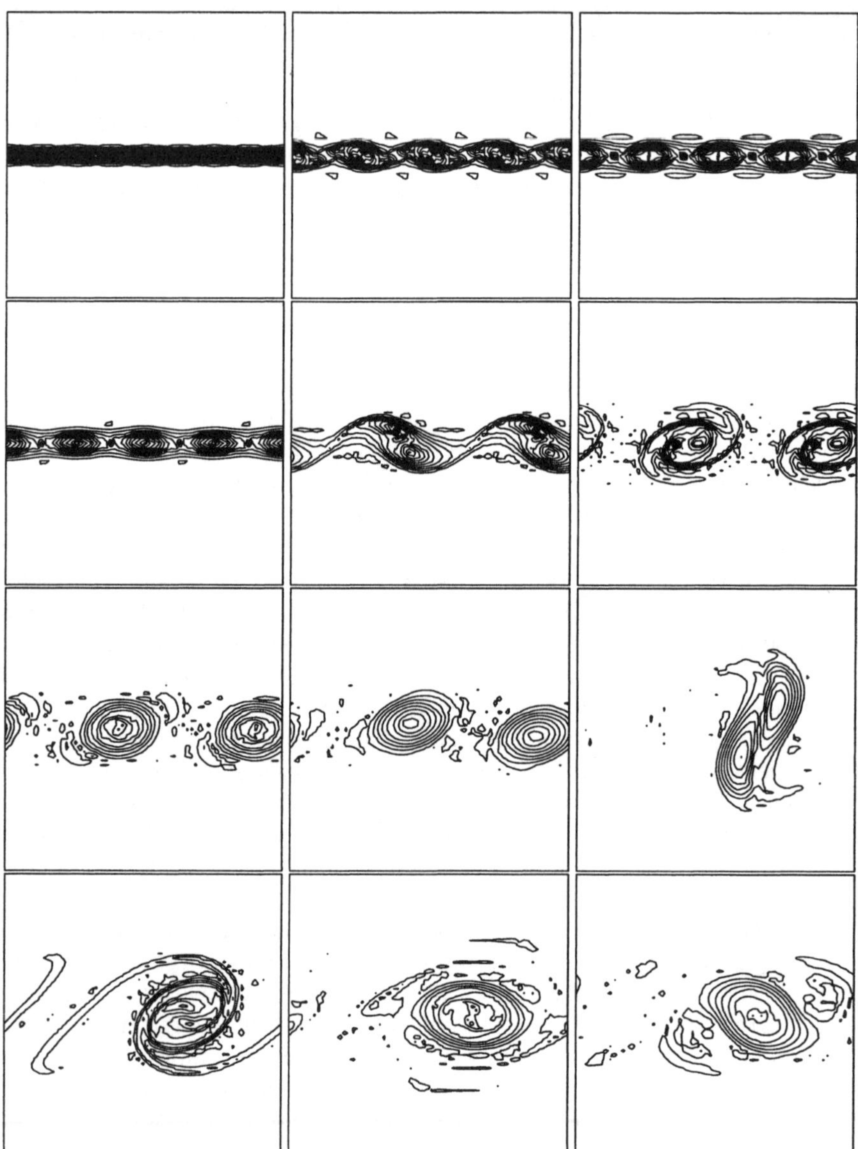

Fig. 11.17. Rational LES model with auxiliary problem, Iliescu-Layton model (4.31) as subgrid scale model with $c_S = 0.5$, vorticity on level 5, at time units 0, 20, 30, 50, 70, 80, 100, 120, 140, 160, 180, 200 (left to right, top to bottom)

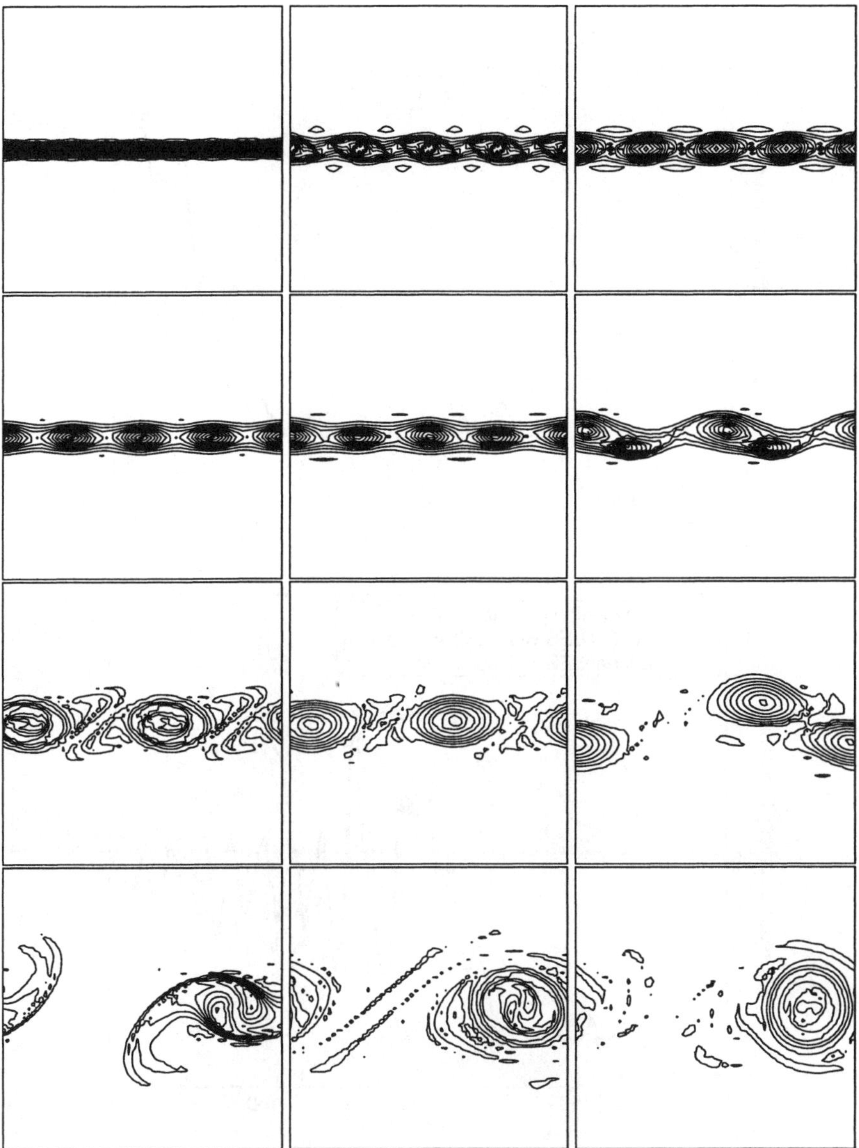

Fig. 11.18. Rational LES model with convolution, Iliescu-Layton model (4.31) as subgrid scale model with $c_S = 0.5$, vorticity on level 5, at time units 0, 20, 30, 50, 70, 80, 100, 120, 140, 160, 180, 200 (left to right, top to bottom)

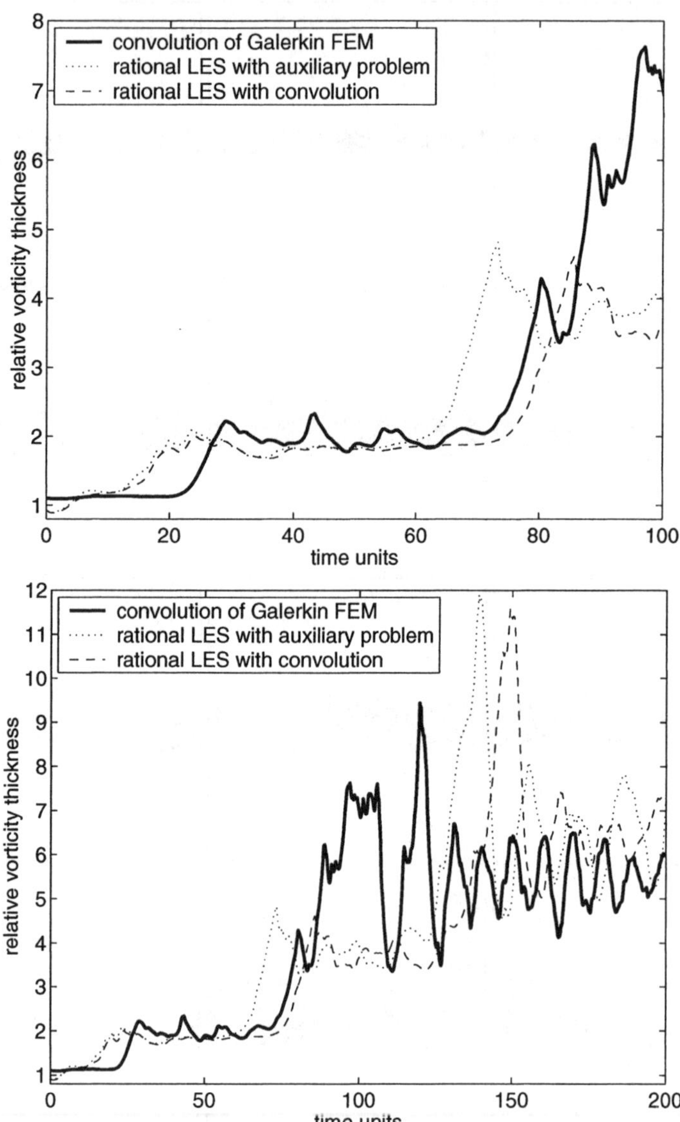

Fig. 11.19. Relative vorticity thickness to σ_0, Iliescu-Layton model (4.31) as subgrid scale model with $c_S = 0.5$, level 5

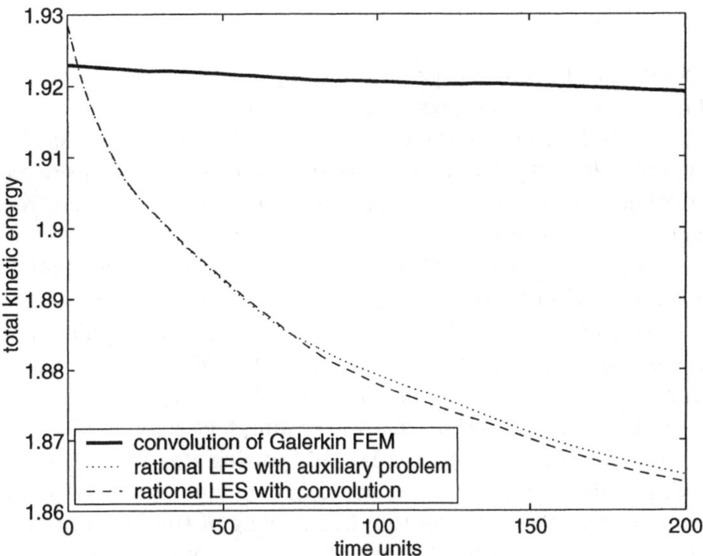

Fig. 11.20. Total kinetic energy, Iliescu-Layton model (4.31) as subgrid scale model with $c_S = 0.5$, level 5

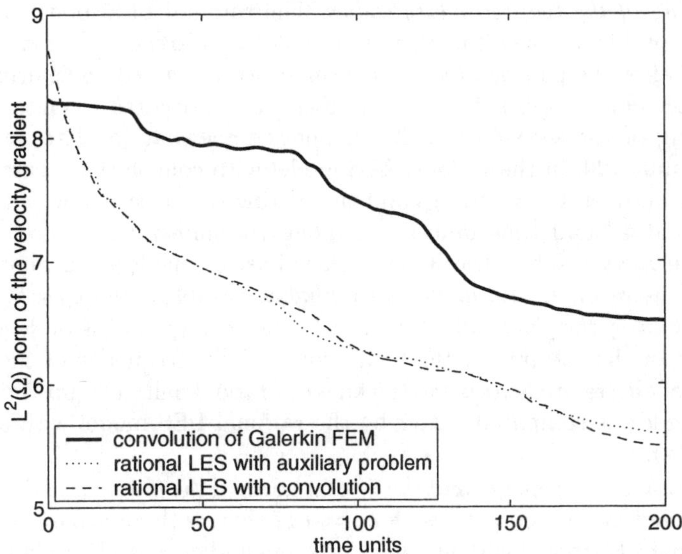

Fig. 11.21. $L^2(\Omega)$ norm of the gradient of the velocity, Iliescu-Layton model (4.31) as subgrid scale model with $c_S = 0.5$, level 5

The Scaling Factor $c_S = 0.17$

The development of the vorticity for the computations with the scaling factor $c_S = 0.17$ is presented in Figures 11.22 and 11.23.

The computed results obtained with $c_S = 0.5$ and $c_S = 0.17$ are also compared here. It will be mentioned only if one result is considered better than the other one. If there is no statement of comparison, the results are considered to have a similar quality.

The computational results obtained with $c_S = 0.17$ are as follows:

- *Development of the four primary eddies.* The four primary eddies can be clearly seen at time unit 30. This development starts earlier than in the reference solution and the relative vorticity thickness increases moderately, see Figure 11.24. The increase of the relative vorticity thickness is computed a little bit better with the scaling factor $c_S = 0.5$.
- *Pairing of the four primary eddies.* This pairing starts somewhat too late, after time unit 80. But for the rational LES model with auxiliary problem, one can observe the non-simultaneous pairing of the four primary eddies, see especially time unit 90 in Figure 11.27. This pairing is computed much better than with all the other considered LES models.
  In the rational LES model with convolution, the non-simultaneous pairing cannot be observed.
- *Pairing of the two secondary eddies.* The rational LES model with auxiliary problem behaves further similar to the reference solution. Immediately after the pairing of the four primary eddies starts the pairing of the two secondary eddies. Since the former pairing occured too late, also the pairing of the secondary eddies happens somewhat too late, at around time unit 120. In the rational LES model with convolution, there is some delay between both pairings and the pairing of the secondary eddy takes place at around time unit 140. Among the numerical studies with the Iliescu-Layton subgrid scale term, this phase of the flow is computed best with the rational LES model with auxiliary problem and $c_S = 0.17$.
- *Rotation of the final eddy.* Like in all other computations on level 5, the speed of the rotation of the final eddy and the frequency of the oscillation of the relative vorticity thickness are too small. The position of the final eddy is computed better by the rational LES model with auxiliary problem.

The total kinetic energy and the $L^2(\Omega)$ norm of the gradient of the velocity are predicted more accurately with $c_S = 0.17$ than with $c_S = 0.5$. Altogether, the rational LES model with auxiliary problem and $c_S = 0.17$ was in nearly all considered aspects superior to the simulations with the rational LES model and the Iliescu-Layton subgrid scale model with $c_S = 0.5$. Especially, the computation of the non-simultaneous pairing of the four primary eddies is remarkable.

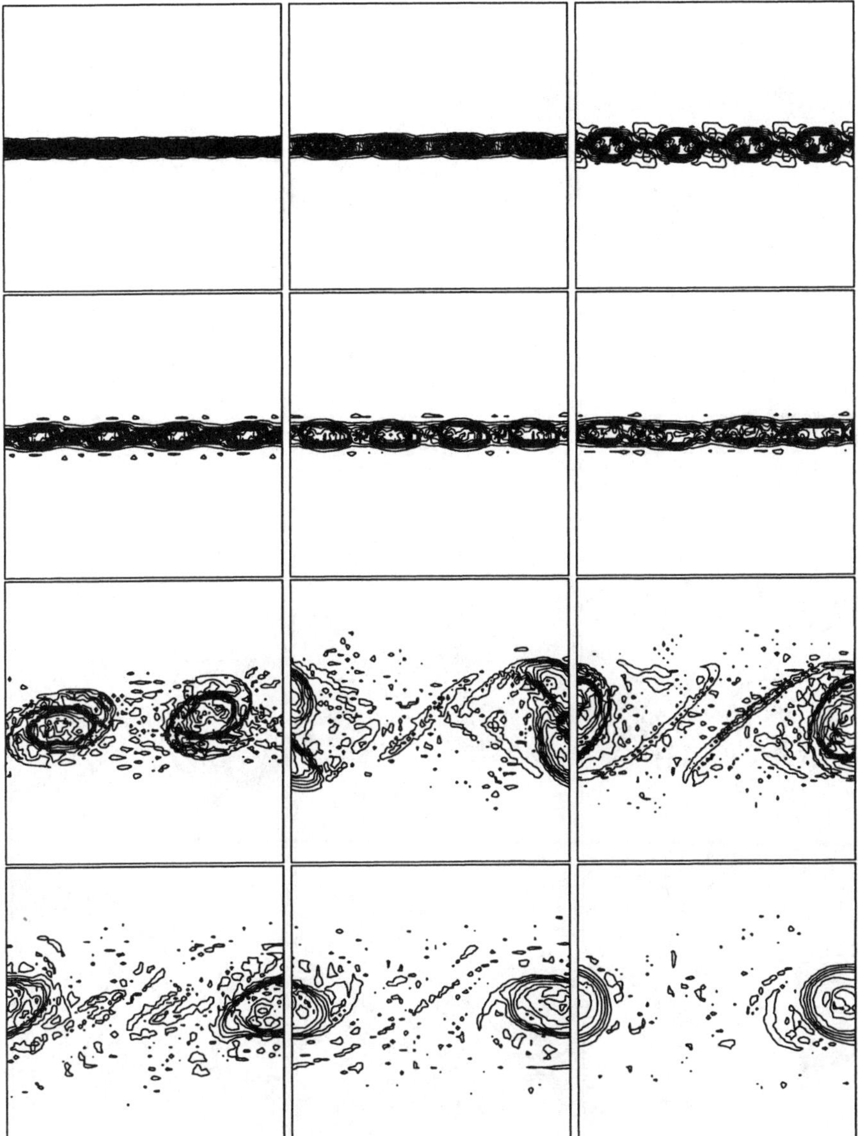

Fig. 11.22. Rational LES model with auxiliary problem, Iliescu-Layton model (4.31) as subgrid scale model with $c_S = 0.17$, vorticity on level 5, at time units 0, 20, 30, 50, 70, 80, 100, 120, 140, 160, 180, 200 (left to right, top to bottom)

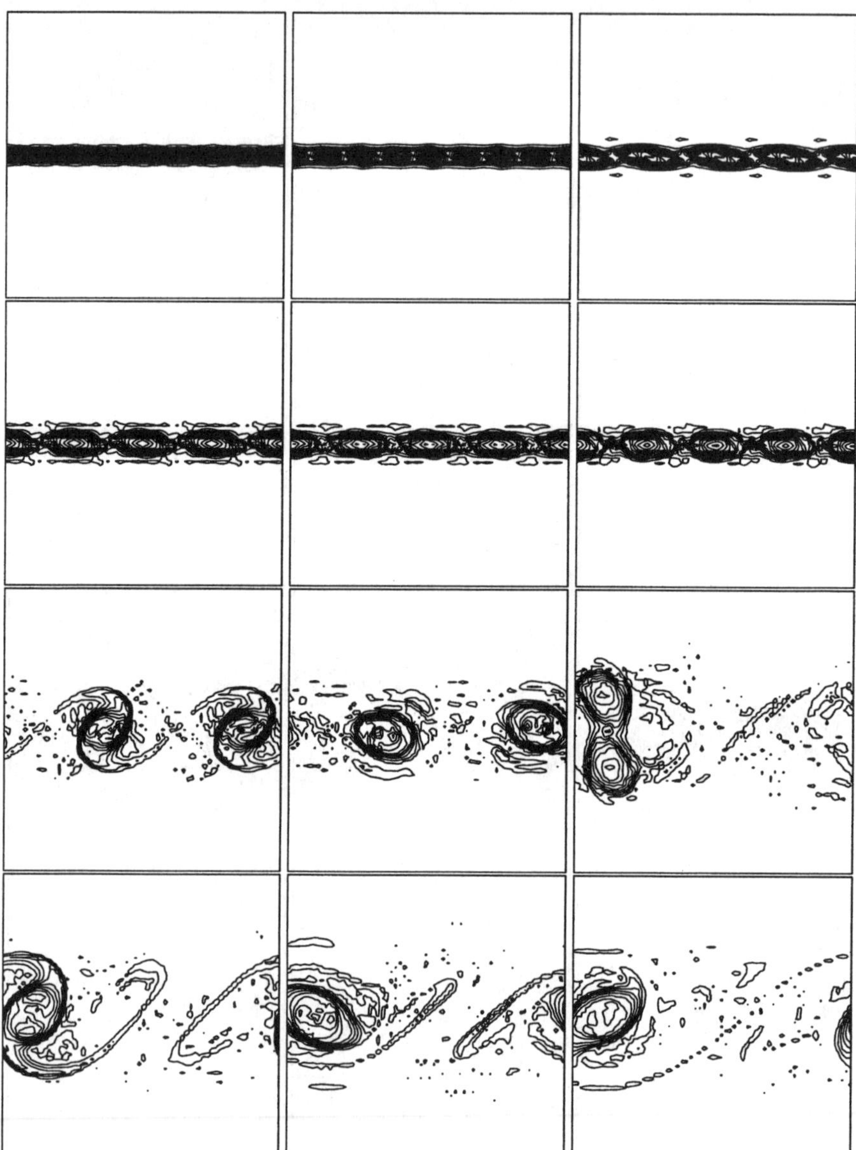

Fig. 11.23. Rational LES model with convolution, Iliescu-Layton model (4.31) as subgrid scale model with $c_S = 0.17$, vorticity on level 5, at time units 0, 20, 30, 50, 70, 80, 100, 120, 140, 160, 180, 200 (left to right, top to bottom)

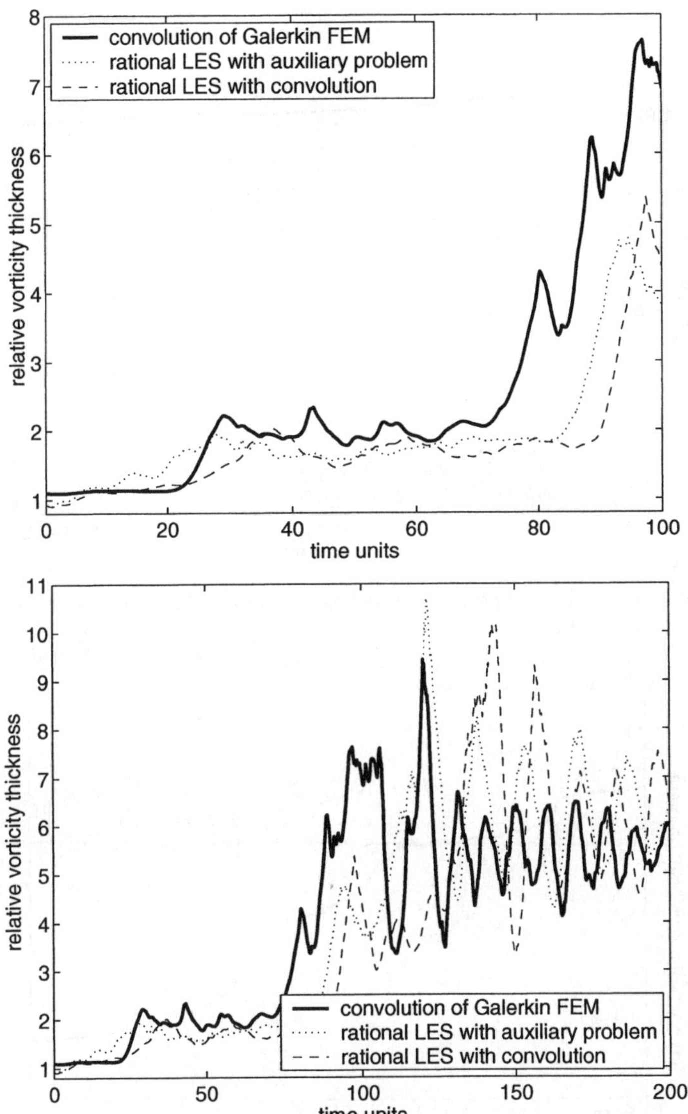

Fig. 11.24. Relative vorticity thickness to σ_0, Iliescu-Layton model (4.31) as subgrid scale model with $c_S = 0.17$, level 5

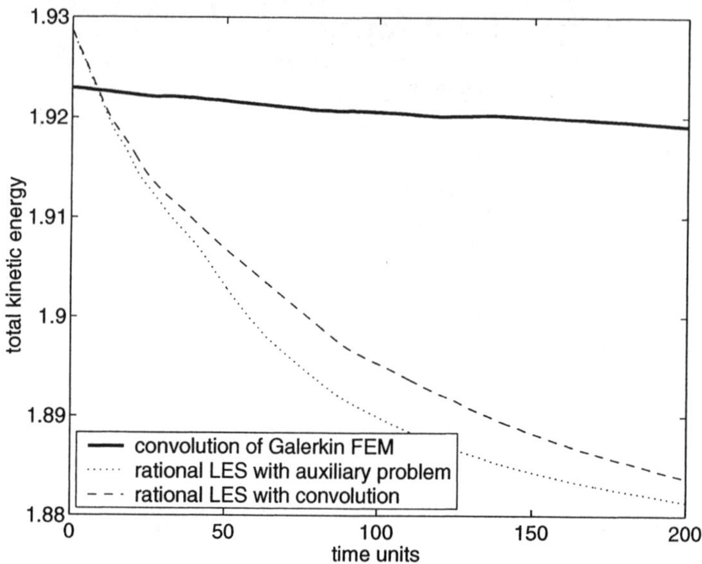

Fig. 11.25. Total kinetic energy, Iliescu-Layton model (4.31) as subgrid scale model with $c_S = 0.17$, level 5

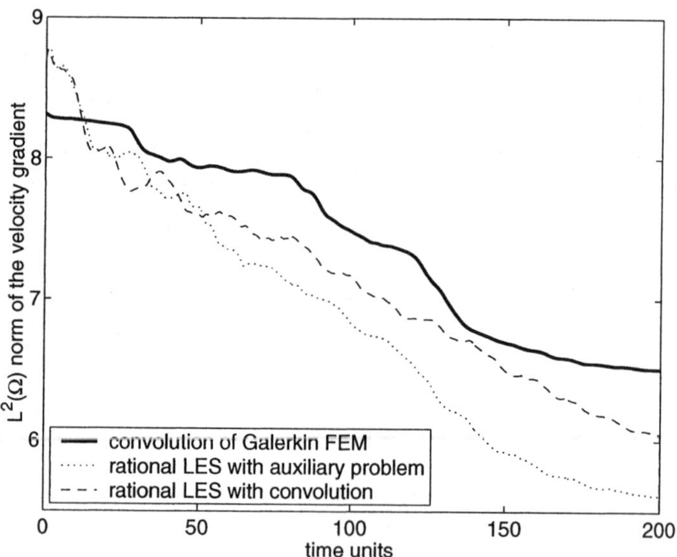

Fig. 11.26. $L^2(\Omega)$ norm of the gradient of the velocity, Iliescu-Layton model (4.31) as subgrid scale model with $c_S = 0.17$, level 5

Fig. 11.27. Rational LES model with auxiliary problem and Iliescu-Layton model (4.31) as subgrid scale model with $c_S = 0.17$, time unit 90

11.1.4 The Rational LES Model Without Model for the Subgrid Scale Term

The rational LES model without model for the subgrid scale term, $\nu_T = 0$, blows up already in forming the four primary eddies, between time unit 12 and 13, see Figure 11.28.

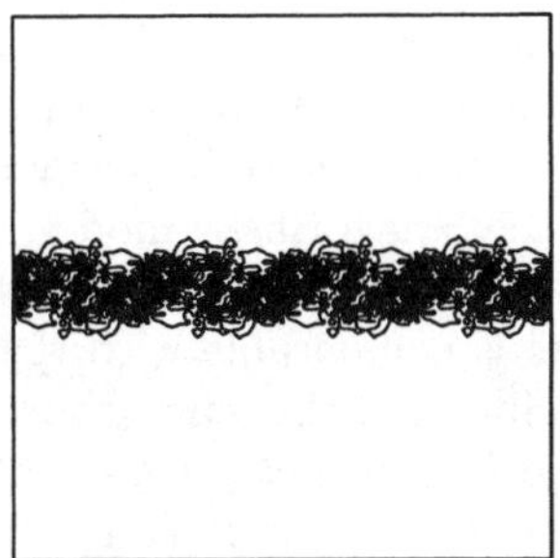

Fig. 11.28. Blow up of rational LES model with auxiliary problem without subgrid scale model, vorticity on level 5, time unit 10

11.1.5 A Comparison of the Smagorinsky Subgrid Scale Term and the Iliescu-Layton Subgrid Scale Term

In this section, both approaches of the subgrid scale term are compared, based on the best solution in each case. The evaluation of the numerical studies with the Smagorinsky subgrid scale model came to the conclusion that the result obtained with the Smagorinsky model and $c_S = 0.005$ belongs to the best results in this class of models. Using the Iliescu-Layton subgrid scale model, the rational LES model with auxiliary problem and $c_S = 0.17$ performed best.

- The results obtained with the Iliescu-Layton subgrid scale term are better in the following respects:
  - The computation of the four primary eddies is somewhat better. They are not as flat as computed with the Smagorinsky model.
  - The pairing of the four primary eddies is computed much better. Especially that a non-simultaneous pairing is observed, like in the reference solution, is very remarkable.
  - There is no delay between the pairings of the four primary eddies and the pairing of the two secondary eddies. Thus, the time of the latter pairing is closer to the reference solution.
- The results obtained with the Smagorinsky subgrid scale term are better in the following respects:
  - The total kinetic energy and the $L^2(\Omega)$ norm of the gradient of the velocity are predicted somewhat more accurately.
  - The position of the final eddy is computed a little bit better.
- The following features of the reference solution are reproduced insufficiently by the solutions obtained with both variants of the subgrid scale term:
  - In general, the pairings of the eddies occur at somewhat different times as in the reference solution.
  - The speed of rotation of the final eddy is too small.

The characteristics of the space averaged flow field which are simulated better by the rational LES model with Iliescu-Layton subgrid scale term and $c_S = 0.17$ are in our opinion much more important than the features which are computed better with the Smagorinsky model.

In addition, the quality of the other results obtained with the Iliescu-Layton subgrid scale model is comparable with the Smagorinsky model and $c_S = 0.005$. All these results have the same shortcomings, in particular, a simultaneous pairing of the four primary eddies and a delay between both pairings. In contrast, many of the results computed with the Smagorinsky subgrid scale model are worse than all results obtained with the Iliescu-Layton subgrid scale model, above all in the initial phase of the flow. Altogether, considering all results, the prediction of the flow was better with the Iliescu-Layton subgrid scale model.

Remark 11.1. Comparison of the rational LES model with auxiliary problem and the rational LES model with convolution. The numerical tests at the 2d mixing layer problem reveal that there are sometimes large differences between the solutions computed with the rational LES model with auxiliary problem and the rational LES model with convolution (variant 1), especially for small scaling factors c_S in the subgrid scale model. Both cases happened, namely that the rational LES model with auxiliary problem computed a better solution and vice versa. From the point of view of efficiency, only the rational LES model with auxiliary problem can be recommended. That's why, the rational

LES model with convolution is not included in the numerical study of the 3d mixing layer problem in Section 11.2. □

11.2 A Mixing Layer Problem in Three Dimensions

The mixing layer problem is also an often used test problem for turbulent flow simulations in three dimensions. Computations with this problem can be found, e.g., in Comte et al. [CLL92], Rogers and Moser [RM94] and Vreman et al. [VGK97].

11.2.1 The Definition of the Problem and the Setup of the Numerical Tests

The Definition of the Problem

We define this problem similarly to the 2d mixing layer problem. The domain of computation is $\Omega = (-1,1)\times(-2,2)\times(0,2)$. It was important to extend the computational domain in y-direction compared to the 2d mixing layer problem in order that the eddies can develop undisturbed by the y-boundary. Free slip boundary conditions are applied at $y = -2$ and $y = 2$. On the other four boundaries, periodic boundary conditions are prescribed. The initial velocity is given by

$$\mathbf{w}_0 = \begin{pmatrix} W_\infty \tanh\left(\dfrac{2y}{\sigma_0}\right) \\ 0 \\ 0 \end{pmatrix} + c_{\text{noise}} W_\infty \begin{pmatrix} \dfrac{\partial\psi}{\partial y} + \dfrac{\partial\psi}{\partial z} \\ -\dfrac{\partial\psi}{\partial x} \\ -\dfrac{\partial\psi}{\partial x} \end{pmatrix} \tag{11.5}$$

with

$$\psi = \exp\left(-(2y/\sigma_0)^2\right)\left(\cos(\pi x) + \cos(2\pi z)\right).$$

The parameters are chosen in the computations as follows:

- $W_\infty = 1$,
- scaling factor $c_{\text{noise}} = 0.01$,
- initial vorticity thickness $\sigma_0 = 1/14$.

The Reynolds number is defined by

$$Re = \frac{\sigma_0 W_\infty}{\nu}. \tag{11.6}$$

We will present computations for $\nu = 10^{-4}$, i.e. for $Re = 10000/14 \approx 714$. For this Reynolds number, the solution of the Navier-Stokes equations with Galerkin finite element discretisation was not possible on refinement level 4 of the initial grid, see Table 11.2 for information to the grids. Even for $Re = 250$, the solution blew up. In the literature, direct numerical simulations with $Re = 100$ can be found, see Vreman et al. [VGK97] or Comte et al. [CLL92]. In these papers, the DNS solution is filtered and then used as reference solution for the LES models in the same fashion as it was done in the 2d mixing layer problem, Section 11.1. Since $Re = 100$ leads to a rather laminar flow, we decided to use a somewhat larger Reynolds number such that, on the one hand, the flow becomes more turbulent but, on the other hand, clear structures of the flow are visible at the initial phase. Since we do not posses a reference solution, the evaluation of the computational results becomes more difficult.

The Momentum Thickness

In the 3d mixing layer problem, it is common to use the momentum thickness instead of the vorticity thickness for evaluating the numerical simulations [RM94, VGK97]. Let $\Omega = (x_0, x_1) \times (-y_0, y_0) \times (z_0, z_1)$ with $y_0 > 0$. The momentum thickness $\mu(t)$ is given by

$$\mu(t) = \int_{-y_0}^{y_0} \left(\frac{1}{4} - \left(\frac{\langle w_1 \rangle (t, y)}{2W_\infty} \right)^2 \right) dy,$$

where w_1 is the first component of the velocity and

$$\langle w_1 \rangle (t, y) = \frac{\int_{x_0}^{x_1} \int_{z_0}^{z_1} w_1(t, x, y, z) dz dx}{\int_{x_0}^{x_1} \int_{z_0}^{z_1} dz dx}.$$

For the undisturbed initial velocity, $c_{\text{noise}} = 0$, a straighforward computation gives

$$\mu(0) = \frac{1}{4} \int_{-y_0}^{y_0} 1 - \tanh^2 \left(\frac{2y}{\sigma_0} \right) dy = \frac{\sigma_0}{8} \tanh \left(\frac{2y}{\sigma_0} \right) \Big|_{y=-y_0}^{y=y_0}.$$

The ratio $2y_0/\sigma_0$ is in general sufficiently large such that

$$\tanh \left(\frac{2y_0}{\sigma_0} \right) \approx 1, \quad \tanh \left(-\frac{2y_0}{\sigma_0} \right) \approx -1,$$

from what follows

$$\mu(0) \approx \frac{\sigma_0}{4}.$$

The momentum thickness is an integral quantity while the vorticity thickness is obtained by differentiation. Thus, the momentum thickness will be smoother and less sensitive to noise in the flow.

The Discretisation in Time and Space

The numerical studies presented in this section were performed with the fractional-step θ-scheme as discretisation in time and the Q_2/P_1^{disc} finite element spatial discretisation. A time unit $\bar{t}$ is defined by $\bar{t} = \sigma_0/W_\infty$. The time discretisation was applied with an equal distant time step of $\Delta t_n = 0.5\bar{t} = 1/28 \approx 3.5714e-2$ and the final time was set to $T = 80\bar{t} \approx 5.714285$. The initial grid (level 0) consists of 16 cubes of edge length one. This grid is refined uniformly. The computations are carried out on level 4. The number of the degrees of freedom and the mesh width are given in Table 11.2.

The non-linear discrete systems in each time step were solved up to an Euclidean norm of the residual vector less than 10^{-6}.

Table 11.2. Degrees of freedom and mesh width on different levels

level	h	velocity	pressure	total
0	$\sqrt{3}$	432	64	494
4	$\sqrt{3}/16$	1 585 152	262 144	1 847 296

The Parameters of the LES Models

We will present computations with the Smagorinsky model and the rational LES model with auxiliary problem. In the rational LES model, the Smagorinsky subgrid scale term as well as the Iliescu-Layton model (4.31) was used. The scaling factor in the Smagorinsky model was chosen to be $c_S = 0.01$ and in the Iliescu-Layton model $c_S = 0.17$ as proposed by Layton and Lewandowski [LL02]. The filter width $\delta = h$ was used in all tests.

The auxiliary problems which have to be solved in the rational LES model with auxiliary problem and in the Iliescu-Layton subgrid scale model are equipped with periodic boundary conditions on all boundaries on which the mixing layer problem possesses such boundary conditions. On the other two boundaries, the usual homogeneous Neumann boundary conditions are prescribed.

11.2.2 Evaluation of the Numerical Results

The evaluation of the computations with the LES models is based on:

- the z-component of the vorticity in the plane $z = 1$, Figures 11.29 - 11.31, where the part $(-1,1) \times (-1,1)$ of the cut plane is presented. The isolines are drawn at the values $\pm 2k+1, k = 0, 1, \ldots$.
- the relative momentum thickness, Figure 11.32. The computed momentum thickness of the disturbed initial velocity on level 4 was $\mu(0) = 1.503835e-2$.

- the total kinetic energy, Figure 11.33.
- the $L^2(\Omega)$ norm of the gradient of the velocity which is an indicator of the amount of noise in the computed solution, Figure 11.34.

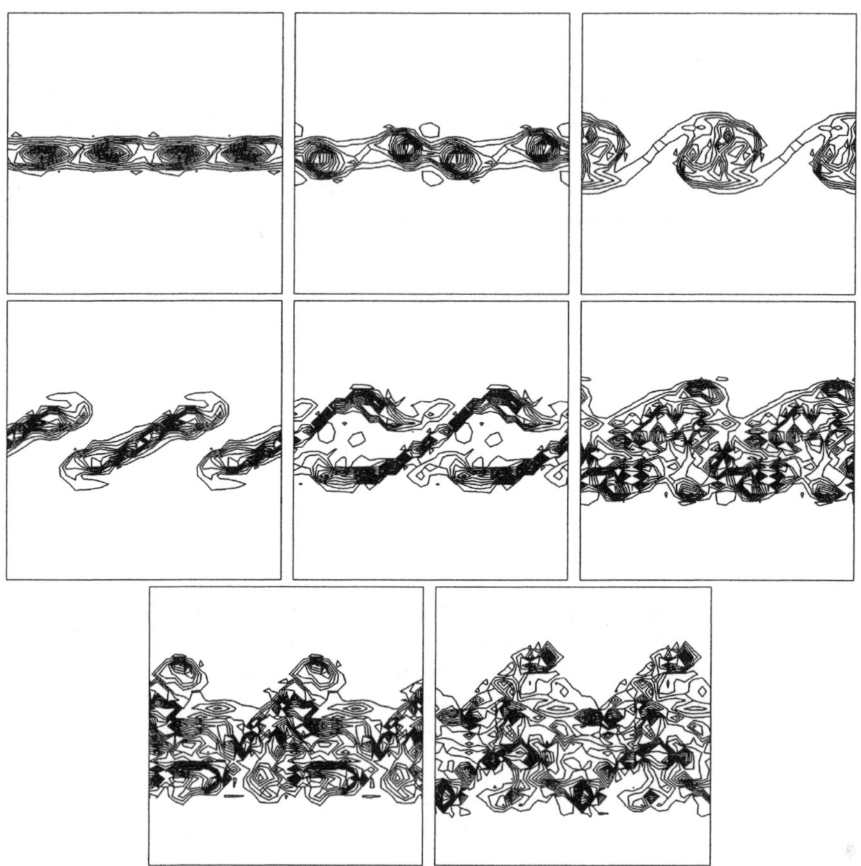

Fig. 11.29. Smagorinsky model (4.3), $c_S = 0.01$, vorticity on level 4 in the plane $z = 1$, at time units 10, 20, 30, 40, 50, 60, 70, 80 (left to right, top to bottom)

At time unit 10, four primary eddies can be seen in $z = 1$ for each computation using a LES model. The relative momentum thickness has approximately doubled in each computation. The evolution of the relative momentum thickness is similar in the rational LES model with both types of the subgrid scale model.

Then, the rational LES model with both types of the subgrid scale term predicts a fast pairing of the primary eddies which is reflected by a rather steep increase in the relative momentum thickness at time unit 20 to a value of 4. At this time unit, the pairing is completed. The Smagorinsky model

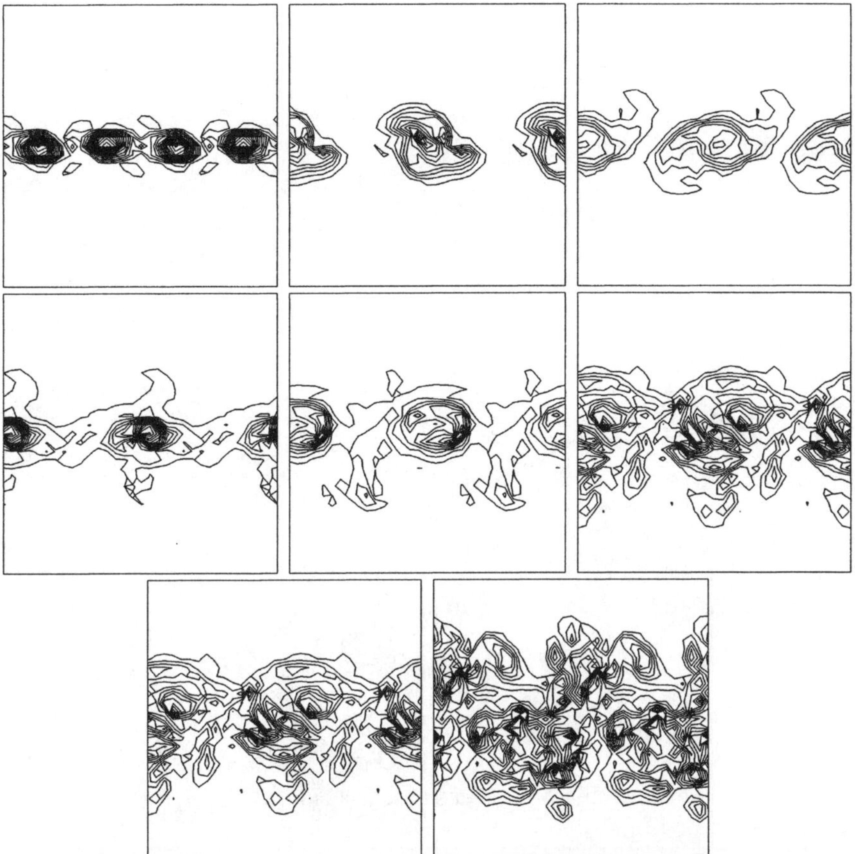

Fig. 11.30. Rational LES model with auxiliary problem, Smagorinsky model (4.3) as subgrid scale model with $c_S = 0.01$, vorticity on level 4 in the plane $z = 1$, at time units 10, 20, 30, 40, 50, 60, 70, 80 (left to right, top to bottom)

shows a slower development of this pairing and a more gentle increase of the relative momentum thickness. This behaviour is similar to the gentle increase of the relative vorticity thickness of the Smagorinsky model with $c_S = 0.01$ in the 2d mixing layer problem, Figure 11.8. In analogy to the 2d mixing layer problem, we expect that the fast pairing with the steep increase of the relative momentum thickness reflects the behaviour of the large eddies better than the slower pairing with the moderate increase of the relative momentum thickness. Thus, in our opinion, the results obtained with the rational LES model are better for the first phase of the flow (up to time unit 25) than the result computed with the Smagorinsky model. At time unit 30, two large eddies can be seen in all computations and the values of the relative momentum thickness are around 4.

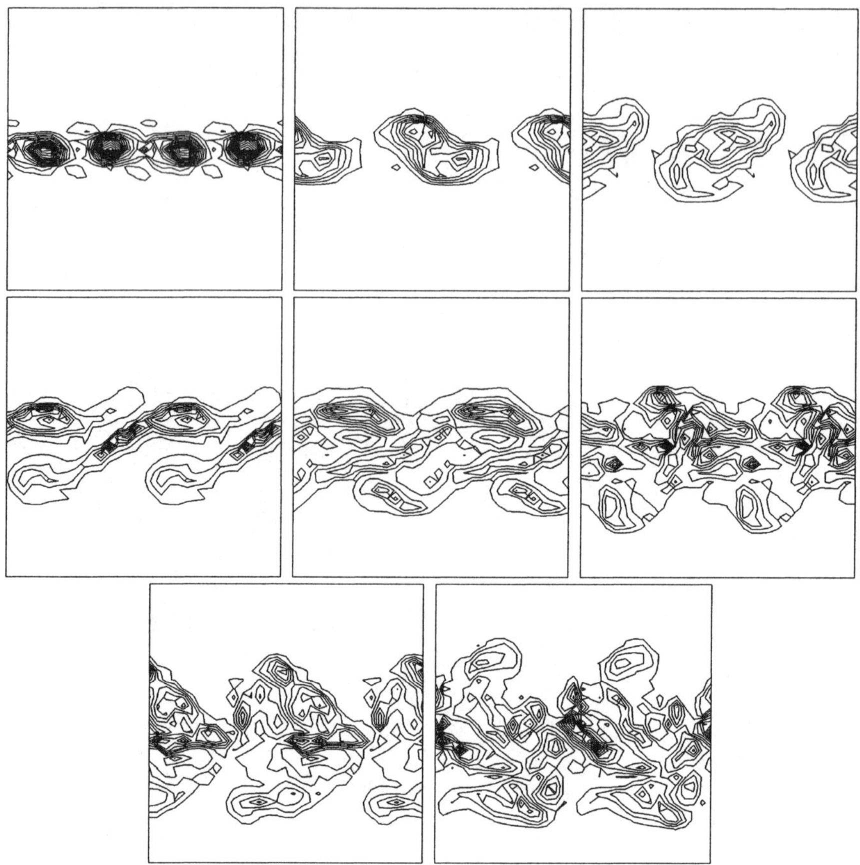

Fig. 11.31. Rational LES model with auxiliary problem, Iliescu-Layton model (4.31) as subgrid scale model with $c_S = 0.17$, vorticity on level 4 in the plane $z = 1$, at time units 10, 20, 30, 40, 50, 60, 70, 80 (left to right, top to bottom)

After time unit 30, the rational LES model with Smagorinsky subgrid scale term starts to behave different from the rational LES model with Iliescu-Layton subgrid scale term. In the rational LES model with Smagorinsky subgrid scale term, two large eddies are formed, see time units 40 and 50. From these eddies, an unstructured flow evolves. In the rational LES model with Iliescu-Layton subgrid scale term, the primary eddies do not really pair. The two large eddies which are present at time unit 30 break up again. In the plane $z = 1$, two larger eddies can be seen above the centre line $y = 0$ and two smaller eddies below the centre line, time units 40 and 50. Then, the flow becomes more and more unstructured. In particular, the two larger eddies above the centre line break up in a number of small eddies.

Also in the Smagorinsky model, the two large eddies at time unit 30 break up again. But the breaking is much more symmetric than in the rational LES model with Iliescu-Layton subgrid scale term. The computed flow with the Smagorinsky model is unstructured at time unit 60.

Despite that the computed flows in time unit 80 consist of many small eddies, one can observe that all solution are still periodic in x-direction with period 1. That means, two periods of the solution are visible in Ω. A certain similarity of the solutions computed with the rational LES model and both types of subgrid scale model can be seen in time unit 70.

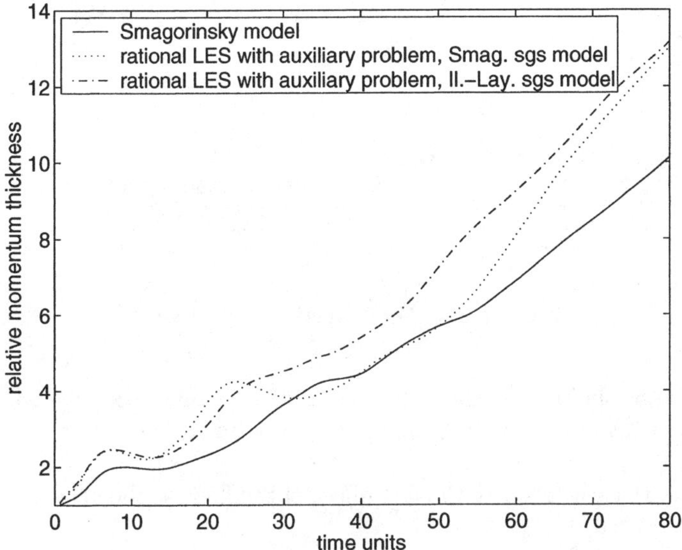

Fig. 11.32. Relative momentum thickness, level 4

Considering the relative momemtum thickness, the following characteristics can be observed:

- It is always larger for the rational LES model with Iliescu-Layton subgrid scale term than for the Smagorinsky model.
- In the initial phase of the flow, up to time unit 25, it is very similar for the rational LES model with both types of the subgrid scale term.
- From time unit 30 - 50, it is almost the same for the Smagorinsky model and the rational LES model with Smagorinsky subgrid scale term.
- At time unit 80, it is again similar for the rational LES model with both types of the subgrid scale term.

The total kinetic energy of the solutions behaves as follows:

- It is always larger for the solution of the Smagorinsky model than for the solution of the rational LES model with Iliescu-Layton subgrid scale term. This is like in the 2d mixing layer problem.

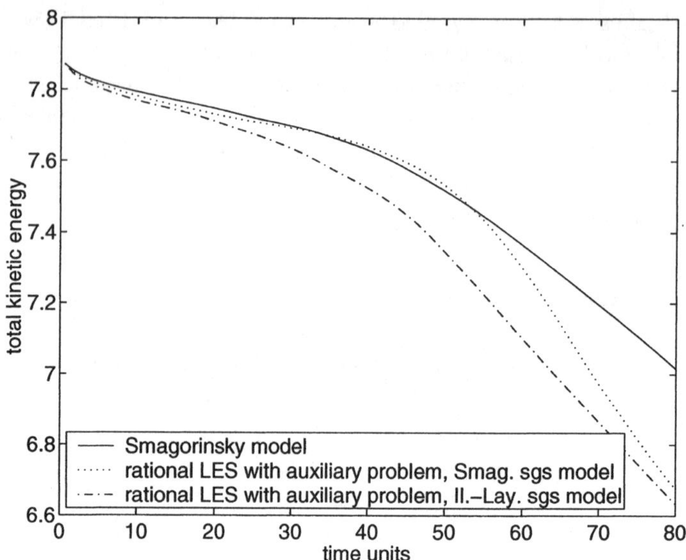

Fig. 11.33. Total kinetic energy, level 4

- It is very similar for the solutions of the Smagorinsky model and the rational LES model with Smagorinsky subgrid scale term up to time unit 50.
- The decrease is faster in the last phase of the flow, as the solutions become unstructured.

The observations for the $L^2(\Omega)$ norm of the gradient of the velocity are as follows:

- It is similar for the solutions of the rational LES model with both types of the subgrid scale term up to time unit 20.
- After time unit 20, the velocity obtained with the Smagorinsky model possesses in general the largest $L^2(\Omega)$ norm of the gradient of the velocity. This is similar to the 2d mixing layer problem.
- The solution obtained with the rational LES model with Iliescu-Layton subgrid scale term is less noisy than the other solutions. This is reflected in the, in general, smallest $L^2(\Omega)$ norm of the gradient of the velocity.

From the similar behaviour of the relative momentum thickness in the 3d mixing layer problem and the relative vorticity thickness in the 2d mixing layer problem, we think that the solutions obtained with the rational LES model and both types of the subgrid scale term are better than the solution of the Smagorinsky model in the initial phase of the flow, up to time unit 25. Then, all solutions become mutually rather different and an assessment of the solutions cannot be made from the available results.

Remark 11.2. Computations with other parameters.

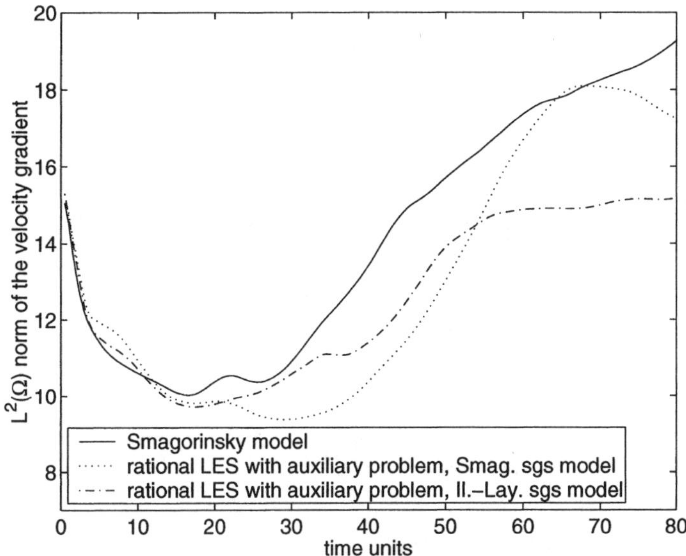

Fig. 11.34. $L^2\,(\Omega)$ norm of the gradient of velocity, level 4

Computations in the domain $(-1,1) \times (-3,3) \times (0,2)$ give very similar results as in Ω. This shows that the extension of Ω in y-direction is sufficiently large.

Numerical results obtained with $\nu = 10^{-4}/4$ ($Re \approx 2856$) are similar to the results computed with $\nu = 10^{-4}$. This indicates that the LES model, especially the model for the subgrid scale term, has already a large influence in the presented tests.

□

12

Problems for Further Investigations

This chapter contains a number of unresolved problems for the LES models considered in this monograph and topics for further investigations. The solution of each of the problems would improve the understanding of either the LES models or of the numerical algorithms used in the computations. The following list does not claim to be complete.

Chapter 3: The Space Averaged Navier-Stokes Equations and the Commutation Error

- In Theorem 3.17 and Remark 3.19, a minimal order of convergence for the weak form of the commutation error is proved in the case $d = 2$. A similar result in the three dimensional case is still missing. Also, the optimality of the two dimensional result requires further investigations.
- It has to be studied if the analysis can be extended to other situations, e.g., to the case that the Navier-Stokes equations are equipped with others than homogeneous Dirichlet boundary conditions. Such boundary conditions require first an appropriate extension of all functions from Ω to $\mathbb{R}^d$.
- An approach for modelling the boundary integral in (3.25) is still missing.
- Other commutation errors, e.g., arising by a non-constant filter width $\delta(\mathbf{x})$, need to be studied.

Chapter 4: LES Models Which are Based on Approximations in Wave Number Space

- An error analysis for the derivation of the Taylor LES model and the rational LES model in a bounded domain is open.
- Homogeneous Neumann boundary conditions have been used for the auxiliary problem on a bounded domain, see Remark 4.7. These boundary conditions are not supported by mathematical arguments. Mathematically supported boundary conditions for the auxiliary problem are still missing.

Likewise, the fourth order rational LES model requires yet the derivation of mathematically supported boundary conditions, see Remark 4.11.

- The numerical tests in Chapters 10 and 11 have been shown that the model for the subgrid scale tensor (Smagorinsky or Iliescu-Layton) can lead to substantially different numerical results. In view of this behaviour, it would be desirable if the model for the subgrid scale tensor is derived in the same way as for the large scale and cross terms, e.g., like in the fourth order rational LES model. The implementation and testing of this model has yet to be done.

Chapter 5: The Variational Formulation of the LES Models

- An error analysis of the modelling error $\|\overline{\mathbf{u}} - \mathbf{w}\|$, $\|\overline{p} - r\|$ in appropriate norms is completely open for all LES models. If such an analysis is possible, its results would be very valuable for the assessment of the quality of a LES model.
- The question of appropriate boundary conditions for LES models is in principle unresolved. Considering the slip with friction boundary conditions as a good approach, the great difficulty consists in the choice of the friction parameter β (linear or non-linear with respect to $\mathbf{w}$, function in space and time, ...).

Chapter 6: Existence and Uniqueness of Solutions of the LES Models

- The existence and uniqueness of a solution of the rational LES model with Smagorinsky subgrid scale term is open. Numerical tests, Chapters 10 and 11, show that this model behaves often similar stable like the Smagorinsky model in the case that c_S is sufficiently large such that one should investigate in this case existence and uniqueness of solutions for large data and long time intervals.
- LES models with the Iliescu-Layton subgrid scale term (4.30) or (4.31) need to be studied analytically. There is only a first result on the existence of a solution for a model with a turbulent viscosity similar to (4.30) obtained by Layton and Lewandowski [LL02], see Remark 6.18. This analytical study has to be continued, it has to be checked if results for the Gaussian filter and the rational LES model can be obtained with a similar analysis.

Chapter 7: Discretisation of the LES Models

- It should be tested if a fully non-linear treatment of the LES model $A\left(\nabla \mathbf{w}_k \nabla \mathbf{w}_k^T\right)$ and the Iliescu-Layton subgrid scale model (7.5) improves the quality of the computed results.

- The assembling of the discrete systems in 3d for higher order finite element discretisations takes approximately the same time as one multigrid cycle. Since the stopping criterion for the linear systems is reached often after one or two multigrid cycles, the time for assembling is a considerable part of the total computing time. It has to be studied if other approaches, which are used, e.g., in p-finite element methods, can reduce the assembling time.

Chapter 8: Numerical Analysis of Finite Element Discretisations of the LES Models

- An analysis of time discretisations for the Smagorinsky model and the Taylor LES model has to be done yet. In connection with the analysis of the finite element discretisation error, an estimate of the complete discretisation error would be available.
- A finite element error analysis of the rational LES model with all kinds of subgrid scale terms is open.

Chapter 9: The Solution of the Linear Systems

- The Vanka smoother does not seem to be an optimal smoother for time dependent problems since it does not profit from small time steps. In general, one hopes that the equations can be solved the faster the smaller the time step is since the well-conditioned mass matrix is becoming dominant. For the Vanka smoother, one can observe this behaviour only to a certain smallness of the time steps. For smaller time steps, the number of iterations per time step stays constant (using the same stopping criteria). Other smoothers, like the smoothers of Braess-Sarazin-type [BS97], take much more advantage from small time steps, see, e.g., the study in John and Tobiska [JT00] for the Navier-Stokes equations. However, the behaviour of the Braess-Sarazin-type smoothers depends heavily on a scaling factor whose optimal choice, which may be a function of time, is not clear. A wrong choice of the scaling factor leads to dramatically longer computing times or even to divergence.
  Within the framework of coupled multigrid methods, smoothers for time dependent problems have to be studied and improved. Also alternative ways of solving the saddle point problem (9.3) have to be tested.
- A large amount of computing time was necessary to perform the numerical simulations presented in this monograph, e.g., each of the 3d simulations presented in Section 11.2 took about two weeks. A parallelisation of the numerical algorithms would be shorten the computing times significantly. All main components of the algorithm, like
  - the assembling of the linear systems,
  - the Krylov subspace method (FGMRES, PCG),
  - the multigrid method with local smoothers (Vanka),

  can be parallelised very efficiently, see, e.g., Haase [Haa99] or John [Joh97].

Chapter 11: A Numerical Study of the Approximation of Space Averaged Flow Fields by the Considered LES Models

Here, open problems in the numerical simulation using the rational LES model are summarised.

- The rational LES model with Smagorinsky subgrid scale term has been tested only in one problem where the turbulence occurs at the boundary of the domain, see Fisher and Iliescu [FI01]. There is the need of much more such studies, also including the Iliescu-Layton subgrid scale term. In [FI01], some modifications of the model near the boundary were necessary to obtain good results. A systematic study of possible modifications which can be applied in the rational LES model is necessary.
- The rational LES model has to be compared numerically with other LES models or turbulence models, especially with those which are considered currently as most successful. Such models have yet to be implemented in the used code. The numerical studies have to be performed as well for academic tests problems as for real turbulent flows for which experimental data are available.
- The dynamic subgrid scale model by Germano et al. [GPMC91] and Lilly [Lil92], Section 4.1.2, has been proved very successful in recent years although this model is based on the Smagorinsky model which possesses a number of drawbacks, see Remark 4.3. The numerical tests with the mixing layer problems have shown that the quality of the computed solution depends very much on a good choice of the scaling factor c_S. A dynamic choice of c_S in a similar fashion as in the dynamic subgrid scale model should be possible also for the subgrid scale term in the rational LES model and it will certainly improve the computational results. However, a rigorous mathematical analysis of such dynamic approaches seems to be far beyond the present state of art.
- In Chapter 7, a number of discretisation schemes in time and space are presented. A brief study of the effects of using different discretisations on the LES models is presented in Section 10.2. The results show that the influence of the chosen discretisation on the computed results may be considerably. There are much more studies necessary to understand advantages and drawbacks of discretisations applied to the LES models and to find conditions for which a discretisation is sufficiently accurate or not. Since in applications the computing time is also a very important factor, the accuracy of discretisations has to be considered always in connection with the possibility of solving the arising discrete systems efficiently.
- The solvers presented in Chapter 9 posses a large number of parameters which influence their efficiency. Parameter studies for the solvers, used in numerical simulations with LES models, are important to quantify the differences in efficiency (computing times) between different choices of parameters. These studies have to be done yet.

13

Notations

symbol	meaning	section
$a_0(\delta)$	artificial viscosity in the Smagorinsky model	8.1
A	operator which determines the LES model	5.1
A	linear system matrix	9.1
A	block in linear saddle point problem	9.3
$\mathcal{A}$	system matrix of the linear saddle point problem	9.3
$\mathcal{A}_j$	local saddle point matrix	9.3.2
$A_\delta(\mathbb{S}(\mathbf{u},p))$	commutation error	3.3
$A(t)$	area of cross sections	3.5
A^h	discrete version of the operator A	7.3
$\mathbf{A}$	non-linear operator in the LES models	6.1
$b(\cdot,\cdot,\cdot)$	trilinear form (convective form)	5.1
$b_s(\cdot,\cdot,\cdot)$	trilinear form (skew-symmetric form)	5.1
r	right hand side of linear system	9.1
B	block in linear saddle point problem	7.3
$B(\mathbf{x},R)$	open ball with centre $\mathbf{x}\in\mathbb{R}^d$ and radius R	3.5
$B_\delta(\mathbf{x})$	function to estimate in the proof of Theorem 3.13	3.5
c_S	scaling factor in turbulent viscosity ν_T	4.1, 4.3
c_{noise}	scaling of the initial noise	10.1,11.1
C	block in linear saddle point problem	7.3
$C_\delta(\mathbf{x})$	function to estimate in the proof of Theorem 3.13	3.5
$C^m(\Omega)$	space of m times continuous differentiable functions on Ω	2.1
$C^m(\overline{\Omega})$	subspace of $C^m(\Omega)$ whose functions with all partial derivatives can be extended continuously to $\partial\Omega$	2.1
$C_0^m(\Omega)$	space of m times continuous differentiable functions with compact support	2.1
$C^{0,\beta}(\overline{\Omega})$	space of Hölder continuous functions	2.1

$C^m(I;V)$	space of m times continuous differentiable functions from the interval I to the Banach space V	2.1
d	dimension	3.1
$d(\mathbf{x},\mathbf{y})$	distance of $\mathbf{x},\mathbf{y} \in \mathbb{R}^d$	3.5
$\mathcal{D}_j$	local saddle point matrix	9.3.2
$\mathbb{D}(\mathbf{u})$	deformation tensor of $\mathbf{u}$	3.1
$\mathbf{e}$	error between the solution of the continuous problem and the finite element solution	8.1.2
$\mathbf{e}_i$	vector of the canonical basis of $\mathbb{R}^d$	5.2.6
$E_{\text{kin}}(\mathbf{w}^h)$	total kinetic energy of $\mathbf{w}^h$	10
f	vector of discrete right hand side	7.3
$\mathbf{f}$	right hand side of the Navier-Stokes equations and of the LES models	3.1
$\mathcal{F}$	right hand sides in the finite element error estimates	8.1
F_K	reference map from $\hat{K}$ onto K	7.4
$\mathcal{F}(f)$	Fourier transform of a function f	2.3
g	filter function	3.3
g_δ	Gaussian filter	3.4
h	characteristic mesh width	7.4
h_K	diameter of a mesh cell K	7.4
$H^m(\Omega)$	Sobolev space $W^{m,2}(\Omega)$	2.1
$H_0^1(\Omega)$	Sobolev space $W_0^{1,2}(\Omega)$	2.1
$H^{-1}(\Omega)$	dual space of $H_0^1(\Omega)$	2.1
$\mathbb{I}$	unit tensor	3.1
$I_l\left(K,V_l^h\right)$	set of local nodes with respect to K	9.3
$J_K(\hat{\mathbf{x}})$	Jacobian of F_K in $\hat{\mathbf{x}}$	7.9
K	mesh cell	7.4
$\hat{K}$	reference mesh cell	7.4
$\mathbb{K}$	tensor in the dynamic subgrid scale model	4.1.2
$\mathbb{L}$	tensor in the dynamic subgrid scale model	4.1.2
l	order in Gaussian quadrature rules	7.9
$L^p(\Omega)$	Lebesgue space	2.1
$L_0^2(\Omega)$	Lebesgue space whose functions have mean value zero	2.1
$L^p(a,b;V)$	Lebesgue space of functions from the interval (a,b) to the Banach space V	2.1
M	preconditioner for linear system	9.1
$\mathbb{M}$	tensor in the dynamic subgrid scale model	4.1.2
$\mathbf{n}_{\partial\Omega}$	outward pointing unit normal on $\partial\Omega$	3.3
$N_{l,j}^K$	dual basis of local nodal functionals in mesh cell K	9.3.1
N_p	number of degrees of freedom of the pressure	7.3

N_v	number of degrees of freedom for one component of the velocity	7.3
p	pressure of Navier-Stokes equations	3.1
$\overline{p}$	filtered or space averaged pressure	3.2
p'	subgrid scale component of the pressure	3.2
$P(\hat{K})$	finite dimensional space on $\hat{K}$	7.4
$P_m(\hat{K})$	finite element space on the reference mesh cell	7.4
$P_m(K)$	finite element space on the mesh cell K	7.4
P_m	finite element space on simplicial meshes	7.4
P_1^{nc}	non-conforming finite element space on simplicial meshes	7.4
P_m^{disc}	discontinuous finite element space on quadrilateral / hexahedral meshes	7.4
P_{l-1}^{l}	prolongation operator	9.3
$\mathcal{P}(K)$	parent mesh cell	9.3
q_i^a	basis functions of Q_a^h	7.3
Q, Q_a	ansatz space for the continuous pressure	5.1
Q^h	ansatz space for the discrete pressure	7.4
$\mathcal{Q}^h, \mathcal{Q}_j^h$	sets of pressure degrees of freedom	9.3.2
$Q_m(\hat{K})$	finite element space on the reference square / cube	7.4
$Q_m(K)$	finite element space on the quadrilateral / hexahedron K	7.4
Q_m	finite element space on quadrilateral / hexahedral meshes	7.4
Q_t	test space for the continuous pressure	5.1
Q_1^{rot}	non-conforming finite element space on quadrilateral / hexahedral meshes	7.4
r	pressure of LES models	5.1
$\mathbf{r}$	vector of discrete pressure coefficients	7.3
$R_l^{*,l-1}$	defect restriction	9.3
R_l^{l-1}	function restriction	9.3
Re	Reynolds number	3.1
$\mathbb{S}(\mathbf{u}, p)$	stress tensor (Navier-Stokes equations)	3.3
$\mathbb{S}(\mathbf{u}, p, A)$	stress tensor (LES models)	5.1
$\text{supp}\,(v)$	support of a function v	2.1
t	time	3.1
$\bar{t}$	time unit in mixing layer problems	11.1, 11.2
t_k, t_n	discrete time	7.1
T	final time	3.1
$\mathbb{T}$	Reynolds stress tensor	4.1
$\mathcal{T}_h$	triangulation of Ω	7.4
$\text{tr}(\cdot)$	trace of a matrix or tensor	2.4
$\mathbf{u}$	velocity of Navier-Stokes equations	3.1
$\overline{\mathbf{u}}$	filtered or space averaged velocity	3.2

$\mathbf{u}'$	subgrid scale component of the velocity	3.2
$\mathbf{v}_i^a$	basis function of V_a^h	7.3
V, V_a	ansatz space for the continuous velocity	5.1
V_a^h	discrete ansatz space	7.3
V_{div}	subspace of V with weakly divergence free functions	8.1
V^h	ansatz space for the discrete velocity	7.4
V_{div}^h	subspace of V^h with discrete divergence free functions	8.1
$\mathcal{V}^h, \mathcal{V}_j^h$	sets of velocity degrees of freedom	9.3.2
V_{conv}^h	space for the discrete convolution	7.8
V_{aux}^h	finite element space for the auxiliary problem	7.7
V_t	test space for the continuous velocity	5.1
V_t^h	discrete test space	7.3
V^*	dual space of V	8.1
w	vector of discrete velocity coefficients	7.3
$\mathbf{w}$	velocity of LES models	5.1
$\mathbf{w}_{k-1}^h$	finite element approximation of $\mathbf{w}$ at the previous discrete time t_{k-1}	7.3
$\mathbf{w}_{\text{old}}^h$	current finite element approximation of $\mathbf{w}$ in fixed point iteration	7.3
$\mathbf{w}^h$	discrete velocity of LES models	8.1
W_∞	characteristic velocity	11.1, 11.2
$W^{m,p}(\Omega)$	Sobolev space	2.1
$W_0^{1,p}(\Omega)$	Sobolev space of functions which vanish on $\partial\Omega$	2.1
$W^{-1,q}(\Omega)$	dual space of $W_0^{1,p}(\Omega)$, $p \in (1,\infty)$, $p^{-1}+q^{-1}=1$	2.1
$\mathcal{W}^h, \mathcal{W}_j^h$	sets of velocity and pressure degrees of freedom	9.3.2
x	solution of linear system	9.1
$\mathbb{X}, \mathbb{X}^h$	solution tensor for auxiliary problem	7.7
$\mathbb{Y}, \mathbb{Y}^h$	test tensor for auxiliary problem	7.7
y_c	cut-off wave number	3.4
α	resistance parameter for boundary penetration	5.2.5
α_s	parameter for least squares stabilisation	8.1.1
β	friction parameter	5.2.4
γ	constant in the Gaussian filter $= 6$	3.4
Γ	boundary of Ω, $\Gamma = \partial\Omega$	3.1
Γ_0	boundary with homogeneous Dirichlet conditions	8.1
Γ_j	$1 \le j \le J$, boundary with slip with linear friction conditions	8.1
Γ_{diri}	boundary with Dirichlet conditions	5.2.1
$\Gamma_{\text{diri,hom}}$	boundary with no slip conditions	5.2.1
Γ_{in}	inflow boundary	10.1
Γ_{out}	boundary with outflow conditions	5.2.2

Γ_{sfpr}	boundary with slip with linear friction and penetration conditions	5.2.5
Γ_{slfr}	boundary with slip with linear friction conditions	5.2.4
Γ_{slip}	boundary with free slip conditions	5.2.3
δ	characteristic filter width	1.3
$\overline{\delta}$	radius for integration in the discrete convolution	7.8
δ_{ij}	Kronecker delta	9.3
Δt_n	time step	7.1
η	interpolation error	8.1
θ	parameter in the estimates of the commutation error	3.5
$\theta_1, \ldots, \theta_4$	weights in the time stepping schemes	7.1
θ_l	weight in quadrature rules	7.9
κ	parameter in the estimates of the commutation error	3.5
$\mu(t)$	momentum thickness	11.2
ν	viscosity parameter of Navier-Stokes equations	3.1
ν_S	scaling factor in Smagorinsky model, usually $\nu_S = c_S\delta^2$	4.1.1
ν_T	turbulent viscosity parameter	4.1
σ_0, $\sigma(t)$	vorticity thickness	11.1
$\Sigma(P)$	set of linear functionals in definition of finite elements	7.4
τ_k	$1 \leq k \leq d-1$, orthonormal systems of tangential vectors on the boundary of a domain	5.2
ϕ^h	finite element error	8.1
$\Phi(t)$	parameter function in upwind stabilisation	7.5
$\psi_{l,j}^h\vert_K$	function of local finite element basis	9.3.1
$\boldsymbol{\omega}$	vorticity	11.1
ω_i	control volume in upwind stabilisation	7.5
Ω	domain where the equations have to be solved	3.1
$\partial\Omega$	boundary of Ω	3.1
$(\cdot,\cdot)$	inner product in $\left(L^2(\Omega)\right)^d$	2.1
$\Vert\cdot\Vert_2$	Euclidean norm in $\mathbb{R}^d$	2.1
$\Vert\cdot\Vert_F$	Frobenius norm of a matrix	2.4

References

[Ada75] R.A. Adams. *Sobolev spaces.* Academic Press, New York, 1975.

[Ald90] A.A. Aldama. *Filtering Techniques for Turbulent Flow Simulation*, volume 56 of *Springer Lecture Notes in Eng.* Springer, Berlin, 1990.

[Bau97] M. Bause. *Optimale Konvergenzraten für voll diskretisierte Navier-Stokes-Approximationen höherer Ordnung in Gebieten mit Lipschitz-Rand.* PhD thesis, Universität-Gesamthochschule Paderborn, 1997.

[Bec92] M. Beckers. *Numerical integration in high dimensions.* PhD thesis, Katholieke Universiteit Leuven, 1992.

[Bey95] J. Bey. Tetrahedral grid refinement. *Computing*, 55:355 - 378, 1995.

[BF91] F. Brezzi and M. Fortin. *Mixed and Hybrid Finite Element Methods*, volume 15 of *Springer Series in Computational Mathematics.* Springer-Verlag, 1991.

[BGLI02] L.G. Berselli, G.P. Galdi, W.J. Layton, and T. Iliescu. Mathematical analysis for the rational large eddy simulation model. *Math. Models and Meth. in Appl. Sciences*, 12:1131 - 1152, 2002.

[BHM00] W.L. Briggs, E. van Henson, and S.F. McCormick. *A multigrid tutorial.* SIAM, Philadelphia, PA, 2nd edition, 2000.

[BKNW97] B.J. Boersma, M.N. Kooper, F.T.M. Neuwstadt, and P. Wesseling. Local grid refinement in large-eddy simulations. *J. Engineering Mathematics*, 32:161 - 175, 1997.

[BR94] R. Becker and R. Rannacher. Finite element discretization of the Stokes and Navier-Stokes equations on anisotropic grids. In G. Wittum and W. Hackbusch, editors, *Proceedings of 10th GAMM-Seminar, Kiel, January 14-16*, volume 49 of *Notes on Numerical Fluid Mechanics*, pages 52 - 61. Vieweg, Braunschweig, 1994.

[Bre98] M. Breuer. Large eddy simulation of the subcritical flow past a circular cylinder: numerical and modeling aspects. *Int. Jour. Num. Meth. Fluids*, 28:1281 - 1302, 1998.

[Bro65] F.E. Browder. Existence and uniqueness theorems for solutions of nonlinear boundary value problems. *Proc. Sympos. Appl. Math.*, 17:24 - 49, 1965.

[BS94] S.C. Brenner and L.R. Scott. *The Mathematical Theory of Finite Element Methods*, volume 15 of *Texts in Applied Mathematics.* Springer-Verlag New York, ..., 1994.

[BS97] D. Braess and R. Sarazin. An efficient smoother for the Stokes problem. *Applied Numer. Math.*, 23(1):3 – 19, 1997.

[CFR79] R.A. Clark, J. H. Ferziger, and W.C. Reynolds. Evaluation of subgrid-scale models using an accurately simulated turbulent flow. *J. Fluid Mech.*, 91:1 – 16, 1979.

[CH87] R. Cools and A. Haegemans. Construction of minimal cubature formulae for the square and the triangle using invariant theory. Report TW 96, Dept. of Computer Science, K.U. Leuven, 1987.

[Cho68] A.J. Chorin. Numerical solution for the Navier-Stokes equations. *Math. Comp.*, 22:745 – 762, 1968.

[Cia78] P.G. Ciarlet. *The finite element method for elliptic problems.* North-Holland Publishing Company, Amsterdam - New York - Oxford, 1978.

[Cia91] P.G. Ciarlet. Basic error estimates for elliptic problems. In P.G. Ciarlet and J.L. Lions, editors, *Handbook of Numerical Analysis II*, pages 19–351. North-Holland Amsterdam, New York, Oxford, Tokyo, 1991.

[CLL92] P. Comte, M. Lesieur, and E. Lamballais. Large- and small-scale stirring of vorticity and a passive scalar in a 3-d temporal mixing layer. *Phys. Fluids A*, 4(12):2761 – 2778, 1992.

[Col97] P. Coletti. A global existence theorem for large eddy simulation turbulence model. *Mathematical Models and Methods in Applied Sciences*, 7:579 – 591, 1997.

[Col98] P. Coletti. *Analytic and Numerical Results for $k - \varepsilon$ and Large Eddy Simulation Turbulence Models.* PhD thesis, University of Trento, 1998.

[Coo99] R. Cools. Monomial cubature rules since "Stroud": a compilation - part 2. *J. Comput. Appl. Math.*, 112:21 – 27, 1999.

[CR73] M. Crouzeix and P.-A. Raviart. Conforming and nonconforming finite element methods for solving the stationary Stokes equations I. *R.A.I.R.O. Anal. Numér.*, 7:33–76, 1973.

[CR93] R. Cools and P. Rabinowitz. Monomial cubature rules since "Stroud". *J. Comput. Appl. Math.*, 48:309 – 326, 1993.

[CT87] M. Crouzeix and V. Thomée. The stability in L^p and $W^{1,p}$ of the L^2-projection onto finite element function spaces. *Math. Comp.*, 48:521 – 523, 1987.

[CWL94] M.E. Cantekin, J.J. Westerink, and R.A. Luettich Jr. Low and moderate Reynolds number transient flow simulations using space filtered Navier-Stokes equations. *Num. Meth. Part. Diff. Equ.*, 10:491 – 524, 1994.

[DG90] Q. Du and M.D. Gunzburger. Finite-element approximations of a Ladyzhenskaya model for stationary incompressible viscous flow. *SIAM J. Numer. Anal.*, 27:1 – 19, 1990.

[DG91] Q. Du and M.D. Gunzburger. Analysis of a Ladyzhenskaya model for incompressible viscous flow. *J. Math. Anal. Appl.*, 155:21 – 45, 1991.

[DJL03a] A. Dunca, V. John, and W.J. Layton. Approximating local averages of fluid velocities: the equilibrium Navier-Stokes equations. *Appl. Numer. Math.*, 2003. accepted for publication.

[DJL03b] A. Dunca, V. John, and W.J. Layton. The commutation error of the space averaged Navier-Stokes equations on a bounded domain. *J. Math. Fluid Mech.*, 2003. accepted for publication.

[Emm99] E. Emmrich. Discrete versions of Gronwall's lemma and their application to the numerical analysis of parabolic problems. Preprint 637, Technische Universität Berlin, Fachbereich Mathematik, 1999.

[Emm01] E. Emmrich. *Analysis von Zeitdiskretisierungen des inkompressiblen Navier-Stokes-Problems.* PhD thesis, Technische Universität Berlin, 2001. appeared also as book from Cuvillier Verlag Göttingen.

[FI01] P. Fischer and T. Iliescu. A 3d channel flow simulation at $Re_\tau = 180$ using a rational LES model. In C. Liu, L. Sakell, and T. Beutner, editors, *DNS/LES - Progress and Challenges (Proceedings of the Third AFOSR International Conference on DNS/LES)*, pages 283 – 290. Greyden Press, Columbus, 2001.

[Fol95] G.B. Folland. *Introduction to Partial Differential Equations*, volume 17 of *Mathematical Notes.* Princeton University Press, 2nd edition, 1995.

[For93] M. Fortin. Finite element solution of the Navier-Stokes equations. In A. Iserles, editor, *Acta Numerica*, pages 239 – 284. Cambridge University Press, 1993.

[FP99] J.H. Ferziger and M. Perić. *Computational Methods for Fluid Dynamics.* Springer Berlin, 2nd edition, 1999.

[Fri95] U. Frisch. *Turbulence, the Legacy of A.N. Kolmogorov.* Cambridge, 1995.

[FT97] C. Fureby and G. Tabor. Mathematical and physical constraints on large-eddy simulations. *Theoret. Comput. Fluid Dynamics*, 9:85 – 102, 1997.

[Gal94] G.P. Galdi. *An Introduction to the Mathematical Theory of the Navier-Stokes Equations, Vol. I: Linearized Steady Problems*, volume 38 of *Springer Tracts in Natural Philosophy.* Springer, 1994.

[Gal00] G.P. Galdi. An introduction to the Navier-Stokes initial-boundary value problem. In G.P. Galdi, J.G. Heywood, and R. Rannacher, editors, *Fundamental Directions in Mathematical Fluid Dynamics*, pages 1 – 70. Birkhäuser, 2000.

[GK00] M. Griebel and F. Koster. Adaptive wavelet solvers for the unsteady incompressible Navier-Stokes equations. In J. Malek, J. Nečas, and M. Rokyta, editors, *Advances in Mathematical Fluid Mechanics*, pages 67 – 118. Springer Verlag, 2000.

[GL00] G.P. Galdi and W.J. Layton. Approximation of the larger eddies in fluid motion II: A model for space filtered flow. *Math. Models and Meth. in Appl. Sciences*, 10(3):343 – 350, 2000.

[GM95] S. Ghosal and P. Moin. The basic equations for large eddy simulation of turbulent flows in complex geometries. *Journal of Computational Physics*, 118:24 – 37, 1995.

[GPMC91] M. Germano, U. Piomelli, P. Moin, and W. Cabot. A dynamic subgrid-scale eddy viscosity model. *Phys. Fluids A*, 3:1760 – 1765, 1991.

[GR86] V. Girault and P.-A. Raviart. *Finite Element Methods for Navier-Stokes equations.* Springer-Verlag, Berlin-Heidelberg-New York, 1986.

[Gri92] P. Grisvard. *Singularities in Boundary Value Problems*, volume 22 of *Research Notes in Applied Mathematics.* Masson, Springer-Verlag, 1992.

[GS00] P.M. Gresho and R.L. Sani. *Incompressible Flow and the Finite Element Method.* Wiley, 2000.

[Gun89] M.D. Gunzburger. *Finite Element Methods for Viscous Incompressible Flows.* Academic Press, Boston, 1989.

[Haa99] G. Haase. *Parallelisierung numerischer Algorithmen für partielle Differentialgleichungen. (Parallelization of numerical algorithms for partial differential equations).* Teubner Verlag Stuttgart, 1999.

[Hac85] W. Hackbusch. *Multi-Grid Methods and Applications.* Springer Series in Computational Mathematics, 4. Springer-Verlag, Berlin-Heidelberg-New York, 1985.

[HMJ00] T.J. Hughes, L. Mazzei, and K.E. Jansen. Large eddy simulation and the variational multiscale method. *Comput. Visual. Sci.*, 3:47 – 59, 2000.

[Hör90] L. Hörmander. *The Analysis of Partial Differential Operators I.* Springer - Verlag, Berlin, ..., 2nd edition, 1990.

[HR82] J.G. Heywood and R. Rannacher. Finite element approximation of the nonstationary Navier-Stokes problem I: Regularity of solutions and second order error estimates for spatial discretization. *SIAM J. Numer. Anal.*, 19:275 – 311, 1982.

[HR90] J.G. Heywood and R. Rannacher. Finite element approximation of the nonstationary Navier-Stokes problem iv: Error analysis for second order time discretizations. *SIAM J. Num. Anal.*, 27:353 – 384, 1990.

[HRT96] J.G. Heywood, R. Rannacher, and S. Turek. Artificial boundaries and flux and pressure conditions for the incompressible Navier-Stokes equations. *Int. J. Numer. Methods Fluids*, 22(5):325–352, 1996.

[HS52] M.R. Hestenes and E. Stiefel. Methods of conjugate gradients for solving linear systems. *J. Res. Nat. Bur. Standards*, 49:409–436, 1952.

[IJL02] T. Iliescu, V. John, and W.J. Layton. Convergence of finite element approximations of large eddy motion. *Num. Meth. Part. Diff. Equ.*, 18:689 – 710, 2002.

[IJL+03] T. Iliescu, V. John, W.J. Layton, G. Matthies, and L. Tobiska. A numerical study of a class of LES models. *Int. J. Comput. Fluid Dyn.*, 17:75 – 85, 2003.

[IL98] T. Iliescu and W.J. Layton. Approximating the larger eddies in fluid motion III: the Boussinesq model for turbulent fluctuations. *Analele Ştiinţifice ale Universităţii "Al. I. Cuza" Iaşi, Tomul XLIV, s.I.a, Matematică*, 44:245–261, 1998.

[JKMT02] V. John, P. Knobloch, G. Matthies, and L. Tobiska. Non-nested multilevel solvers for finite element discretizations of mixed problems. *Computing*, 68:313 – 341, 2002.

[JL01] V. John and W.J. Layton. Approximating local averages of fluid velocities: The Stokes problem. *Computing*, 66:269 – 287, 2001.

[JL02] V. John and W.J. Layton. Analysis of numerical errors in large eddy simulation. *SIAM J. Numer. Anal.*, 40:995 – 1020, 2002.

[JLS03] V. John, W.J. Layton, and N. Sahin. Derivation and analysis of near wall models for channel and recirculating flows. *Comput. Math. Appl.*, 2003. accepted for publication.

[JM01] V. John and G. Matthies. Higher order finite element discretizations in a benchmark problem for incompressible flows. *Int. J. Num. Meth. Fluids*, 37:885 – 903, 2001.

[JM03] V. John and G. Matthies. MooNMD - a program package based on mapped finite element methods. *Comput. Visual. Sci.*, 2003. accepted for publication.

[JMM+00] V. John, G. Matthies, T.I. Mitkova, L.Tobiska, and P.S. Vassilevski. A comparison of three solvers for the incompressible Navier-Stokes equations. In M. Griebel, S.D. Margenov, and P. Yalamov, editors, *Large-Scale Scientific Computations of Engineering and Environmental Prob-*

lems, volume 73 of *Notes on Numerical Fluid Mechanics*, pages 215 – 222. Vieweg, 2000.

[JMST98] V. John, G. Matthies, F. Schieweck, and L. Tobiska. A streamline-diffusion method for nonconforming finite element approximations applied to convection-diffusion problems. *Comp. Meth. Appl. Mech. Engrg.*, 166:85 – 97, 1998.

[JMT97] V. John, J.M. Maubach, and L. Tobiska. Nonconforming streamline-diffusion-finite-element-methods for convection-diffusion problems. *Numer. Math.*, 78:165 – 188, 1997.

[Joh97] V. John. *Parallele Lösung der inkompressiblen Navier-Stokes Gleichungen auf adaptiv verfeinerten Gittern.* PhD thesis, Otto-von-Guericke-Universität Magdeburg, Fakultät für Mathematik, 1997.

[Joh99] V. John. A comparison of parallel solvers for the incompressible Navier-Stokes equations. *Comput. Visual. Sci.*, 4(1):193 – 200, 1999.

[Joh02a] V. John. Higher order finite element methods and multigrid solvers in a benchmark problem for the 3D Navier-Stokes equations. *Int. J. Num. Meth. Fluids*, 40:775 – 798, 2002.

[Joh02b] V. John. Slip with friction and penetration with resistance boundary conditions for the Navier-Stokes equations - numerical tests and aspects of the implementation. *J. Comp. Appl. Math.*, 147:287 – 300, 2002.

[JT00] V. John and L. Tobiska. Smoothers in coupled multigrid methods for the parallel solution of the incompressible Navier-Stokes equations. *Int. J. Num. Meth. Fluids*, 33:453 – 473, 2000.

[Kol42] A. Kolmogorov. The equations of turbulent motion in an incompressible fluid. *Isv. Acad. Sci. USSR Phys.*, 6:56 – 58, 1942.

[Kos97] B. Kosović. Subgrid-scale modelling for the large-eddy simulation of high-Reynolds-number boundary layers. *J. Fluid Mech.*, 336:151 – 182, 1997.

[KR94] P. Klouček and F.S. Rys. Stability of the fractional step θ-scheme for the nonstationary Navier-Stokes equations. *SIAM Num. Anal.*, 31:1312 – 1335, 1994.

[Kra67] R.H. Kraichnan. Inertial ranges in two dimensional turbulence. *Phys. Fluids*, 10:1417 – 1423, 1967.

[Lad67] O.A. Ladyzhenskaya. New equations for the description of motion of viscous incompressible fluids and solvability in the large of boundary value problems for them. *Proc. Steklov Inst. Math.*, 102:95 – 118, 1967.

[Lad69] O.A. Ladyzhenskaya. *The Mathematical Theory of Viscous Incompressible Flow.* Gordon and Breach, 2nd edition, 1969.

[Leo74] A. Leonard. Energy cascade in large eddy simulation of turbulent fluid flows. *Adv. in Geophysics*, 18A:237 – 248, 1974.

[Les97] M. Lesieur. *Turbulence in Fluids*, volume 40 of *Fluid Mechanics and its Applications.* Kluwer Academic Publishers, 3rd edition, 1997.

[Lil92] D.K. Lilly. A proposed modification of the Germano subgrid-scale closure method. *Phys. Fluids A*, 4:633 – 635, 1992.

[LL02] W.J. Layton and R. Lewandowski. Analysis of an eddy viscosity model for large eddy simulation of turbulent flows. *J. Math. Fluid Mechanics*, 4:374 – 399, 2002.

[LSLC88] M. Lesieur, C. Staquet, P. Le Roy, and P. Comte. The mixing layer and its coherence examined from the point of view of two-dimensional turbulence. *J. Fluid Mech.*, 192:511 – 534, 1988.

[Mat01] G. Matthies. Mapped finite elements on hexahedra. Necessary and sufficient conditions for optimal interpolation errors. *Numer. Alg.*, 27:317 – 327, 2001.

[Max79] J.C. Maxwell, 1879. Phil. Trans. Royal Society.

[Mic64] A. Michalke. On the inviscid instability of the hyperbolic tangent velocity profile. *J. Fluid Mech.*, 19:543 – 556, 1964.

[Min62] G. Minty. Monotone (nonlinear) operators in Hilbert space. *Duke Math. J.*, 29:341 – 346, 1962.

[MP94] B. Mohammadi and O. Pironneau. *Analysis of the K-Epsilon Turbulence Model.* John Wiley & Sons, 1994.

[MT98] M. Marion and R. Temam. Navier-Stokes equations: Theory and approximation. In P.G. Ciarlet and J.L. Lions, editors, *Handbook of Numerical Analysis VI*, pages 503–688. Elsevier, Amsterdam, 1998.

[MU94] S. Müller-Urbaniak. Eine Analyse des Zwischenschritt-θ-Verfahrens zur Lösung der instationären Navier-Stokes-Gleichungen. Preprint 94-01, Universität Heidelberg, Interdisziplinäres Zentrum für wissenschaftliches Rechnen, 1994.

[Nav23] C.L.M.H. Navier. Mémoire sur les lois du mouvement des fluiales. *Mém. Acad. Royal Society*, 6:389–440, 1823.

[NW03] S. Nägele and G. Wittum. Large-eddy simulation and multigrid methods. *Electon. Trans. Numer. Anal.*, 15:152 – 164, 2003.

[OU84] K. Ohmori and T. Ushijima. A technique of upstream type applied to a linear nonconforming finite element approximation of convective diffusion equations. *R.A.I.R.O. Anal. Numer.*, 18(3):309–332, 1984.

[Par92] C. Parés. Existence, uniqueness and regularity of solutions of the equations of a turbulence model for incompressible fluids. *Applicable Analysis*, 43:245 – 296, 1992.

[Pio99] U. Piomelli. Large-eddy simulation: achievements and challenges. *Progress in Aerospace Sciences*, 35:335 – 362, 1999.

[Pop00] S. B. Pope. *Turbulent flows.* Cambridge University Press, 2000.

[RM94] M.M. Rogers and R.D. Moser. Direct simulation of a self-similar turbulent flow. *Phys. Fluids*, 6:903 – 923, 1994.

[RST96] H.-G. Roos, M. Stynes, and L. Tobiska. *Numerical Methods for Singularly Perturbed Differential Equations.* Springer, 1996.

[RT92] R. Rannacher and S. Turek. Simple nonconforming quadrilateral Stokes element. *Numer. Meth. Part. Diff. Equ.*, 8:97 – 111, 1992.

[Rud91] W. Rudin. *Functional Analysis.* International Series in Pure and Applied Mathematics. McGraw-Hill, Inc., New York, 2nd edition, 1991.

[Saa93] Y. Saad. A flexible inner-outer preconditioned GMRES algorithm. *SIAM J. Sci. Comput.*, 14(2):461–469, 1993.

[Saa96] Y. Saad. *Iterative Methods for Sparse Linear Systems.* PWS Publishing Company, 1996.

[Sag01] P. Sagaut. *Large Eddy Simulation for Incompressible Flows.* Springer-Verlag Berlin Heidelberg New York, 2001.

[Sch97] F. Schieweck. *Parallele Lösung der stationären inkompressiblen Navier-Stokes Gleichungen.* Otto-von-Guericke-Universität Magdeburg, Fakultät für Mathematik, 1997. Habilitation.

[Sch00] F. Schieweck. A general transfer operator for arbitrary finite element spaces. Preprint 00-25, Fakultät für Mathematik, Otto-von-Guericke-Universität Magdeburg, 2000.

[Sma63] J.S. Smagorinsky. General circulation experiments with the primitive equations. *Mon. Weather Review*, 91:99 – 164, 1963.

[Soh01] H. Sohr. *The Navier-Stokes Equations, An Elementary Functional Analytic Approach.* Birhäuser Advanced Texts. Birkhäuser Verlag Basel, Boston, Berlin, 2001.

[SS86] Y. Saad and M.H. Schultz. GMRES: A generalized minimal residual algorithm for solving nonsymmetric linear systems. *SIAM J. Sci. Stat. Comput.*, 7(3):856 – 869, 1986.

[ST96a] M. Schäfer and S. Turek. The benchmark problem "Flow around a cylinder". In E.H. Hirschel, editor, *Flow Simulation with High-Performance Computers II*, volume 52 of *Notes on Numerical Fluid Mechanics*, pages 547 – 566. Vieweg, 1996.

[ST96b] F. Schieweck and L. Tobiska. An optimal order error estimate for an upwind discretization of the Navier-Stokes equations. *Num. Meth. Part. Diff. Equ.*, 12:407 – 421, 1996.

[Str71] A.H. Stroud. *Approximate calculation of multiple integrals.* Prentice-Hall, Englewood Cliffs, N.J., 1971.

[Tem77] R. Temam. *Navier-Stokes Equations. Theory and Numerical Analysis*, volume 2 of *Studies in Mathematics and Its Applications.* North-Holland Publishing Company, Amsterdam, New York, Oxford, 1977.

[Tem95] R. Temam. *Navier-Stokes Equations and Nonlinear Functional Analysis*, volume 66 of *CBMS-NSF Regional Conference Series in Applied Mathematics.* SIAM, Philadelphia, 2nd edition, 1995.

[Tur99] S. Turek. *Efficient Solvers for Incompressible Flow Problems: An Algorithmic and Computational Approach*, volume 6 of *Lecture Notes in Computational Science and Engineering.* Springer, 1999.

[Van86] S. Vanka. Block-implicit multigrid calculation of two-dimensional recirculating flows. *Comp. Meth. Appl. Mech. Eng.*, 59(1):29–48, 1986.

[VGK97] B. Vreman, B. Guerts, and H. Kuerten. Large-eddy simulation of turbulent mixing layer. *J. Fluid Mech.*, 339:357 – 390, 1997.

[VLM98] O.V. Vasilyev, T.S. Lund, and P. Moin. A general class of commutative filters for LES in complex geometries. *Journal of Computational Physics*, 146:82 – 104, 1998.

[Zei86] E. Zeidler. *Nonlinear Functional Analysis and its Applications, I: Fixed-Point Theorems.* Springer-Verlag New York, Berlin, Heidelberg, Tokyo, 1986.

[Zei90] E. Zeidler. *Nonlinear functional analysis and its applications. II/B: Nonlinear monotone operators.* Springer-Verlag, New York, ..., 1990.

[ZSK93] Y. Zang, R.L. Street, and J.R. Koseff. A dynamic mixed subgrid-scale model and its application to turbulent recirculating flows. *Phys. Fluids A*, 5:3186 – 3196, 1993.

Index

Editorial Policy

§1. Volumes in the following three categories will be published in LNCSE:

i) Research monographs
ii) Lecture and seminar notes
iii) Conference proceedings

Those considering a book which might be suitable for the series are strongly advised to contact the publisher or the series editors at an early stage.

§2. Categories i) and ii). These categories will be emphasized by Lecture Notes in Computational Science and Engineering. **Submissions by interdisciplinary teams of authors are encouraged.** The goal is to report new developments – quickly, informally, and in a way that will make them accessible to non-specialists. In the evaluation of submissions timeliness of the work is an important criterion. Texts should be well-rounded, well-written and reasonably self-contained. In most cases the work will contain results of others as well as those of the author(s). In each case the author(s) should provide sufficient motivation, examples, and applications. In this respect, Ph.D. theses will usually be deemed unsuitable for the Lecture Notes series. Proposals for volumes in these categories should be submitted either to one of the series editors or to Springer-Verlag, Heidelberg, and will be refereed. A provisional judgment on the acceptability of a project can be based on partial information about the work: a detailed outline describing the contents of each chapter, the estimated length, a bibliography, and one or two sample chapters – or a first draft. A final decision whether to accept will rest on an evaluation of the completed work which should include

- at least 100 pages of text;
- a table of contents;
- an informative introduction perhaps with some historical remarks which should be accessible to readers unfamiliar with the topic treated;
- a subject index.

§3. Category iii). Conference proceedings will be considered for publication provided that they are both of exceptional interest and devoted to a single topic. One (or more) expert participants will act as the scientific editor(s) of the volume. They select the papers which are suitable for inclusion and have them individually refereed as for a journal. Papers not closely related to the central topic are to be excluded. Organizers should contact Lecture Notes in Computational Science and Engineering at the planning stage.

In exceptional cases some other multi-author-volumes may be considered in this category.

§4. Format. Only works in English are considered. They should be submitted in camera-ready form according to Springer-Verlag's specifications. Electronic material can be included if appropriate. Please contact the publisher. Technical instructions and/or TeX macros are available via http://www.springer.de/math/authors/help-momu.html. The macros can also be sent on request.

General Remarks

Lecture Notes are printed by photo-offset from the master-copy delivered in camera-ready form by the authors. For this purpose Springer-Verlag provides technical instructions for the preparation of manuscripts. See also *Editorial Policy.*

Careful preparation of manuscripts will help keep production time short and ensure a satisfactory appearance of the finished book.

The following terms and conditions hold:

Categories i), ii), and iii):
Authors receive 50 free copies of their book. No royalty is paid. Commitment to publish is made by letter of intent rather than by signing a formal contract. Springer-Verlag secures the copyright for each volume.

For conference proceedings, editors receive a total of 50 free copies of their volume for distribution to the contributing authors.

All categories:
Authors are entitled to purchase further copies of their book and other Springer mathematics books for their personal use, at a discount of 33,3 % directly from Springer-Verlag.

Addresses:

Timothy J. Barth
NASA Ames Research Center
NAS Division
Moffett Field, CA 94035, USA
e-mail: barth@nas.nasa.gov

Michael Griebel
Institut für Angewandte Mathematik
der Universität Bonn
Wegelerstr. 6
53115 Bonn, Germany
e-mail: griebel@iam.uni-bonn.de

David E. Keyes
Department of Applied Physics
and Applied Mathematics
Columbia University
200 S. W. Mudd Building
500 W. 120th Street
New York, NY 10027, USA
e-mail: david.keyes@columbia.edu

Risto M. Nieminen
Laboratory of Physics
Helsinki University of Technology
02150 Espoo, Finland
e-mail: rni@fyslab.hut.fi

Dirk Roose
Department of Computer Science
Katholieke Universiteit Leuven
Celestijnenlaan 200A
3001 Leuven-Heverlee, Belgium
e-mail: dirk.roose@cs.kuleuven.ac.be

Tamar Schlick
Department of Chemistry
Courant Institute of Mathematical
Sciences
New York University
and Howard Hughes Medical Institute
251 Mercer Street
New York, NY 10012, USA
e-mail: schlick@nyu.edu

Springer-Verlag, Mathematics Editorial IV
Tiergartenstrasse 17
69121 Heidelberg, Germany
Tel.: *49 (6221) 487-8185
e-mail: peters@springer.de
http://www.springer.de/math/
peters.html

Lecture Notes in Computational Science and Engineering

Vol. 1 D. Funaro, *Spectral Elements for Transport-Dominated Equations.* 1997. X, 211 pp. Softcover. ISBN 3-540-62649-2

Vol. 2 H. P. Langtangen, *Computational Partial Differential Equations.* Numerical Methods and Diffpack Programming. 1999. XXIII, 682 pp. Hardcover. ISBN 3-540-65274-4

Vol. 3 W. Hackbusch, G. Wittum (eds.), *Multigrid Methods V.* Proceedings of the Fifth European Multigrid Conference held in Stuttgart, Germany, October 1-4, 1996. 1998. VIII, 334 pp. Softcover. ISBN 3-540-63133-X

Vol. 4 P. Deuflhard, J. Hermans, B. Leimkuhler, A. E. Mark, S. Reich, R. D. Skeel (eds.), *Computational Molecular Dynamics: Challenges, Methods, Ideas.* Proceedings of the 2nd International Symposium on Algorithms for Macromolecular Modelling, Berlin, May 21-24, 1997. 1998. XI, 489 pp. Softcover. ISBN 3-540-63242-5

Vol. 5 D. Kröner, M. Ohlberger, C. Rohde (eds.), *An Introduction to Recent Developments in Theory and Numerics for Conservation Laws.* Proceedings of the International School on Theory and Numerics for Conservation Laws, Freiburg / Littenweiler, October 20-24, 1997. 1998. VII, 285 pp. Softcover. ISBN 3-540-65081-4

Vol. 6 S. Turek, *Efficient Solvers for Incompressible Flow Problems.* An Algorithmic and Computational Approach. 1999. XVII, 352 pp, with CD-ROM. Hardcover. ISBN 3-540-65433-X

Vol. 7 R. von Schwerin, *Multi Body System SIMulation.* Numerical Methods, Algorithms, and Software. 1999. XX, 338 pp. Softcover. ISBN 3-540-65662-6

Vol. 8 H.-J. Bungartz, F. Durst, C. Zenger (eds.), *High Performance Scientific and Engineering Computing.* Proceedings of the International FORTWIHR Conference on HPSEC, Munich, March 16-18, 1998. 1999. X, 471 pp. Softcover. 3-540-65730-4

Vol. 9 T. J. Barth, H. Deconinck (eds.), *High-Order Methods for Computational Physics.* 1999. VII, 582 pp. Hardcover. 3-540-65893-9

Vol. 10 H. P. Langtangen, A. M. Bruaset, E. Quak (eds.), *Advances in Software Tools for Scientific Computing.* 2000. X, 357 pp. Softcover. 3-540-66557-9

Vol. 11 B. Cockburn, G. E. Karniadakis, C.-W. Shu (eds.), *Discontinuous Galerkin Methods.* Theory, Computation and Applications. 2000. XI, 470 pp. Hardcover. 3-540-66787-3

Vol. 12 U. van Rienen, *Numerical Methods in Computational Electrodynamics.* Linear Systems in Practical Applications. 2000. XIII, 375 pp. Softcover. 3-540-67629-5

Vol. 13 B. Engquist, L. Johnsson, M. Hammill, F. Short (eds.), *Simulation and Visualization on the Grid.* Parallelldatorcentrum Seventh Annual Conference, Stockholm, December 1999, Proceedings. 2000. XIII, 301 pp. Softcover. 3-540-67264-8

Vol. 14 E. Dick, K. Riemslagh, J. Vierendeels (eds.), *Multigrid Methods VI.* Proceedings of the Sixth European Multigrid Conference Held in Gent, Belgium, September 27-30, 1999. 2000. IX, 293 pp. Softcover. 3-540-67157-9

Vol. 15 A. Frommer, T. Lippert, B. Medeke, K. Schilling (eds.), *Numerical Challenges in Lattice Quantum Chromodynamics.* Joint Interdisciplinary Workshop of John von Neumann Institute for Computing, Jülich and Institute of Applied Computer Science, Wuppertal University, August 1999. 2000. VIII, 184 pp. Softcover. 3-540-67732-1

Vol. 16 J. Lang, *Adaptive Multilevel Solution of Nonlinear Parabolic PDE Systems.* Theory, Algorithm, and Applications. 2001. XII, 157 pp. Softcover. 3-540-67900-6

Vol. 17 B. I. Wohlmuth, *Discretization Methods and Iterative Solvers Based on Domain Decomposition.* 2001. X, 197 pp. Softcover. 3-540-41083-X

Vol. 18 U. van Rienen, M. Günther, D. Hecht (eds.), *Scientific Computing in Electrical Engineering.* Proceedings of the 3rd International Workshop, August 20-23, 2000, Warnemünde, Germany. 2001. XII, 428 pp. Softcover. 3-540-42173-4

Vol. 19 I. Babuška, P. G. Ciarlet, T. Miyoshi (eds.), *Mathematical Modeling and Numerical Simulation in Continuum Mechanics.* Proceedings of the International Symposium on Mathematical Modeling and Numerical Simulation in Continuum Mechanics, September 29 - October 3, 2000, Yamaguchi, Japan. 2002. VIII, 301 pp. Softcover. 3-540-42399-0

Vol. 20 T. J. Barth, T. Chan, R. Haimes (eds.), *Multiscale and Multiresolution Methods.* Theory and Applications. 2002. X, 389 pp. Softcover. 3-540-42420-2

Vol. 21 M. Breuer, F. Durst, C. Zenger (eds.), *High Performance Scientific and Engineering Computing.* Proceedings of the 3rd International FORTWIHR Conference on HPSEC, Erlangen, March 12-14, 2001. 2002. XIII, 408 pp. Softcover. 3-540-42946-8

Vol. 22 K. Urban, *Wavelets in Numerical Simulation.* Problem Adapted Construction and Applications. 2002. XV, 181 pp. Softcover. 3-540-43055-5

Vol. 23 L. F. Pavarino, A. Toselli (eds.), *Recent Developments in Domain Decomposition Methods.* 2002. XII, 243 pp. Softcover. 3-540-43413-5

Vol. 24 T. Schlick, H. H. Gan (eds.), *Computational Methods for Macromolecules: Challenges and Applications.* Proceedings of the 3rd International Workshop on Algorithms for Macromolecular Modeling, New York, October 12-14, 2000. 2002. IX, 504 pp. Softcover. 3-540-43756-8

Vol. 25 T. J. Barth, H. Deconinck (eds.), *Error Estimation and Adaptive Discretization Methods in Computational Fluid Dynamics.* 2003. VII, 344 pp. Hardcover. 3-540-43758-4

Vol. 26 M. Griebel, M. A. Schweitzer (eds.), *Meshfree Methods for Partial Differential Equations.* 2003. IX, 466 pp. Softcover. 3-540-43891-2

Vol. 27 S. Müller, *Adaptive Multiscale Schemes for Conservation Laws.* 2003. XIV, 181 pp. Softcover. 3-540-44325-8

Vol. 28 C. Carstensen, S. Funken, W. Hackbusch, R. H. W. Hoppe, P. Monk (eds.), *Computational Electromagnetics.* Proceedings of the GAMM Workshop on "Computational Electromagnetics", Kiel, Germany, January 26-28, 2001. 2003. X, 209 pp. Softcover. 3-540-44392-4

Vol. 29 M. A. Schweitzer, *A Parallel Multilevel Partition of Unity Method for Elliptic Partial Differential Equations.* 2003. V, 194 pp. Softcover. 3-540-00351-7

Vol. 30 T. Biegler, O. Ghattas, M. Heinkenschloss, B. van Bloemen Waanders (eds.), *Large-Scale PDE-Constrained Optimization.* 2003. VI, 349 pp. Softcover. 3-540-05045-0

Vol. 31 M. Ainsworth, P. Davies, D. Duncan, P. Martin, B. Rynne (eds.) *Topics in Computational Wave Propagation.* Direct and Inverse Problems. 2003. VIII, 399 pp. Softcover. 3-540-00744-X

Vol. 32 H. Emmerich, B. Nestler, M. Schreckenberg (eds.) *Interface and Transport Dynamics.* Computational Modelling. 2003. XV, 432 pp. Hardcover. 3-540-40367-1

Vol. 33 H. P. Langtangen, A. Tveito (eds.) *Advanced Topics in Computational Partial Differential Equations.* Numerical Methods and Diffpack Programming. 2003. XIX, 659 pp. Softcover. 3-540-01438-1

Vol. 34 V. John, *Large Eddy Simulation of Turbulent Incompressible Flows.* Analytical and Numerical Results for a Class of LES Models. 2004. XII, 261 pp. Softcover. 3-540-40643-3

Texts in Computational Science and Engineering

Vol. 1 H. P. Langtangen, *Computational Partial Differential Equations.* Numerical Methods and Diffpack Programming. 2nd Edition 2003. XXVI, 855 pp. Hardcover. ISBN 3-540-43416-X

Vol. 2 A. Quarteroni, F. Saleri, *Scientific Computing with MATLAB.* 2003. IX, 257 pp. Hardcover. ISBN 3-540-44363-0

For further information on these books please have a look at our mathematics catalogue at the following URL: `http://www.springer.de/math/index.html`

Druck und Bindung: Strauss Offsetdruck GmbH